INTRODUCTION
TO
KINESIOLOGY

THE SCIENCE OF HUMAN PHYSICAL ACTIVITY

Second Revised First Edition

Marilyn Mitchell

cognella™
San Diego, CA

Bassim Hamadeh, CEO and Publisher
Michael Simpson, Vice President of Acquisitions
Jamie Giganti, Managing Editor
Jess Busch, Graphic Design Supervisor
Jessica Knott, Senior Project Editor
Luiz Ferreira, Licensing Associate

First published in the United States of America in 2013 by Cognella, Inc.

Printed in the United States of America

Cover Image: Copyright © by Michael Slobodian. Reprinted with permission

ISBN: 978-1-62661-486-4

Cover photograph, "Makaila Wallace," by Michael Slobodian.

www.cognella.com 800.200.3908

CONTENTS

Preface

For humans and other animals, movement is critical for survival. Without movement, our ability to communicate through speech and gesturing to satisfy basic needs would be lost. Even a lack of adequate movement or physical activity negatively impacts our physiological and psychological states. Humans incarcerated for long time periods without the opportunity for physical activity suffer physiological and psychological deterioration. On the other hand, optimal amounts of physical activity provide benefits to our physical and mental conditions. In addition, how we actually produce movement is of great importance to coaches and teachers of aspiring athletes, to physical therapists who aid in the rehabilitation of physically impaired individuals, to engineers and ergonomists who build machines and construct work environments to match human abilities, and to scientists trying to unravel the great mysteries of how skilled movement is acquired. Given the importance of movement to our very survival and to our quality of life, the scientific study of human movement is clearly warranted. In short, movement matters.

Kinesiology, derived from the Greek terms *kinesis* (movement) and *logos* (the study of), is a field dedicated to the study of human physical activity. The earlier use of the term *kinesiology* was restricted to applied biomechanics, but recently it is becoming more common to use the term for representing the entire scope of the study of human movement and physical activity.

What is this scope? All movements, large and small, simple and elaborate, are the result of complex interactions of anatomical (bones, muscles, tendons, etc.) and neurophysiological components. The millions of nerve cells in our nervous system allow us to formulate intentional behavior, react to stimuli and maintain balance and posture. Our anatomy serves as the "foundation" for all movement. Thus, human anatomy and physiology are essential elements of the field of study we call kinesiology. Biomechanical principles, adapted from physics, allow us to characterize the forces produced by the body and those that act on the body, such as gravity, as well as provide detailed description and efficiency of movement. Concepts in motor control and motor learning are required to help us understand how we coordinate movements and acquire new motor skills. Psychological factors, such as motivation and stress, are known to influence the quality of performance and determine our desire or aversion to participate in physical activity. Our physical and mental conditions are also affected by the aging process. A "life-span" developmental kinesiology perspective is required to understand reflexive movement, the development of intentional skilled movement, and the decline in physiological and behavioral functions with age. Understanding of the sociocultural aspects of movement and physical activity is needed to answer questions related to differences in physical activity patterns of boys and girls, of different racial groups and older adults, and the influence of sport on the adoption of a physically active lifestyle. Another important element of any field of study is to have an understanding of its history. Identifying the important people and events that have influenced the development of kinesiology is useful in acquiring a perspective of how the field got to where it is, as well as trying to understand where it is going. Related to the historical understanding of kinesiology is the philosophical understanding of kinesiology as a field of study. In what ways can we view kinesiology as a field of study? For example, is kinesiology simply a collection of separate areas of study (biomechanics, exercise physiology, etc.), or does kinesiology provide a home for all the various subfields, such that integration of the contents of the subfields can occur?

Thus, an understanding of how we move, what motivates our actions, and the physiological and psychological benefits of physical activity requires a firm "grounding" in the areas of human anatomy and physiology, biomechanics, exercise physiology, motor learning and motor control, sport and exercise psychology, motor development, and the sociocultural aspects of human movement. It is also important to have a historical and philosophical perspective on human physical activity. The field of study called kinesiology provides a scholarly home for these areas. Today, many major universities in the country have a kinesiology department consisting of scholars in most or all of the above areas. Other related department names are used to describe this field, such as Exercise and Sport Science, Human Movement Studies, and so forth.

As the field of kinesiology continues to develop, I believe it is important to provide an introductory text that reviews the major concepts, principles, and experimental findings for the curious yet serious student. Most major fields of study have a text(s) that provides this type of information in a readable yet scientifically rigorous format for the student. However, there are only a few texts today that specifically focus on the field of kinesiology as described above or that provide suitable introductory information. This text attempts to address this need.

This text is primarily designed for the student who either intends to enter the field of kinesiology or is simply curious about the field. It is my hope that this text will provide a scholarly introduction to the field of kinesiology for those students interested in careers in physical education, coaching, ergonomics, physical therapy, sport psychology, medicine and other health-related professions, for example. The text is also designed to help prepare students for future core courses taken in kinesiology such as biomechanics, exercise physiology, etc.

The ten chapters in the text are arranged in the order that I have taught an Introduction to Kinesiology class in the past. The order of the chapters could be changed to fit the needs of the instructor. Chapter One, **Kinesiology: The Emergence of a New Field of Study**, discusses the focus of kinesiology, defines the various subfields, and describes the cross- and interdisciplinary nature of kinesiological inquiry. In addition, Chapter One identifies the links to professional and performance areas, outlines a typical kinesiology degree, and describes career options for the undergraduate kinesiology student.

Chapter Two, **The History of Kinesiology**, provides an account of the promotion of physical activity in different civilizations and cultures throughout history. This chapter also identifies important names and events of scientific contributions to our understanding of the human engaged in physical activity, and summarizes the development of kinesiology as a field of study. Chapter Two is presented in a chronological (as opposed to a thematic) framework and is an attempt to provide some useful information on the development of inquiry into the study of and participation in physical activity throughout history.

Chapter Three, **Anatomical and Physiological Systems**, provides background information for the beginning student on several physiological systems of the human body. The knowledgeable reader will notice that not all physiological systems are discussed. Only those systems I thought would be helpful in understanding the subsequent chapters were included.

Chapter Four, **Exercise Physiology Foundation**, provides information on energy utilization during movement and exercise, cardiovascular changes with exercise, and ventilatory changes with exercise. This chapter also provides some discussion of the importance of physical activity and exercise to health and fitness.

Chapter Five, **Biomechanical Foundation**, describes types of human motions, as well as linear and angular kinetics and kinematics. The chapter also discusses the nature of Newton's three laws of motion and how they may be applied to human motion.

Chapter Six, **Motor Control and Motor Learning Foundation**, gives an overview of the neuromuscular system and briefly describes some major models of motor control. The chapter also provides an account of variables affecting the speed and accuracy of movement. Finally, the chapter discusses some important factors affecting the learning of motor skills.

Chapter Seven, **Psychological Foundations**, describes major components within the subfield of sport and exercise psychology. This chapter overviews important psychological factors in human performance, cognitive strategies for enhancing performance, and some issues in exercise psychology related to physical activity.

Chapter Eight, **Developmental Foundations**, details some important developmental concepts and terms, anatomical and physiological developmental changes, and characteristics of motor development in the child. In addition, the influence of aging on health and performance is discussed.

Chapter Nine, **Sociocultural Foundations**, discusses the relevance of sociological and cultural factors to kinesiology, different sociology theories, and the influence of rationalization on participation in physical activity. A variety of sociocultural issues related to participation in sports and other physical activities is covered.

Finally, Chapter Ten, **Epilogue**, provides a rationale for an integrative approach to research and teaching in the field of kinesiology. In addition, Chapter Ten discusses some philosophical issues related to the nature of inquiry in kinesiology and provides some closing remarks on the future of kinesiology.

Because of the large content in kinesiology, it was not possible to cover every aspect, issue, or scientific problem within each subfield. I am sure I will be criticized by some who may feel I have left out some important areas.

To improve the readability and hopefully the retention of the material, important words and names are highlighted in each chapter. Also, brief summaries after each section are provided in every chapter, and all chapters finish with a chapter summary. In addition, most of the chapters contain one or more highlighted boxes and several figures to enhance the material in the text. Important terms are listed at the end of each chapter. To challenge the reader's understanding of the material, I have included a section at the end of each chapter called "Integrating Kinesiology: Putting It All Together." These sections contain questions and exercises on the material that occasionally require the reader to refer to material in previous chapters. Finally, I have included some websites at the end of each chapter that provide additional information about the discussion material and important links to related information that will hopefully be interesting for the reader.

There are a number of people I must thank who have helped me in different ways throughout the duration of this writing project. Dennis Poremski was my teaching assistant at the University of Colorado at Boulder where I first began writing the text. Dennis provided considerable feedback as I was writing the text and teaching an introductory course in kinesiology that I developed. I want to thank Upen Patil, Elena Lazaretnik, Tamara Hellen Mull, and Claire Furlotte, students at San Francisco State University, who helped with some of the graphics and figures contained in the book. Brian Maraj and Bob Brustad provided helpful editorial comments and encouragement. Penny McCullagh gave me several suggestions and editorial comments. I would also like to thank Melissa Accornero and Brent Hannify at University Readers for all their help and support in the final phases of this project. Finally, this book is dedicated to my parents, Ed and Jean, for their love and support throughout my career, and to my daughter, Makaila, a professional ballet dancer who has reinforced my belief that movement certainly does matter.

Marilyn Mitchell, Ph.D.
May 2013

Chapter One
Kinesiology: The Emergence of a New Field of Study

Important Terms

Integrating Kinesiology: Putting It All Together

Kinesiology on the Web

References

CHAPTER ONE

Kinesiology: The Emergence of a New Field of Study

STUDENT OBJECTIVES

1. To appreciate the focus of inquiry in kinesiology.
2. To identify the various subfields or foundations in kinesiology.
3. To know the difference between cross- and interdisciplinary fields of study.
4. To understand the nature of a degree in kinesiology.
5. To identify a number of career options following a degree in kinesiology.

THE PHYSICAL ACTIVITY FOCUS

Since ancient times, there has always been an interest in how and why humans engage in physical activity. Recently, there have been serious efforts to embody this interest within a coherent, scholarly field of study. The field of study we call today kinesiology is the result of these efforts. The term kinesiology derives from the Greek words *kinesis* (movement) and *logos* (the study of). The emphasis in kinesiology inquiry is on human movement, although animal movements are studied and used to help us better understand the human condition.

In the preface, a rationale for understanding human movement from several viewpoints was discussed. In Chapter One, this rationale is further developed. A part of this rationale deals with the focus of inquiry in kinesiology. Should kinesiology as a field of study be restricted to a certain type of movement or physical activity? Should kinesiologists study sport? Is the study of dance within the focus of kinesiology? What about activities of daily living, such as tying one's shoes, doing the dishes, or driving a car? Is physical exercise within the scope of kinesiology? The basic answer to these questions is essentially *yes*. As argued by Newell (1990), kinesiological inquiry can be used to study all types of physical activity, from spontaneous play, exercise, rehabilitation, to recreational and competitive sports. Charles (1994) suggested that there are three major types of movement that can be scrutinized by kinesiology researchers as shown in Figure 1.1:

3 main types

- *Sportive movement*—refers to skill-related physical activity
- *Symbolic movement*—is physical activity that expresses thoughts and feelings through the symbolic medium of the body; the main emphasis is on physical expressiveness such as in dance, gestures, and speech
- *Supportive movement*—is physical activity of a functional nature necessary to support a certain lifestyle and the major emphasis is health-related physical activity, such as the activities of daily living, rehabilitation from injury, disease prevention, and work.

This distinction by no means implies that the categories of movement are mutually exclusive (Charles, 1994, p. 16). For example, a ballet dancer's movements are clearly symbolic but they are also sportive, in the sense that they reflect the skill of the dancer. In addition, ballet movements also may be supportive if the dancer considers their production as contributing to the dancer's health. Thus, any observed movement could fit into one or more categories.

Knowledge gained from the study of one type of movement may also apply to our understanding of the other types. For example, many supportive movements have to be learned first before they can be used. That is, whether one engages in supportive movement may depend on how well the movement can be executed. For example, a person with amputation, who could functionally benefit from the use of an artificial hand, often rejects using it because of the lack of sufficient motor control. As pointed out by McKensie (1970), "Learning to use an arm prosthesis never comes instinctively, and its effective use is an acquired skill, so much so that no worthwhile return in the way of function is apparent to the user, and rejection may result." This is just one example of how an understanding of one type of movement (sportive or skillful) can facilitate the understanding of another (supportive).

In addition to the three major types of movement studied by kinesiologists, there also are a variety of settings *where* they may be examined. We can expect the qualities of human movement to be at least partly dependent on the situation. For example, sportive, symbolic, and supportive movements all can be studied in the laboratory where the researcher can better control the environmental conditions. Much kinesiology research has been investigated in this setting. However, further insight into physical activity can be gained by studying human movement outside of the laboratory in more real-life conditions, such as in the workplace, the clinic, the gymnasium, the dance studio, the swimming pool, a city park, or on a mountain. There is much to learn about human physical activity in natural settings, where the difficulties of controlling environmental influences can be offset somewhat by the richness of the information gathered. In addition, some physiological and psychological contributions to movement will vary from one physical activity setting to another, while others will not. How we experience and ascribe meaning to physical activity may also likely depend on the physical activity setting.

Differences and similarities in human physical activity can be examined across cultures, racial and ethnic groups, socioeconomic status, between the sexes, and across different developmental ages. While the focus of kinesiology is on *human* physical activity, this does not imply a restriction to *only* human inquiry. Often, insight into human physical activity can be gained by studying other animals before applying this knowledge to the human. In addition, a growing amount of physical activity research utilizes computer technology, where the researcher relies on computer models and simulations of human and animal movement. Here, human or animal movement is investigated through the use of a set of derived equations! The researcher derives such equations to mimic a theoretical model of movement or factors affecting movement, and tests the model using computer simulations before applying the model to actual human behavior. Thus, some scientific inquiries of physical activity are not performed in the experimental laboratory or in the natural environment, but on the computer! In summary, restricting our inquiry to a certain type of movement, setting, type of individual, or model will only serve to limit our understanding of the various dimensions of human physical activity.

Figure 1.1
Types of movement, (adapted from Charles, 1994).

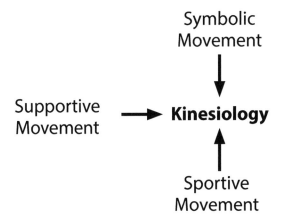

SUBFIELDS IN KINESIOLOGY

Figure 1.2 illustrates a conceptualization of kinesiology made up of various **subfields** and dimensions. In this conceptualization, the major subfields of kinesiology are biomechanics, exercise physiology, motor control and motor learning, motor development, sport and exercise psychology, and sociology of physical activity. Furthermore, there is a historical dimension of kinesiology that includes important people and events that have shaped the development of the field. There is a philosophical dimension to kinesiology dealing with how and why we view the field of kinesiology as a field of study. The circles representing the subfields are purposely placed close to one another and all connect to the field of kinesiology. As a result, the content of each subfield can be thought to be connected both to the field of kinesiology and to the other subfields. In this way, the field of kinesiology can be thought of as an "integrative" field of study.

In my view, no subfield is more important than another because without information from a subfield, a more complete understanding of how and why humans move, work, and perform in the environment is not possible. As we will see in subsequent chapters, each subfield focuses on a certain aspect of human physical activity. The subfields identified in Figure 1.2 have been well researched for a number of years and contain a number of facts, models, theories, and phenomena. Because each subfield has been thoroughly researched and has contributed to our understanding of kinesiology as a whole, each of them will be referred to as a **foundation**. This is not to suggest that a complete understanding of each foundation has been achieved because the body of knowledge within each is constantly expanding as research continues.

The chapter entitled **History of Kinesiology** documents, in more or less chronological order, the relative importance of physical activity to society from ancient to modern times. This chapter also identifies important people in history who have either studied physical activity or have affected the promotion of physical activity. Finally, the history of kinesiology as a developing field of study is examined.

The **Anatomical and Physiological Systems** chapter is an overview of some of the major anatomical structures and important physiological systems of the human body. This chapter has been included to give the reader some useful background before going into the various subfields within kinesiology.

The **Exercise Physiology Foundation** chapter is the study of how the structures and the various physiological systems in the body adapt acutely and chronically to the stress of exercise. The **Biomechanical Foundation** chapter provides detailed description of the muscular forces produced by the human body and the mechanics of human motion. How movements are controlled, coordinated, and learned are investigated within the **Motor Control and Motor Learning Foundation** chapter. The **Psychological Foundation** chapter explores the various cognitive, emotional, and social factors that affect both how and why we move and perform in a number of physical activity environments. Changes in both the structure and function of the human body over the life span are investigated within the **Developmental Foundation** chapter. Some may argue that development is not a separate foundation, but rather a dimension of the other subfields. I would not disagree. However, there is sufficient accumulated knowledge of the human development and aging process to consider development as a unique foundation. The **Sociocultural Foundation** chapter examines the many social and cultural factors that influence human physical activity. Finally, the last chapter examines some of the philosophical elements of kinesiology.

Some time ago, Jerry Barham (1963; 1966) conceptualized the various subfields in a unique way by using different adjectives in front of the noun kinesiology. For example, biomechanics was called *Mechanical* Kinesiology, and exercise physiology was called *Physiological* Kinesiology. Using Barham's conceptualization, it is possible to characterize the other subfields in a similar manner, as shown in Figure 1.2. For example, sport and exercise psychology could be called *Psychological* Kinesiology and sociology of physical activity could be called *Sociological* Kinesiology. Some kinesiology departments use these types of descriptions to describe subfield courses within their kinesiology curriculum. While the traditional names of the subfields are acceptable, using Barham's approach brings can bring some unity to the subfields by directly associating them with the overarching field of study, kinesiology.

One could argue that specific names for each subfield should be dropped altogether and replaced by a thematic area within kinesiology. For example, in 1990 Newell suggested the following five thematic areas that attempt to cover the entire scope of kinesiological inquiry: coordination, control, and skill; growth, development, and form; energy, work, and efficiency; involvement, achievement, and enculturation; aesthetics, meanings, and values.

There is merit to Newell's proposal because a thematic approach attempts to link the various subfields and helps promote cross-disciplinary knowledge (see below, and Chapter 10). The adjective descriptions of the subfields also bring a sense of unity to kinesiology inquiry. The traditional names of the subfields are probably adequate because they accurately represent the content within kinesiology and they are familiar to most people in the field. However, both the Barham and Newell proposals for describing the developing field of kinesiology have merit and are worthy of further consideration.

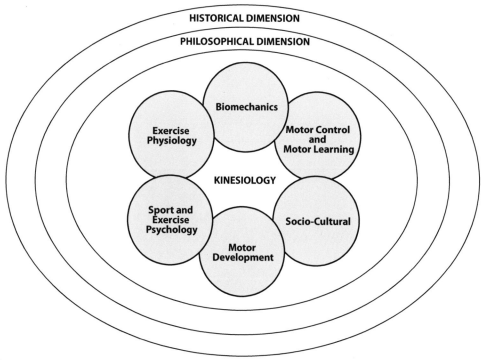

Figure 1.2

The Sub-Fields in Kinesiology

LEVELS OF ANALYSIS

levels of analysis /

The antecedents (causes) and the consequences (outcomes) of human movement can be examined at different levels of analysis. A **level of analysis** can be thought of as representing the size of the unit under investigation. In describing the human engaged in physical activity, there are several levels of analysis: molecular, cellular, systems, behavioral, psychological, and sociocultural. Each subfield in kinesiology focuses on certain level(s) of analysis to evaluate the antecedents and consequences of physical activity.[1]

The unit of study at the **molecular** level of analysis is extremely small. Observing and measuring events at this level of analysis require sophisticated tools such the electron microscope. For example, the molecular level of analysis has been used by exercise physiologists to examine how muscle contraction occurs. Magnetic resonance imaging (MRI) measures the amount of certain atoms at different locations in the body. This technique has been used by motor control researchers to measure brain activity during movement. It also has been used by biomechanics and exercise physiology researchers to study a variety of anatomical and physiological functions.

At the **cellular** level of analysis, cellular events occurring in the nerve, bone, muscle, and other tissues are examined. The cellular level of analysis is used primarily by the biomechanics, exercise physiology, and motor learning and control subfields.

At the **systems** level of analysis, activity within the various physiological systems is evaluated, such as the nervous, skeletal, muscular, cardiovascular, and respiratory systems, to name a few. Once again, the biomechanics, exercise physiology, and motor learning and control subfields extensively examine this level of analysis. There are also some researchers in sport and exercise psychology who measure events occurring in certain physiological systems.

At the **behavioral** level of analysis, movements that are readily observable by the naked eye are measured or evaluated. At the behavioral level of analysis, we can examine the causes of antecedents of movement using the principles of **kinetics**, a major area of biomechanics that studies the forces produced by or exerted on the body. Kinetic measures such as force can be examined through the use of devices called strain gauges, force transducers, and force platforms. **Kinematics**, another major area with biomechanics, examines the various biomechanical descriptions of movement such as displacement, velocity, and acceleration of the limbs or body.[2] An example of a kinematic measurement tool is motion analysis technology, such as videotape analysis. Certain types of markers are strategically placed on the body, the motion is videotaped, and the changes in markers' positions during the movement are digitized and entered into a computer for analysis purposes. There also are other sophisticated technologies today that allow for kinematic measurement (see Magill, 2011).

Some years ago, Higgins (1977) proposed that the events at the behavioral level also could be examined using kinesic and subjective-description measurement. **Kinesic** measurement is a qualitative method to evaluate nonverbal communicative behavior of an individual or between individuals. It can be used to measure nonverbal communication of a performer, teacher, or coach, for example. **Subjective description** is another way to qualitatively evaluate movement based upon the opinion of the evaluator. A golf instructor evaluating a student's swing or an audience watching a dance performance are examples of subjective description of movement. The outcome of the movement, involving the consequence, efficiency, and effectiveness of the goal accomplishment of the movement, can also be evaluated (Higgins, 1977). Did the basketball go in the basket, did the dart hit the bull's-eye, is the amputee improving in the use of a new prosthesis, are examples of questions that can be raised about movement outcome. How efficient a movement is produced can also be examined. Thus, at the behavioral level of analysis, the causes, description, and outcome of movement can be evaluated. The behavioral level of analysis has been used by all the subfields in kinesiology.

At the **psychological** level of analysis, cognitive and emotional events of the human are evaluated. Measurement of cognitive and emotional events may take the form of validated questionnaires or interviews. These types of measurements assess the mental events associated with participating in physical activity such as the motivation of the individual. The psychological level of analysis is used primarily in the sport and exercise psychology subfield, but some of the other subfields also take measures at this level of analysis (Duda, 1998).

The **sociocultural** level of analysis examines the largest unit of study in kinesiology: the behavior of groups of people or societies. One example of a question addressed at this level of analysis is what sociocultural factors influence the participation of an individual or a large group of individuals in physical activity? Does the racial or ethnic background of a group of people affect their participation in exercise? Does the socioeconomic status of a group of people have an effect on whether they choose to live a physically active lifestyle or not?

Other than the size of the unit, another distinguishing characteristic of the various levels of analysis are time-scale differences. **Time scales** can be thought of as the elapsed time between significant events occurring at a given level of analysis. The fastest time scale occurs at the molecular level of analysis (not including the atomic or subatomic levels) and the slowest time scale occurs at the sociocultural level of analysis. In other words, important changes in events occur at a much faster rate at the molecular level, sometimes on the order of nanoseconds (i.e., one billionth of a second), compared to the sociocultural level, where significant events may be separated by years, decades, or even centuries! One of the great challenges for kinesiologists is determining the proper time scale between important events at any given level of analysis and to make measurements that encompass that time scale. For example, the time scale for a given muscle contraction, say for flexing the elbow, might be no longer than a second or two. However, what is the time scale for understanding why a person chooses not to engage in a physically active lifestyle? The relevant time scale is certainly not as short as two seconds! But what *is* the relevant time scale in this case? Answering the question depends on the definition of what is meant by "a significant event" at a given level of analysis. The events that occur at a particular level are numerous. The researcher must determine which are the relevant and irrelevant events before a time scale can be determined. Determining what are relevant and irrelevant events at any given level of analysis is a difficult but nonetheless important challenge facing research in kinesiology as well as in other fields of study.

All the levels of analysis operate as an individual is engaged in physical activity. When we perform a golf swing, molecular events are occurring at unbelievable rates, contributing to both the physical and cognitive aspects of the performance. But superimposed on molecular events are all the other levels of analysis that contribute in certain ways to the swing. As a result of the many important processes occurring at different time scales of the various levels of analysis, the golf swing occurs! It must be the case that there is some relationship between what is happening at the molecular level and the other levels. Indeed, yet another major challenge to researchers in kinesiology (as well as other fields of study) is how the events at any one level of analysis relate to events at the other levels.

The last point I want to make in this section is that because all behavioral movements are influenced by all the levels of analysis, no one level of analysis is more important than another. Ignoring one or more levels of analysis results in an incomplete description of the movement. If we ignore the molecular, cellular or systems level of analysis, the physiological description of the movement is lost. If the psychological level is ignored, we fail to appreciate the motivation for performing the movement, for example. The sociocultural level of analysis provides the overall context of the movement, and probably helps define the meaning of the movement to the individual.

While all levels of analysis are required to fully describe a movement, a particular level of analysis might be more appropriate for certain situations. For example, the three types of movement, (i.e., sportive, symbolic, and supportive) can be investigated from several levels of analysis. In some cases, a particular level of analysis may be necessary to measure for a given type of movement or movement situation. For example, if a golfer is slicing the ball and wishes to improve the swing, the instructor may wish to analyze the behavioral level of analysis through videotaping. In this way, the student may be able to actually see what is wrong with the swing. In this situation, the student's heart rate, one measure of the cardiovascular system, is probably irrelevant in terms of providing helpful information that could be used to correct the student's slice. This example appears rather straightforward; however, determining the most useful level of analysis for understanding a certain movement or movement situation is another important challenge faced by kinesiologists. The reasoning is that movements, and their antecedents and consequences, are complex, and complex phenomena are difficult to describe and understand.

CROSS- AND INTERDISCIPLINARY FIELDS OF STUDY

Each foundation of kinesiology has been investigated by scholars both within and outside of kinesiology. For example, medical researchers have contributed to the development of the Exercise Physiology Foundation and researchers in the field of psychology have helped in laying some of the groundwork in the Motor Learning/ Control and Psychological Foundations. To some extent, each of the Foundations could be part of other disciplines, such as physiology or psychology. However, the *major* focus of kinesiology is human movement and physical activity from a number of perspectives. As such, it is different from the major focus of other academic disciplines.

If research in each subfield or foundation in kinesiology were to continue in an isolated fashion without any integration from the other subfields, then kinesiology would become an **interdisciplinary** field of study (see Henry, 1978, and Lawson and Morford, 1979, for excellent discussions). This type of academic discipline develops subfields that typically become highly specialized. Researchers in each subfield develop unique research skills and techniques, and speak a "different language" in describing the content within their specialized area. In this scenario, each subfield becomes so specialized that it could be encompassed within a traditional or parent discipline (e.g., biomechanics into physics, exercise physiology into biology, sport and exercise psychology into psychology). In some present-day departments of kinesiology, it is common for researchers in exercise physiology, for example, to have little interaction with their colleagues in sport and exercise psychology, in spite of the fact that there are many areas of potential overlap and issues of common interest. I suspect similar interdisciplinary trends are evident in many other scientific fields of study.

Cross-disciplinary means that the focus of the field of study is the communication and interaction *among* the subfields and not just the isolated advancement of each subfield. Many academic fields of study, such as psychology, physics, and chemistry, started out being cross-disciplinary but as time went on, tended to become more interdisciplinary in nature due to specialization and fragmentation. Suffice it to say, interdisciplinary research within each subfield of kinesiology is important, but a more comprehensive understanding of how and why we engage in physical activity is likely to be better achieved through cross-disciplinary research. With cross-disciplinary research, thematic problems within the field of kinesiology can be explored using expertise and knowledge from more than one subfield. For example, the question "How does one improve cardiovascular fitness?" can be addressed from several subfield perspectives. From an exercise physiology point of view, specific physiological training techniques can be evaluated. From a sport and exercise psychology perspective, the proper motivational orientation of the individual can be assessed. From a developmental viewpoint, the age of an individual can affect the type of training technique used and may have an influence on the individual's motivation. Whether the individual initiates and adheres to a cardiovascular fitness program may depend on his or her socioeconomic background, information emphasized from a sociocultural point of view. Whether kinesiology as a field of study develops into an interdisciplinary or cross-disciplinary field of study is an open question.

TYPES OF KNOWLEDGE IN KINESIOLOGY

As with other fields of study, the content of kinesiology consists of declarative and procedural knowledge (Ryle, 1949). One type of **declarative knowledge** is the theoretical and empirical information generated within a field of study as a result of scientific research. After decades of research in the various subfields, kinesiology as a field of study contains considerable declarative knowledge about the human engaged in physical activity. The views of Franklin Henry, expressed in two important papers (1964; 1978), had much to do with the emphasis on declarative knowledge in the kinesiology curriculum today in most universities.[3]

Another type of declarative knowledge is that generated by the practitioner, such as the physical education teacher, sports coach, and exercise trainer (Newell, 1990). We have all probably met practitioners in the field such as physical education teachers or coaches who have declarative knowledge of their profession of a practical nature. The declarative knowledge they possess is about the real-world setting of their profession. For example, it is one thing to know about the biomechanics of a soccer kick, but it is quite another to be able to teach soccer to elementary school children. Any beginning student physical education teacher will confirm this point! There are many "tricks of the trade" in any profession, some gained with theoretical knowledge, but much usually gained from years of practical experience on the job. One reason for bringing up this distinction between types of knowledge is to emphasize that this book is primarily about declarative knowledge of the theoretical and empirical types within kinesiology. However, real-life examples of how theoretical information can be used in practical settings will be discussed throughout the book.

Procedural knowledge entails knowing *how* to do something. In the case of kinesiology, procedural knowledge may pertain to the ability to execute a movement or an action. Having procedural knowledge of a golf swing means being able to actually execute the golf swing. In many physical education degrees and some kinesiology degrees, it is required for students to demonstrate their ability to perform certain skills or exercises or be required to take physical activity classes. The emphasis on procedural knowledge is common in most dance degrees. Procedural knowledge also can be demonstrated in laboratories associated with lecture classes in kinesiology classes. The ability to perform certain measurement techniques at a given level of analysis indicates some mastery of procedural knowledge. Student demonstrations of a movement, skill, or exercise in a lecture or laboratory class are other examples of procedural knowledge. It is my view that declarative and procedural knowledge about humans engaged in physical activity should be part of the kinesiology curriculum.

THREE ASPECTS OF KINESIOLOGY PROGRAMS

According to Newell (1990), a kinesiology program at the university level can consist of three emphases: disciplinary, professional and performance. The **disciplinary emphasis** relates primarily to the declarative knowledge within kinesiology: that is, the theoretical information within the various subfields. In most universities, a disciplinary emphasis is required in all fields of study. The **professional emphasis** focuses on preparing the student for a specific career or profession. In physical education departments and some kinesiology departments, students may major in physical education that will prepare them to become a physical education teacher in elementary or secondary school, a sports coach, or sports administrator, for example. In other kinesiology departments, the professional emphasis is not a part of the curriculum. The **performance emphasis** relates primarily to certain aspects of procedural knowledge within kinesiology—for example, the demonstration of skill for competitive or aesthetic purpose. According to Newell (1990), this emphasis greatly decreased in most physical education and kinesiology programs and probably has continued to do so. In fact, the field of dance, which used to be an inherent part of physical education before kinesiology evolved, moved out of physical education and formed its own degree program in many universities. The performance emphasis is typically less emphasized in most kinesiology curriculums across North America.

In my view, whether the three program emphases are a part of a given degree or not, people in a given emphasis can benefit from the knowledge generated in the other two. For instance, the theoretical knowledge generated by researchers in the disciplinary emphasis may be useful to both the physical activity practitioner and the performer. Examples of physical activity practitioners are physical education teachers, coaches, exercise leaders, physical therapists and other health professionals. Physical activity performers may include dancers and other athletes. While debated (Best, 1978; Newell, 1990), it is possible for the three types of emphases (theoretical, practitioner, and performance) to inform each other. Practitioners such as physical education teachers can apply principles of biomechanics, exercise physiology, and the other subfields to their teaching of exercise habits and motor skill development. Basic knowledge of kinesiology can help physical therapists develop preventive injury or rehabilitation programs for their clients.

It is also possible, I believe, for the researcher to gain insight into theoretical physical activity problems by studying or interacting with professionals, clinicians, or performers. "Real-world" problems faced by these professionals, clinicians, and performers offer the researcher a steady reminder of the complexity of behaviors outside the laboratory. The kinesiology researcher can potentially gain great insight into a physical activity problem by conducting research outside of the laboratory. Field research is a common strategy used by researchers in the psychological and sociocultural areas, but also practiced by some researchers in the other subfields. Performers also can benefit from the other two areas. A dancer who interacts with a physical activity researcher may discover a better method to improve his or her leg strength. If the dancer becomes injured, knowledge gained by communication with a physical therapist may improve rehabilitation and help in preventing future injury.

THE DEGREE IN KINESIOLOGY

Regardless of the desired career, it is important for kinesiology students to receive a strong liberal arts education. A liberal arts education exposes the student to a wide variety of areas of study outside of the student's specific major. A liberal arts education is designed to provide the student with a greater perspective of the world and an appreciation of the importance of different fields of study, regardless of one's specific interest. In addition, many liberal arts courses can provide important background for understanding the content in the various subfields of kinesiology. For example, general psychology provides an excellent foundation for the study of sport and exercise psychology. Chemistry and biology are important prerequisites for exercise physiology. Physics is important for a better understanding of biomechanics.

In my opinion, it is important that students take necessary prerequisite coursework that adequately prepares them for kinesiology. Table 1.1 illustrates one possible model of a kinesiology degree. While the model is not meant to represent the best or the only structure, it does serve to illustrate some of the important content necessary for a strong academic degree in kinesiology.

For some careers, more coursework may be necessary. For example, many physical therapy programs across North America require an entire year of physics or additional classes in chemistry.

The kinesiology core classes are designed to introduce students to the various subfields within kinesiology. Following these core classes, a number of elective classes in kinesiology are offered to further prepare students for specific careers in kinesiology or to expand on their interests.

Table 1.1
An Example Undergraduate Degree in Kinesiology

Pre-Kinesiology Requirements	
General Biology	1 year
General Chemistry	1 year
General Physics	1 year
Algebra and/or Calculus	1 semester minimum
Introductory Psychology	1 semester
College Writing	1 semester
Kinesiology Major Requirements*	
Human Anatomy **	
Human Physiology **	
Physical Activity Classes (emphasizing procedural knowledge)	1 or more semesters
Introduction to Kinesiology	1 semester
Introduction to Statistics and Research Methods	1 semester
Exercise Physiology	1 semester
Biomechanics	1 semester
Motor Learning and Motor Control	1 semester
Sport and Exercise Psychology	1 semester
Motor Development	1 semester
History/Philosophy of Kinesiology	1 or more semesters
Sociocultural Kinesiology	1 semester
Senior Seminar—Integrative Kinesiology (see Ch. 10 for discussion)	1 semester
Ideally these courses should be accompanied by laboratory experiences	
**These courses could be offered in other departments (e.g., Biology)*	
A Sample of Kinesiology Electives	
Aging and Performance	
Athletic Training	
Human Factors	
Honors Thesis in Kinesiology	
Independent Study in Kinesiology	
Internship in Kinesiology	
Management of Exercise and Wellness Programs	
Nutrition and Performance	
Physical Education Methods	
Prevention and Treatment of Performance and Sports Injuries	
Selected Topics in Kinesiology	
Sports Medicine	

HIGHLIGHT

What can I do with a degree in kinesiology?

First of all, countless studies have concluded that a college degree in any major is an asset in many ways in later life. Most liberal arts degrees (e.g., psychology, biology, and mathematics) are not designed to lead to a specific employment opportunity, but rather to provide a well-rounded education and to develop the basic knowledge and skills inherent in almost any occupation. One of the requirements of a liberal arts degree is a concentration of work in one specific area known as a major. In most cases, students naturally select an area of interest in which to major in and, of course, it makes sense that this might lead to some related employment situation in the future. Thus the question, "What can I do with a degree in kinesiology?"

In an attempt to answer this question, the following occupations and fields related to kinesiology were compiled:

Adapted Physical Education
Aerospace Medicine
Allied Health
Anatomical Science
Anatomy
Anesthesiology Education
Applied Physiology
Aquatics Director
Athletic Clubs
Athletic Director
Athletic Training
Athletic or Sports Administration
Biobehavioral Science
Biodynamics
Biomedical Engineering
Biomedical Science
Buying and Selling Equipment
Camp Director
Cardiac Rehabilitation
Cardiopulmonary Technology
Cardiorespiratory Science
Cell Biology
Chiropody
Chiropractic
Clinical Biology
Clinical Medicine
Coaching
Corporate Fitness/Wellness
Health Administration

Health Care Management
Health Club/Management
Health Education
Health Fitness Management
Health Promotion
Health Records
Health Science Admin.
Health Services
Health Services
Health Spa Management
Health and Wellness
Home Economics/Nutrition
Hotel Programs
Human Biology
Human Factors
Human Nutrition
Industrial Fitness Programs
Industrial Hygiene and Safety
Leisure Management
Leisure Studies
Manager–Recreational
Medical Assistant
Medical Biology
Medical Laboratory Technician
Medical Records Administration
Medical Sciences
Medical Technology
Medicine
Mental Health

Microbiology
Naturopathic Medicine
Nuclear Medicine Technology
Nursing
Nutrition
Nutrition Biochemistry
Nutrition Science
Occupational Therapy
Ophthalmology
Optometry
Orthopedic Assistant
Orthopedic Medicine
Osteopathic Therapy
Osteopathy
Outdoor Education
Paramedic
Park and Recreation Resources
Pharmacology
Physical Ed/Special Population
Physical Education
Physical Therapist
Physical Therapy/Aid
Physical Therapy Assistant
Physician's Assistant
Physiology
Podiatry
Preschool Program
Preventive Medicine
Private Sports/Rec Clubs
Promotional Manager—Sports Equipment
Psychiatric Medicine
Psychomotor Therapy
Public Health
Radiation Technology

Radiation Therapy
Radiological Technology
Recreation
Recreation Administration
Recreation Leadership
Recreational Therapy
Recreation and Leisure
Recreation and Parks
Rehabilitation Specialist
Research Assistant
Resort Programs
Sport Information
Sport Psychologist
Sport Sociologist
Sports Clubs
Sports Communication
Sports Health
Sports Journalism
Sports Management
Sports Medicine
Sports Sciences
Sports Studies
Strength and Conditioning Director
Stress Management
Surgeon Assistant
Surgical Technician
Therapeutic Recreation
Therapy Pool Manager
Underwater EMT
Veterinary Medicine
Veterinary Technology
Vision Science
Wellness Centers
YMCA

List originally compiled by Dale Mood, Department of Kinesiology, University of Colorado at Boulder.

As one can see, there are a number of possible career opportunities awaiting kinesiology graduates. In the next section, six career opportunities are briefly discussed to provide additional detail for the interested student: physical education, physical therapy, physician assistant, allopathic and osteopathic medicine, chiropractic, and university teaching and research.

Physical Education

As we will see in the next chapter, the field of kinesiology evolved from physical education, the traditional area of study for those interested in teaching physical education or coaching sports. One of the goals in teaching physical education in the schools is to improve the quality of children's lives through participation in various types of physical activity, including sports (Lumpkin, 2011). The National Association for Sport and Physical Education (NASPE) has been given responsibility from the National Council for Accreditation of Teaching Education (NCATE) to ensure that physical education undergraduate and graduate programs meet the minimum standards for accreditation. The general competencies required of physical education and sport teachers are the following:

- understanding the scientific and philosophical bases of physical education and sport
- developing a comprehensive knowledge about analyzing movement
- developing a wide range of motor skills, especially those related to the area of teaching
- studying the teaching-learning processes specifically related to the area of physical education and sport
- becoming knowledgeable about planning, organizing, administering, supervising, evaluating, and interpreting various aspects of a balanced physical education and sport program (from Lumpkin, 2011).

There are more specific competencies for elementary, middle and secondary physical education and sport teachers.

An undergraduate degree in either kinesiology or physical education is required to teach at the elementary, middle, or secondary levels. Students also are required to teach in a nearby school for at least one semester before graduation. In some programs, an additional year is required for teaching certification. In addition, some degrees in kinesiology or physical education offer concentrations or minors allowing the student to specialize in a given area. Some of the various concentrations within physical education are coaching, fitness, sport management, exercise and sport science, adapted physical education, therapeutic recreation, and teaching (Lumpkin, 2011). It should be noted that many departments of physical education have dropped the teaching professional focus of their programs to concentrate more on the disciplinary emphasis or the academic orientation, as some have called it (e.g., Siedentop, 2009). Thus, students interested in the teaching and coaching profession should make sure departments of their interest have accredited teaching certification programs.

Athletic Training

A student interested in preventing, recognizing, managing, and rehabilitating sports injuries may very well wish to pursue a career in athletic training. In actuality, it is possible to specialize in a number of areas within the profession of athletic training, including prevention of athletic injuries, recognition, evaluation and immediate care of athletic injuries, rehabilitation and reconditioning of athletic injuries, health care administration, and finally, education and counseling. The American Medical Association recognizes athletic training as an allied health care profession. Athletic training is listed by the Bureau of Labor Statistics as one of the fast growing professions.

The athletic trainer typically works under the direction of a physician and within a team of allied health professionals. Athletic trainers are consulted to develop physical conditioning and injury rehabilitation programs, prepare athletes for competition and practice using recognized procedures for taping, bandaging, and bracing injured areas, and to determine whether an athlete requires further medical treatment. Athletic trainers may find employment in secondary schools, colleges and universities, professional sports and sports medicine clinics, corporate health programs, health clubs, clinical and industrial health care programs, and athletic training curriculum programs.

To become a certified athletic trainer, it is necessary to graduate from an accredited athletic training program with either an undergraduate or graduate degree or meet internship requirements established by the National Athletic Trainers' Association Board of Certification. Many kinesiology degrees offered at universities also have certified programs in athletic training.

Physical Therapy

One of the more popular careers open for kinesiology graduates is the field of Physical Therapy. It should be noted that physical therapy is a much-desired profession, and entrance into a physical therapy program is a highly competitive process, not unlike medical school. Nonetheless, highly qualified kinesiology graduates have had success getting accepted into physical therapy schools.

Physical therapy is a type of health care profession involved in the promotion of optimal health using scientific principles to prevent, assess, and correct movement dysfunction. Physical therapists commonly work with other health care providers, such as physicians, occupational therapists, rehabilitation nurses, psychologists, social workers, dentists, podiatrists, and speech pathologists and audiologists (American Physical Therapy Association). A large number of physical therapists work in hospitals, but more than 70 percent can also be found in private physical therapy offices, community health centers, corporate or industrial health centers, sports facilities, research institutions, rehabilitation centers, nursing homes, home health agencies, schools, pediatric centers, and colleges and universities across North America. In Canada and Europe, physical therapists are called *physiotherapists*.

Physical therapists or physiotherapists often are involved in evaluating and assessing patients recovering from injury, surgery, or disease, and they help develop and implement treatment programs. They can teach patients how to use artificial limbs and other assistive devices. Physical therapists also can provide instruction and home programs to patients and their families to continue the recovery process once the patient is free from the physical therapist's direct care.

According to a 2008 report by the U.S. Department of Labor's Bureau of Labor Statistics, physical therapy (and physical therapy assistants) is one of the fastest growing health care occupations. There are over 185,000 practicing physical therapists in the United States alone. Today, graduates from physical therapy programs receive either master's or doctoral degrees in physical therapy.

Another related employment opportunity is a physical therapy assistant. Physical therapy assistants work under the supervision of the physical therapist and assist the therapist in a wide range of activities such as implementing treatment and rehabilitation programs, training patients in exercises and activities of daily living, and monitoring and reporting patient progress. Training in physical therapy usually involves two years of academic and clinical experiences following the completion of a four-year undergraduate degree (such as in kinesiology). Physical therapy assistants must complete a two-year education program, usually offered in a community or junior college. Training consists of one year of general education and one year of technical courses on physical therapy procedures and clinical experience (American Physical Therapy Association).

Allopathic and Osteopathic Medicine

Kinesiology students who wish to receive a medical degree actually have a choice between the standard medical degree (MD) and a degree in osteopathy (DO). The degrees are comparable in that medical and osteopathic students both complete four years of medical college, complete a residency program, pass similar state boards, practice in fully accredited and licensed hospitals and medical centers, can become specialists, and can prescribe drugs and perform surgery. The difference between them is their philosophy of health care. Allopathic medicine, ascribed to by the conventional medical community, centers on the treatment of the disease, primarily through medication and surgery.

Osteopathic physicians also utilize medication and surgery, but in addition receive 300–500 hours of extensive training in body manipulation, including the spine, joints, and connective tissue. The osteopath's philosophy is not to treat the disease but the whole person, and the focus is on prevention as well as curing disease or injury. Osteopathic medicine takes a more holistic approach, emphasizing the interconnectedness of various physiological systems (e.g., muscular, nervous, skeletal) to the psychological and even spiritual state of the patient. This holistic approach may be appealing especially to kinesiology graduates (see Chapter 10 for a discussion). There are approximately 40,000 practicing DOs in the United States, compared to 600,000 allopathic physicians. As of this writing, there are 28 osteopathy schools compared with 130 traditional medical schools in the United States.

Physician Assistants

Another rapidly growing health care profession is physician assistant (PA). These are formally trained professionals who assist physicians in routine tasks such as taking medical histories, performing physical examinations, ordering laboratory tests, ordering X-rays, making preliminary diagnoses, and giving inoculations. In 30 states and the District of Columbia, PAs can prescribe medication. They can perform about 80 percent of physicians' duties but they are not medical doctors and must be under a physician's supervision. Most assistants are in hospitals and doctors' offices, but demand is rising for their help in family practice and health maintenance organizations. There are about 75,000 practicing PAs in the United States and the U.S. Department of Labor expects a nearly 30 percent increase by 2018. There are 142 PA programs in the United States that usually involve two years of academic and clinical training and one year of internship experience. According to the American Academy of Physician Assistants, there are many jobs available for every graduating PA. Of course, while it is not possible to predict with certainty how long this demand will continue, a career as a physician assistant appears to be an attractive one going into the 21st century.

Chiropractic

There was a time when chiropractic medicine was considered quackery and even illegal, but this view is changing. Today, all 50 states in the United States, the District of Columbia, and Puerto Rico have statutes recognizing and regulating the practice of chiropractic. The word chiropractic, taken from the Greek words *cheri* (hand) and *prato* (I do), means "done by hand." The chiropractic profession is an approach to health that 1) utilizes the body's inherent and natural recuperative powers; that 2) places emphasis on maintaining the structural integrity of the body; and which 3) does not utilize drugs or surgery. Chiropractors, like DOs, ascribe to a more holistic approach to health and make use of massage and manipulation techniques to help patients with pain and disabilities. However, chiropractors are not licensed physicians. There are over 53,000 chiropractics in the United States alone. About 1 in 20 Americans sees a chiropractor during the course of a year (Purvis, 1991). As such, chiropractic is the second largest of the three primary health care providers in the United States (medical, chiropractic, and osteopathy, Peterson, 1989). To become a doctor of chiropractic (DC), students must complete general college-level studies (in some states, an undergraduate degree is required), complete four years of training at an accredited chiropractic college, pass required National Board or other exams, and meet individual state requirements for licensure. There are 18 accredited programs in the United States, 2 in Canada, 6 in Australia, and 5 in Europe.

University Teaching and Research

Another career option for kinesiology graduates is teaching and research at the college or university levels. Every state and province in the United States and Canada, for example, has at least one or more higher education institutions that offer physical education or kinesiology degrees. According to the American Kinesiology Association, there are well over 800 university programs in kinesiology (or related names) in the United States alone.

These programs need trained scholars with either a master's or doctoral degree in physical education or kinesiology to teach and possibly conduct research in the subfields within kinesiology. Some physical education programs focus on the training of physical education teachers and coaches who may require a teaching certification beyond the four-year undergraduate degree.

The opportunities to work in a physical education or kinesiology department at a college or university are good, with the competition for jobs generally increasing as one moves from a two-year community college to a four-year major university. At the junior college and smaller university, the emphasis is primarily on teaching the subject matter within physical education or kinesiology, and a master's degree usually is required. At the four-year major university level, research is more emphasized and a doctoral degree usually is required for employment. There is a trend at many major universities to require an additional year or more of post-doctoral experience in teaching or research. In other words, following completion of the doctoral degree in kinesiology, the individual gains additional experience working in a different research laboratory or teaching at another institution before applying for a major university faculty position. This additional experience is common for exercise physiology positions at major universities and in some of the other subfields as well. A master's degree in kinesiology usually requires one to two years of additional academic training beyond the undergraduate degree, and may involve the writing of a thesis or independent research project. The doctoral degree in kinesiology typically involves two to three years of academic training (for full-time students) and the writing of a doctoral dissertation of a major independent research project requiring an additional one to two years, depending upon the scope of the project.

Obviously, there are many career options available to kinesiology graduates. In general, professions that involve the teaching, research, or service aspects of physical activity and health care are excellent options. Many of these professions require additional certification, experience, or degrees beyond the undergraduate degree. Some professions are extremely competitive to enter, such as physical therapy, osteopathy, and traditional medical school. However, with a solid foundation in physical activity coursework and a strong liberal education, I believe the future is bright for the motivated and dedicated kinesiology student.

CHAPTER SUMMARY

- Kinesiology is the study of movement. The emphasis in this field of study is the human engaged in physical activity even though animal movement may be examined to help understand the human condition.
- Kinesiology inquiry examines different types of movements in different settings with different models.
- The human engaged in physical activity can be measured using a variety of tools at several levels of analysis.
- There are several subfields or foundations of knowledge within kinesiology.
- Both interdisciplinary and cross-disciplinary knowledge contributes to these foundations.
- A cross-disciplinary approach is suggested for developing a more comprehensive understanding of how and why we engage in physical activity.
- Theoretical knowledge in kinesiology can contribute to and can be influenced by professional and performance related areas of physical activity.
- The degree in kinesiology should consist of pre-kinesiology preparation, kinesiology core, elective classes, and experiences to complete a strong liberal arts education.
- There are many possible career options in the physical activity and health care related fields for the kinesiology graduate.

IMPORTANT TERMS

Career opportunities – There are many career opportunities for kinesiology students discussed in the chapter

Cross-disciplinary – The communication and interaction among the sub-fields of any field of study including kinesiology

Declarative knowledge – Theoretical and empirical knowledge generated within any field of study

Degrees in kinesiology – The degree in kinesiology should allow for a a strong liberal education, include necessary prerequisites for the kinesiology, core classes and elective classes that help prepare the student for a future career or expands their interests

Disciplinary emphasis – Focuses on the theoretical information within a field of study

Interdisciplinary – A type of specialized development of a sub-field within a discipline having little communication with other sub-fields

Levels of analysis- The size of the unit of investigation

Performance emphasis – Relates to procedural knowledge within a kinesiology degree

Procedural knowledge – Is knowing how to do something such as being able to execute an action

Professional emphasis – Focuses on preparing students within a kinesiology degree for a career or profession

Sportive movement – Refers to skill related physical activity

Sub-fields of kinesiology – These are the various areas of kinesiology that investigate physical activity from different levels of analysis

Supportive movement – Physical activity of a functional nature necessary to support a certain lifestyle

Symbolic movement – Physical activity that expresses thoughts and feelings through the symbolic medium of the body

Time scales – The elapsed time between significant events occurring at a given level of analysis

FOOTNOTES

1. Below the molecular level are the atomic and subatomic levels. But these levels are not widely examined in kinesiology, compared to the field of nuclear physics, for example.
2. Kinetic and kinematic measures can also be used at the systems and cellular levels of analysis.
3. Actually, as pointed out by Tulving (1972), declarative knowledge can be divided into semantic and episodic. Semantic, declarative knowledge is that theoretical and empirical information generated by scientific research referred to here. However, episodic, declarative knowledge is that information gained through personal experiences, a type of knowledge not emphasized in this book.

INTEGRATING KINESIOLOGY: PUTTING IT ALL TOGETHER

1. Charles (1994) identified three types of movements that can be studied by kinesiology researchers: sportive, symbolic, and supportive. Can you think of other ways to categorize movements?
2. Which levels of analysis are related to the various subfields of kinesiology?
3. Provide some examples of declarative and procedural knowledge about kinesiology.
4. Should every kinesiology degree at the university contain a disciplinary, professional, and performance emphasis? Defend your answer.
5. After receiving an undergraduate degree in kinesiology, which career path seems most interesting to you?

KINESIOLOGY ON THE WEB

- http://www.americankinesiology.org This Web site contains various kinesiology links to physical education and kinesiology home pages and organizations and institutes related to kinesiology.

REFERENCES

Best, D. (1978). Degree studies in human movement and physical education. *Journal of Human Movement Studies, 4,* 119–128.

Charles, J. (1994). *Contemporary Kinesiology: An Introduction to the Study of Human Movement in Higher Education.* Englewood, CO: Morton.

Duda, J. L. (1998), *Advances in Sport and Exercise Psychology Measurement*, pp. 471–83. Morgantown, WV: Fitness Information Technology

Henry, F. M. (1964). Physical education: An academic discipline. *Journal of Health, Physical Education and Recreation.* September, pp. 32, 33, 69.

Henry, F.M. (1978). The academic discipline of physical education. *Quest*, Monograph 29, 13–29.

Higgins, J. R. (1977). *Human Movement: An Integrated Approach.* St. Louis, MO: CV Mosby.

Lawson, H. A., and Morford, R. (1979). The cross-disciplinary structure of
kinesiology and sports studies: Distinctions, implications and advantages. *Quest, 31,* 222–30.

Magill, R. A. (2011). *Motor Learning and Control: Concepts and Applications.* **New York: McGraw-Hill, 9th Edition,**

Lumpkin, A. (2011). *Introduction to Physical Education, Exercise Science, and Sport Studies*, New York: McGraw-Hill, 8th Edition.

Newell, K. M. (1990). Physical activity, knowledge types, and degree programs. *Quest, 42,* 243–68.

Peterson, R. G. (1989). Chiropractic: The prognosis for the profession. *New Realities*, September/October, 32–37.

Purvis, A. (1991). Is there a method to manipulation? *Time*, September 23, 60–61.

Ryle, G. (1949). *The Concept of Mind.* Middlesex, England: Penguin.

Siedentop, D. (2009). *Introduction to Physical Education, Fitness, and Sport.* New York: McGraw-Hill, 7th Edition.

Tulving, E. (1972). Episodic and semantic memory. In E. Tulving and W. Donaldson (eds.), *Organization of Memory*, pp. 382–403. New York: Academic Press.

Chapter Two
The History of Kinesiology

Physical Activity and Early Humans

Physical Activity in Ancient Civilizations

Egypt

Ancient Civilizations in the Far East

Ancient Greece

Ancient Rome

The Middle Ages

The Renaissance Period and the Birth of Scientific Inquiry

Promotion of Physical Activity and Exercise

Humanism

Moralism

Realism

The 17th, 18th and 19th Centuries and the Age of Enlightenment

The Promotion of Physical Activity and Exercise in Europe

The Promotion of Physical Activity and Exercise in the United States

The Beginning of Physical Education in the Schools

Physical Education as a Profession

Other Promoters of Physical Activity

Scientific Contributions to the Understanding of Human Movement

The 20th Century

1900 to 1930 and the Rise of Physical Education

Other Promoters of Physical Activity

Sports and Athletics

Intramurals

Organizations

Recreation

Dance

Physical Therapy

Scientific Contribution to the Understanding of Movement and Physical Activity (1900–1930)

Physical Education Research

Movement Research in Other Disciplines

1930 to 1950

Continued Development of Physical Education

Promotion of Physical Activity

Sports

Dance

Physical Therapy and Adapted Physical Education

Scientific Contributions to the Understanding of Movement and Physical Activity (1930–1950)

Physical Education Research

CHAPTER TWO

The History of Kinesiology

The history of kinesiology is a fascinating account of 1) the promotion of physical activity throughout the ages; 2) the many scientific contributions to our understanding of the human engaged in physical activity; and 3) the development of kinesiology as a field of study.

STUDENT OBJECTIVES

1. To appreciate how kinesiology evolved into the present field of study.
2. To understand the various contributions to the promotion of physical activity throughout history.
3. The ability to identify important scientific contributions to our understanding of the structure and function of the human body, from ancient civilizations to the present modern era.

Every field of study has a history that has been shaped by important events and people from ancient to modern times. Because a large share of the declarative knowledge in kinesiology is based upon the parent disciplines of biology, psychology, physics, and chemistry, the early history of kinesiology, from ancient times to the 20th century, shares some of the same important events and people that influenced the development of the parent disciplines. As far back as the ancients, there has always been a keen interest in human physical activity. Many of the events and people we will discuss in this chapter have had an impact on the development of the parent disciplines as well as on the development of kinesiology.

I will also trace the interest in physical activity within certain selected early civilizations in Europe, Asia, and North Africa, into the Middle Ages, the Renaissance, and finally into modern times. The civilizations selected represent examples, and are not meant to be all inclusive of the civilizations throughout history that have contributed to knowledge about physical activity. Of particular interest will be the importance of physical activity to the individual and the social fabric of these selected early cultures. We will see that the importance of physical activity in the eyes of civilization has waxed and waned throughout the history of mankind.

Today, physical activity is deemed by many as critical for both our mental and physiological health, but debate and controversy are prevalent. Early and modern scientific contributions to our understanding of the structure and functions of the human body will be highlighted. In this chapter, I will discuss the scientific contributions to the development of the kinesiology field of study from the Renaissance period to around the middle of the 20th century. Much of the specific scientific contributions from 1950 to the present will be incorporated into the subsequent chapters. The emergence of physical education and finally that of kinesiology will also be discussed. At the end of the chapter, the current state of affairs in the field of kinesiology will be commented on and speculations will be made on some of the major issues confronting the future of kinesiology.

Figure 2.1
The mimicking of animals by a baboon dancer of the Bambara tribe of Mali, (adapted from Lonsdale, 1981, Figure 17).

PHYSICAL ACTIVITY AND EARLY HUMANS

All humans and other animals engage in physical activity. The movements of animals are usually associated with various types of activity such as food gathering, chasing prey, or escaping predators and sexual activity. Even in animals, dance-like behavior can be observed. Many insects, birds, animals, and fish display what might be considered ritualized movement patterns resembling dance, and a close analysis of this behavior reveals some underlying purpose. For example, it has been observed that bees perform dance to help direct their coworkers to sites for a new hive. The better the quality of the location, the more prolonged and intense the dance (Meerloo, 1960). Chimpanzees display elaborate dance-like movement during friendly play. Apes have been seen joining hands and moving rhythmically in circles, showering themselves with leaves (Langer, 1951).

The physical activity of early humans centered on food gathering, hunting, obtaining clothes, and shelter and protection from enemies (other humans or animals). Early humans performed ceremonial games and dances related to primitive religious rights and procreational activity. In figure 2.1 a Mali tribal dancer mimics the appearance of the baboon. Like early humans they draw inspiration from local wildlife. Early humans often attributed special characteristics to animals, the baboon, for example, has long been linked to the virtue of wisdom. Another example of our connection with animals has been the adoption of animal mascots for sports teams like the Chicago Bears and the Florida Gators. Displays of physical feats also occurred at these ceremonies and were performed to enhance the selection of mates. Clearly, the *teaching* of certain motor skills such as hunting and fishing must have occurred, but very little physical evidence of this exists. It is believed that early humans observed animal movement very closely.

This assumption is supported by the many paintings and sculptures of animals found in caves. Because early humans were thought to attribute great powers to animals, and even worshipped them, it is not surprising that our early ancestors also danced like them and mimicked their movements (Kraus, Hilsendager, and Dixon, 1990).

PHYSICAL ACTIVITY IN ANCIENT CIVILIZATIONS

Egypt. The sun and water were plentiful, and the soil was rich. The Nile River valley was a perfect place to start one of the world's earliest and influential civilizations. Indeed, ancient Egypt was one of the first cultures to emerge from barbarism into a highly organized and complex society. The Egyptians reached their peak in achievements and influence around 1500 B.C. and gradually declined until 700 B.C., when they were conquered by Assyria. But the contributions of the Egyptians in science, agriculture, engineering, and the arts were extensive.

The Egyptians were also one of the first civilizations to formalize their religious practices into worshipping a host of different gods. Many of these practices were complemented by dance. Trained and professional dancers were common at religious ceremonies. Dance flourished in ancient Egypt, a society that has lasted over 4,000 years. The Nile region is generally considered one of the great centers of ancient dancing (Ellis, 1923).

The ancient Egyptians also practiced other forms of physical activity. Acrobatics were performed mostly by women at feasts to entertain guests. Swimming, symbolized by its own hieroglyphic symbol, was also a popular activity among the Egyptian people. Wrestling developed as part of military training, but it also served as a source of entertainment. Additionally, there is evidence of ball games and bullfighting. Most of these activities were practiced by the lower classes. Interestingly, there is no strong evidence of organized competitions in the above activities. As a population, the ancient Egyptians were not considered particularly athletic people. Later civilizations reaped the many cultural benefits from ancient Egypt. However, athletics and an appreciation of the importance of physical activity to the development of the individual was not one of them (Gardiner, 1930).

Ancient Civilizations in the Far East. Another successful culture and civilization lasting for thousands of years was China. In the early years of the Chou dynasty (1122–249 B.C.), physical training was considered important to the individual's well-being, but subsequent systems and religions such as Confucianism, Taoism, and Buddhism all emphasized a contemplative life. Physical activity was not strongly promoted (Van Dalen and Bennett, 1971). The ancient Chinese developed **kung fu**, involving self-defense, hunting techniques, and military training. The term kung fu literally translates into "martial art." Additionally, dancing was an inherent part of the educational instruction of wealthy Chinese youths. A number of sports were also popular such as boxing, football (ts'u chu), and wrestling. Recreational activities included swimming, hunting, fishing, and flying kites. The ancient Japanese are credited with the development of **jujitsu**, a different type of self-defense system.

The religions of Hinduism and Buddhism were interwoven in the development of India. Hinduism, probably the world's oldest religion, and Buddhism emphasized nonviolence and passivism. However, some religious exercises were practiced such as **yoga**, which was probably the first example of the power of psychological over physical processes. The art of **massage** was practiced in several ancient civilizations of the Far East, including China, Japan, and India. Vigorous manipulation and palpation of the muscles were used to stimulate circulation in the belief that massage maintained and improved one's health (Van Dalen and Bennett, 1971). Today, massage is still used by a number of health care professions such as osteopathic medicine, physical therapy, and chiropractic.

Ancient Greece. Located on the Balkan peninsula of the Mediterranean Sea, ancient Greece became the first European land to be civilized. Partly because of its geography, with many fertile valleys surrounded by high hills and mountains, ancient Greece never became completely unified. Instead, Greece was dotted with several city-states, the most prominent being Athens and Sparta. The antithesis of each other, Athens became a democratic society promoting the full development of the moral, spiritual, intellectual, and physical abilities of the individual, while Sparta became a totalitarian society where citizens were molded from birth into a military caste. The Athenians made significant contributions to the study of art, literature, and science, while the Spartans were mostly preoccupied with the military life.

Much of our understanding of early Greece comes from the two great literary works of Homer, the Iliad and the Odyssey. In these works, the hero Achilles represented the man of action and Odysseus the man of wisdom. One of Homer's motivations was to personify in these two heroes the goals of each Greek citizen, namely, excellence in physical and intellectual prowess. Homer's goals became embedded in later Athenian philosophy that promoted the complete harmony between the physical and the intellect (Harmon, 1936).

Aside from farming, hunting and fishing that were necessary for survival, the Greeks engaged in and vigorously promoted athletics and dancing. It was in ancient Greece that athletic competition became institutionalized, particularly with the establishment of national athletic contests including the Isthmian, Pythian, Nemean, and most well known, the **Olympic festivals**. The Pythian games included contests in poetry, music, and athletics, while the Olympic games were exclusively athletic (Scanlan, 1984). The first recorded beginning of the Olympic games was in 776 B.C. (Slowikowski, 1989). Occurring every four years, the games lasted for five days and were accompanied by ceremonies in honor of the Greek god Zeus. Heavily laden in ritual, the Olympic games were supported by all the city-states in Greece (Slowikowski, 1989). Both poor and aristocratic citizens participated in the Olympics, but year-round training for the games was usually practiced by only the wealthy citizens. Evidence suggests that only young men were allowed to participate in games (Miller, 1985).

Many sporting activities were represented in the Olympics. Appearing in the 18th Olympiad, the pentathlon was established to select the best athlete and consisted of running, jumping, the discus, javelin, and wrestling. Contrary to popular belief, winning events was often accompanied by material rewards (Pleket, 1975). In addition, some of the contests were violent, involving strangulation and the breaking of fingers (Poliakoff, 1987).

Not only were the Olympics and other festivals promotions of physical activity but they also allowed for cultural exchanges between the various city-states. In Athens, great gymnasiums called **palestras** were constructed to help prepare competitors for the games and to provide citizens with exercise facilities. The palestras were also gathering places, where philosophers and citizens could hold conversations. The gymnasium embodied the essence of ancient Greek philosophy of "Healthy mind in a healthy body."

Dance was also extremely popular in Greek culture. Archaeological evidence suggests that the earliest known dances came from the nearby island of Crete between 3000 and 1400 B.C. A wide variety of dances was practiced, such as warlike dances performed with weapons, dances with animal masks and heads, patterned dances for women, fertility dances, and religious dances in honor of the many Greek gods (Lawler, 1964). The Greeks viewed dance not as an isolated event but as a source of entertainment integrated with music and even poetry. The Greeks learned to dance at an early age, apparently from private teachers. Integrated into military training, dance was also used to improve mind-body unity, muscle strength, and discipline.

Many of the great philosophers of ancient Greece stressed the importance of dance and movement and were very interested in the functioning of the human body. **Socrates** (471–399 B.C.) argued that those who honored the gods with beautiful dancing were rewarded on the battlefield. **Hippocrates** (460–370 B.C.), considered to be the father of medicine, wrote at least three works on the joints and bones of the human body. He also described a number of different techniques of joint manipulation such as **succussion**, which was used to help straighten the spinal column, and other manipulation procedures for remedying joint dislocations. These works are thought to serve as a foundation for a number of fields of study involving manipulation of the joints and palpation that are known today as orthopedics, osteopathic medicine, and physical therapy.

Aristotle (384–322 B.C.) believed dance to be an inherent part of general education and that it provided moral training for youths. He also provided several treatises on the analysis of human and animal motion (Chryssafis, 1930). Some of his treatises are thought to have influenced the later thinking of Galileo and Newton (Enoka, 1988).

Erasistratus (304–250 B.C.) developed the theory of pneumatism, which suggested the phenomenon of life was associated with *pneuma*, or spirits, that resided in the body and caused movement. One type of pneuma was called **animal spirits**, thought to be conveyed by the nerves. While his theory was eventually proven wrong, the concept of pneumatism was further elaborated well into the Middle Ages (Singer and Underwood, 1962).

The ancient Greeks were also renowned for their contributions to mathematics, helping lay the foundation of physics and modern-day biomechanics. Including Ptolemy, Euclid, and Pythagoras, **Archimedes** (287–212 B.C.) was considered to be among the greatest mathematicians of ancient times. Besides a number of geometric proofs, he developed the first work on hydrostatic principles that govern floating bodies and the motions used in swimming. Legend has it that Archimedes discovered his principle of buoyancy while taking a bath and then ran naked into the street shouting, "Eureka!" (I have found it!).

In general, dance and physical activity were important elements in the overall Greek philosophy of the interconnectedness of the mind, body, and spirit. Many of the plays written by Sophocles, Euripedes, Aeschylus, and Aristofanes were based on dancing. Groups of performers called the **chorus** danced throughout their plays.

In the later Athenian period, greater emphasis was placed on intellectual development. The preoccupation with year-round physical training preparation for the various festivals disturbed many, including Socrates and his influential student, **Plato** (428–347 B.C.). In the end, Homer's man of action became overshadowed by the rise of intellectualism. However, the fascination with the human body and its capabilities and limitations probably began with the ancient Greeks.

Ancient Rome. To the west of the Balkan peninsula lies the boot-shaped Italian peninsula, the home of the great ancient civilization of Rome. In contrast to the creative and aesthetically minded Greeks, the Romans, who emerged from a small tribal community called Latium near the Tiber River, were a practical, industrious, and ambitious people. After overthrowing the Etruscans to the north in 509 B.C., the Romans unified the entire Italian peninsula by 275 B.C. and conquered the Balkan peninsula in 146 B.C. The Romans established a large empire, encircling all the land and peoples adjacent to Mediterranean Sea, much of Europe, and the island of Britain.

To a large extent, Roman physical training had one purpose—preparation and training for the military. In this sense, Roman philosophy was similar to that of Sparta. In contrast to the Greeks, Romans did not believe in the unity between intellectual and physical development.

There were two general types of athletics in ancient Rome: the spectacula and the palaestra (Scanlan, 1984). The **spectacula** were public athletic contests largely participated in by athletes, sometimes referred to as gladiators, who were either slaves or professionals. These athletic festivals were more for entertainment than for the expression of athletic skill. The **palaestra** was a type of physical education of Roman male youths. The Roman emperor Augustus incorporated these games and contests into the *iuventus*, or youth corps, a program designed for six thousand noble Roman youths. The iuventus involved both formal paramilitary training and athletic contests, consisting of mostly Greek sports such as chariot racing, armed combat, running, javelin and discus throwing, archery, wrestling, boxing, ball playing, and hoop rolling as well as more typical Roman sports of swimming and horseback riding. Augustus was called the *principes iuventutis*, leader of the youth corps (Scanlan, 1984). To emphasize again, the iuventus was designed for certain youths in the noble class. For the average citizen, more emphasis was placed on training youths to become orators rather than athletes (Van Dalen and Bennett, 1971). With the exception of a brief period when Augustus Caesar ruled, athletic contests were only for slaves and professionals, not citizens. In addition, women were not welcome to participate (Scanlan, 1984).

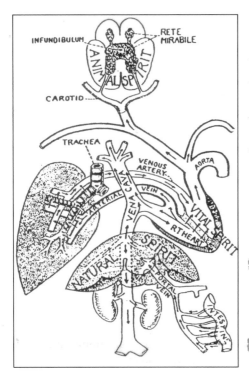

Figure 2.2

Galen's physiological system with three types of "spirits." From Castiglioni, A. (1958). A history of medicine. Translated from the Italian and edited by E. B. Krumbhaar. New York: A. A. Knopf.

Dance was not seriously considered an important physical activity. Probably due to the fact that Rome was filled with such diverse populations and languages, the **pantomime** developed in the theater, with actors dancing and entertaining without the spoken word. As time passed, dance suffered the same fate as the spectacula—it became more of a spectacle to be watched rather than an activity in which to participate. Evidence suggests that dances were even used for gruesome purposes (Lawler, 1964). Some criminals, their clothes treated with chemicals, were forced to dance in arenas before being set afire! After the fall of Rome, the influential Christian church, ruled by the pope in Rome, condemned various Roman practices, including dance. Likewise, dance was not strongly embraced by the church in Rome throughout the subsequent Middle Ages.

In an otherwise dark period for the promotion of physical activity and interest in the functioning of the human body, **Galen** (129–199 A.D.), a Roman physician of Greek origin, strongly advocated physical fitness. He described many ancient muscular strengthening exercises, such as those involving the use of *halteres* or handheld weights (Gardiner, 1955). Galen is considered to be the first known athletic team physician, attending to the wounds and injuries of Roman gladiators. He also served as physician to five Roman emperors over a span of 35 years. Galen made several contributions to our understanding of human anatomy. However, his work was later criticized by Vesalius (and Leonardo da Vinci) during the Renaissance period, partly because Galen's work was based on dissections of monkeys and pigs, not humans (Inglis, 1965). He was apparently the first person to distinguish between motor and sensory nerves and also defined the difference between agonist and antagonist muscles (see Chapter 3). Galen proposed perhaps the first physiological system of the human body that actually included three types of spirits: animals spirits, vital spirits and natural spirits that were located in different internal organs, as shown in Figure 2.2. These spirits had different functions. Elaborating on the earlier views of Erasistratus, Galen believed that muscles became shorter and fatter when activated because they filled up with one of the mysterious substances called animal spirits (Sarton, 1954). We now know that muscles cause movement by contraction of muscular fibers. However, Galen's elaboration of the concept of animal spirits lasted for centuries until it was finally disproved in the 19th century. But Galen's thinking had a great impact on later scientific work during the Renaissance period.

By the 3rd century A.D., the Roman Empire was in decline as a result of famine, internal strife, and war. The empire finally split into two large geographical regions. The eastern or Byzantine Empire was based in the city of Constantinople. The western empire remained centered in Rome. The western part of the Roman Empire collapsed in 476 A.D. with the invasion of the Visigoths, a Germanic people from northern Europe. The eastern Roman Empire continued for several more centuries under the ruling of numerous Christian emperors. In sum, the Romans, like the earlier Spartans, emphasized the importance of physical activity for developing military and combat skills. Outside of this arena, the physical activity involved in athletics or dance was not highly regarded. The Romans tended to view Greek athletics as idle play. Reflecting this notion, Horace, the great Roman poet, felt that the Greek leisure sports were a distraction that could lead youth away from useful military training (Scanlan, 1984).

Recap - Early Humans

SUMMARY

- In addition to movements associated with survival skills such as hunting, fishing, and locomotion, early humans mimicked animals with dancelike behavior.
- Dancing became an art form in ancient Egypt.
- A variety of sports and martial arts originated in ancient Egypt, China, and India.
- Interest in the structure and function of the human body began in ancient Greece. Physical activity in the form of sports and dance was also greatly promoted.
- In ancient Rome, the most highly regarded physical activity was that required for military training.
- Galen, a Greek physician living at the height of the Roman Empire, greatly contributed to our understanding of the human body.

THE MIDDLE AGES

With the collapse of the Roman Empire came the domination of Europe by the Christian church. Particularly in western Europe, which was dominated by the rule of the pope, the Church placed more emphasis on the spiritual and moral fiber of the individual and less on physical development. In the Byzantine Empire, the Eastern Orthodox Christians did not deemphasize the body and physical activity. In Constantinople and many eastern cities, great gymnasiums were built in a tradition and spirit resembling the early Greeks.

The rise in the Christian church in western Europe under the pope's rule was accompanied by the development of two movements that influenced beliefs about the importance of physical activity: asceticism and scholasticism (Wuest and Bucher, 1991). The term asceticism derived from the Greek word *askesis*, which signifies the training required by athletes to prepare for the competition of athletic games. In a sort of ironic twist, some early western Christians applied the term to certain restrictions on daily activities required to lead a virtuous and spiritual life, including limits on certain physical activities! While there were apparently many forms of asceticism (Kaelber, 1998), one version expressed belief that the human body was an instrument of sin, possessed by the devil, and should be punished. In addition to this form of asceticism, **scholasticism** emphasized the importance of facts as the basis for knowledge, truth, and wisdom. The physical development of the body was deemphasized in favor of the mental and spiritual life. The rise of these two movements was largely supported by the Christian church in western Europe, and many physical activities, including several types of dances, were discouraged. The epitome of this view in the early Middle Ages can be summarized in a quote from St. Bernard:

2 types of movements

> *Always in a robust and active body the mind lies more soft and more lukewarm and on the other hand, the spirit flourishes more strongly and more actively in an infirm and weakly body.*
>
> (from Coulton, 1923)

However, recent scholars have disputed the view that the Christian church was totally against the development of the body and physical activity (Twietmeyer, 2008). Later monastic life, particularly in western Europe, embraced more moderate forms of physical activity, particularly in terms of manual labor.

Besides the Church, another set of forces had much to do with life during the Middle Ages. Three systems developed in Europe following the collapse of the Roman Empire. Feudalism was a government and political system based on hierarchical control with the king as the supreme ruler. Manorialism was the agricultural and economical system that supported feudalism. Finally, there was **chivalry**, a moral and social code designed to help the male citizen properly serve and obey his lord or master. The code of chivalry demanded certain laws of etiquette, the protection of women and the poor, and the defense of the Church.

3 systems

The sons of noblemen during this feudal era had essentially two options for their careers: either to join the Church as clergymen, requiring religious and academic training, or to become knights, which involved physical, social, and military training (Wuest and Bucher, 1995). The latter option required training to begin at the age of seven as a *page* under the supervision of women in the lord's castle and involved the learning of social etiquette and household skills. The young page would also practice different types of physical activities such as boxing, running, fencing, jumping, and swimming. At the age of 14, the boy became a *squire* under the supervision of a knight. A variety of physical activity skills were emphasized, including hunting and horseback riding. If the squire proved his worthiness and skill, he became a *knight* at the age of 21 in a ceremony known as the *accolade*.

The development of the knight as a protector of the feudal system parallels, to some degree, the training involved in the earlier Roman *iuventus* or youth corps designed by Augustus to protect the Roman Empire. Eventually, the Church in western Europe heavily influenced the chivalry code to help its expansion throughout Europe, western Asian and northern Africa. Chivalry was perhaps the strongest factor in the rekindling of interest and promotion of physical activity during the Middle Ages. The promotion of tournaments and jousting allowed for the displays of various military skills among the knights such as horsemanship, archery and jousting. Interestingly, these events were for both amusement and display of skill, even though the events themselves were often dangerous.

Dance also became increasingly popular during the Middle Ages, despite not being enthusiastically embraced by the Church in western Europe. Because the Romans had often used dance for perverted entertainment, the Church banned theatrical entertainment, including dance. The first Christian Roman emperors, as early as 300 A.D., denied the rite of baptism to those citizens associated with the circus or pantomime. The Olympic games were banned by the Christian leader Theodosius I in A.D. 393 or his grandson Theodosius II in A.D. 435.

The Church tried to make a distinction between religious and theatrical (or sensual) dancing. For example, the monks participated in many religious dances that evolved from earlier pagan festivals. The festivals influenced many of the customs of the Christian church in western Europe, such as the ringing of bells to help ward off evil spirits. In spite of being generally condemned by the Church, theatrical dancing was too great a force among the masses and continued to flourish during the Middle Ages and into the Renaissance period.

Because it was a reflection of the beliefs of the common citizen, the **Dance of Death** was a particularly strong force in the promotion of physical activity. Certain to have originated with primitive peoples, this dance had to do with the respect people had for the power of death. The Dance of Death was also a reaction against the ascetic preaching of the Church, primarily in western Europe. In its extreme form, the Dance of Death, also called St. John's Dance, evolved into witch dances paying homage to the devil. This dance involved frenzied leaps and turns, uncontrollable screaming, and even foaming at the mouth (Kraus et al. 1991)!

Recap - Middle Ages

SUMMARY

- Chivalry strongly contributed to the promotion of physical activity and rekindled an interest in sports and other forms of recreational activity.
- Throughout the Middle Ages, the Christian church, centered in Rome, vehemently opposed dancing unrelated to religious ceremony. However, dance remained very popular with the masses.

THE RENAISSANCE PERIOD AND THE BIRTH OF SCIENTIFIC INQUIRY

The term renaissance means awakening or rebirth. The Renaissance served as the transition period between the Middle (or Dark) Ages and the modern era. The Renaissance period revived interest in the cultural achievements of Greece and Rome. Freedom of thought and expression and the expansion of scientific achievements eventually led to the establishment of universities all over Europe. Commerce and travel significantly increased throughout Europe during the Renaissance. It is generally believed that the Renaissance first started in Italy around 1350 A.D., moved to France in the middle 1400s A.D., and finally settled into England around 1500 A.D.

Promotion of Physical Activity and Exercise. Three major philosophical movements occurred during the Renaissance period: humanism, moralism, and realism. Each movement had differing views on the promotion of physical activity.

3 major philosophical movements

Humanism

Humanism was an attempt to resurrect the classical literature of the Greeks and Romans. Humanists generally believed in the development of both the physical and the intellect similar to earlier Athenian philosophy. **Mapheus Vegius** (1405–1458) and **Vittorino de Feltre** (1378–1446) both believed that moderate exercise could positively affect the mental state. **Petrus Paulus Vergerius** (1349–1420) promoted games and exercise as a preparation for the military.

While not strongly embraced by the humanist philosophers, dance continued to be popular with peasants as well as with nobility. Beginning in the Middle Ages, a variety of **court dances** developed and were further refined during the Renaissance. The peasant dances tended to have large movement and wide stepping figures, while the court dances of the nobility, developed as part of the chivalric way of life, were typically much less active. Originating in Italy, **ballet** became a highly developed art form in France through the efforts of King Louis XIV. It became known as theatrical storytelling through dance (Van Dalen and Bennett, 1971). Most of the major elements of what is termed today as *classical ballet* were developed by the Frenchman **Jean-Georges Noverre** (1810–1840), considered by some as the father of ballet. Other early influential people in the development of ballet were the Italian **Carlo Blasis** (1787–1878) and the Frenchman **Jules Perrot** (1810–1892).

Moralism

The development of capitalism, the rise in nationalism in Europe, and the development of the Protestant movement against the Catholic church were all significant events during the Renaissance. These events signaled the birth of a new type of individualistic spirit that clearly ran against traditional Christian philosophy in western Europe. A counter-reformist or *moralistic* viewpoint developed in response to Puritanism, which then spread to the New World in North America. It is important at this time to distinguish between two types or classes of physical activity that were either condoned or discouraged during the Renaissance period. Manual labor was a type of physical activity that was strongly embraced by the various Protestant religions and sects. **Martin Luther** (1483–1546), the leader of the Protestant Reformation, saw certain types of physical activity as important in counteracting the negative effects of gambling and drinking, as well as promoting good health (Wuest and Bucher, 1991). However, sports, games, and exercise (other than that related to manual labor) had become associated with the nobility and generally discouraged by the moralists. Dance was generally regarded a private matter by

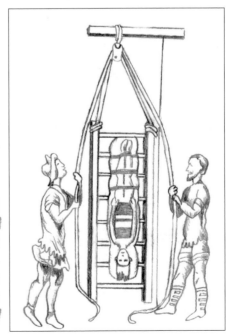

Figure 2.3
The technique of "succussion" for treating spinal dysfunction.

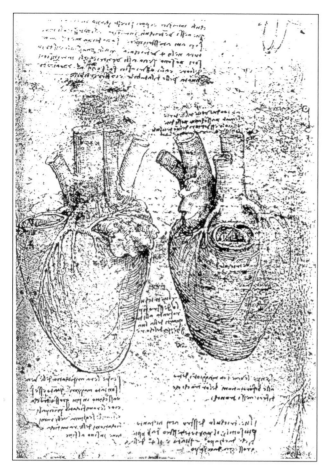

Figure 2.4
Drawings of the heart by Leonardo da Vinci (from Singer and Underwood, 1962, Figure VIII, originally reproduced from Quaderni d'anatomia, II fol. 3V).

Figure 2.5
Drawing of the skeleton by Vesalius (from Singer and Underwood, 1962, Figure X, origin from A. Vesalius, De Humani Corporis Fabrica, Basle, 1543).

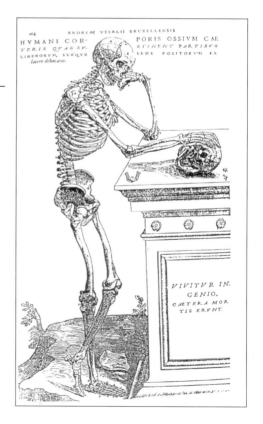

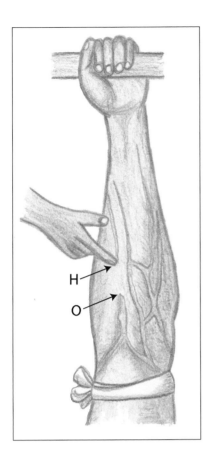

Figure 2.6

One of Harvey's demonstrations of circulation and the workings of the valves in the veins. If the upper arm is bandaged, the valves are shown as nodes on the swollen vein. If the finger is pressed along a vein, in a direction away from the heart from one node to the next (e.g., O to H), the section OH will be emptied of blood. It will remain empty, since the valve at O does not permit blood to flow away from the heart, (adapted from Singer and Underwood, 1962, original from W. Harvey, De Motu Cordis, Frankfurt, 1628).

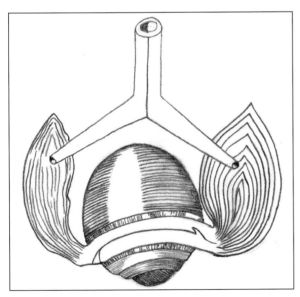

Figure 2.7

According to Descartes (1664), animal spirits flow to and from the eye muscles. Through a myriad of tubes and valves, the animal spirits would be directed in and out of the appropriate muscles, causing them to inflate and deflate. His theory was proven wrong by Swammerdam and other scientists, (adapted from Jeannerod, 1985, original from R. Descartes, Traite de l'Homme, 1664) the Puritans. Mixed dancing between men and women and tavern dancing were prohibited and offenders were severely punished (Kraus et al. 1991). Humanists and moralists were generally preoccupied with earlier classical views or the teachings of either the Catholic or Protestant religions.

Realism

In dramatic response to these traditional teachings was the rise in realism, signaling the birth of scientific inquiry. A tremendous increase in our understanding of the structure and function of the human body resulted from the birth of the realist philosophy.

Most people associate the name **Leonardo da Vinci** (1452–1519) with great artwork, sculpture, and invention, but he also contributed to our understanding of the structure of the human body. Da Vinci provided the first detailed descriptions of how the body moves in relation to the forces of gravity.

Our understanding of human anatomy greatly improved with the work by **Andreas Vesalius** of Brussels (1514–1564), who shook the foundations of Galenian physiology. Vesalius performed dissections on both animal and human bodies. He published a landmark work entitled *The Fabric of the Human Body* in 1543, containing beautiful illustrations of the various structures and organs in the human body. Some have argued that no dramatic changes have occurred in our understanding of the structure of the human body since Vesalius (Singer and Underwood, 1962).

Another profound jolt to traditional beliefs came from **Galileo Galilei** (1564–1642) in Italy, who was convinced that nature could be described in mathematical terms and understood by mortal man. He demonstrated that the acceleration of a falling body was not proportional to its weight. Galileo's work laid the groundwork for the development of classical (Newtonian) mechanics and thus to modern-day biomechanics. One of Galileo's students, **Alfonso Borelli** (1608–1679) greatly contributed to the area of muscular mechanics. Borelli is regarded as the father of modern biomechanics of the locomotor system (Steindler, 1935).

William Harvey (1578–1657) formulated the concept of circulation of blood through the body. As shown in Figure 2.6, he performed many experiments that led to our present day understanding of circulation. **Niels Stensen** (1638–1686) declared that the heart was merely a muscle and not the seat of emotion. He published a very important book on muscle function entitled *Elementorum Myologiae Specimen*, where he distinguished between tendon and muscle tissue and laid out the present-day concept of muscle contraction

René Descartes (1596–1650), a French philosopher, mathematician, and scientist, used Galen's earlier concept of animal spirits to lay out a "hydraulic" explanation of movement control based on activation of agonist and antagonist muscles. While his theory was wrong, it stimulated much research and eventually led to the development of new theories of motor control based on propagation of electrical signals in nerve pathways (Jeannerod, 1985).

The greatest scientist in the Renaissance period was probably the Englishman **Isaac Newton** (1642–1727). The co-inventor of the calculus, Newton formulated his famous three laws of motion in the *Principia*, which formed the foundation of physics and modern biomechanics (see Chapter 5, Biomechanical Foundations). No single scientist has influenced the scientific community and indeed the way we view nature more than Newton (Fauvel, Flood, Shortland, and Wilson, 1988).

There were other important realist philosophers who advocated the importance of physical activity as well as exercise in the schools (see Table 2.1). However, it was during the next two centuries that the advocacy of physical activity in the schools, and indeed, for the general populace, began to be realized in a structured way.

SUMMARY

- Elaborate dance forms evolved, such as ballet; however the Puritan philosophy discouraged recreational dancing by the masses.
- The Renaissance witnessed the birth of scientific inquiry and the renewed interest of the earlier works of the Ancient Greeks and Romans.
- Scientific investigations into the structure and function of the human body were begun by several Europeans.

THE 17TH, 18TH, AND 19TH CENTURIES AND THE AGE OF ENLIGHTENMENT

The French and American revolutions served to stimulate an increase in nationalism all across Europe and the world. This increase in nationalistic spirit impacted scientific communities and educational systems in many countries. During the next 200 years, further promotion of physical activity and exercise as a means of improving health occurred, along with a tremendous increase in the scientific study of human movement.

The promotion of physical activity and exercise in Europe. One of the greatest philosophers in this era was **Jean-Jacques Rousseau** (1712–1778), who argued that people have certain inalienable rights such as freedom from government oppression and slavery. He was also a strong advocate of child education and suggested that children pass through several developmental stages, each requiring certain amounts and kinds of physical activity. For example, in the first period from birth to five years called the *animal stage*, Rousseau believed that the child needed to engage in vigorous physical activity. (Many parents today would probably agree with the name of this stage!) During the next *savage stage*, from ages five to twelve, various games and sports were required, yet academic instruction should be minimized until later in adolescence when the child's power of reasoning was deemed to emerge.

Many of Rousseau's proposals were considered radical but they had a great impact on later educational leaders such as **Johann Guts Muths** (1759–1839) from Germany, who is considered the *grandfather of modern physical education* (Van Dalen and Bennett, 1971). Guts Muths published several books on gymnastics and swimming, some of which were translated into different languages. His physical education classes included both gymnastics and manual labor activities such as cabinetmaking and gardening!

Another educational leader and devout German nationalist, **Friedrich Jahn** (1778–1852) was instrumental in the development of the *Turnplatz (play or exercise ground)*. To Jahn, the physical activities in the Turnplatz were designed to promote both physical development and national pride. Within a large rectangular area, the Turnplatz contained various types of gymnastic equipment and running tracks. Children wore common uniforms to remove distinctions of social class and promote social equality. Youths were also assessed a fee to help maintain the conditions of the exercise area.

While never becoming part of the school systems in Germany, the Turnplatz served as the foundation of the National Union of German Gymnastic Societies, a state-supported organization. By the turn of the century, membership grew to over a million participants and over 11,000 local societies were formed within Germany (Van Dalen and Bennett, 1971). National exhibitions became extremely popular as thousands of participants displayed their gymnastic talents in front of huge audiences. Supporters of the so-called Turner system emigrated to the United States and other parts of the world to develop similar societies. Thus, the Turner system became one of the most important and influential promoters of physical activity and exercise for the masses.

A pioneer in the development of physical education on the European continent was **Franz Nachtegall** (1777–1847), who introduced compulsory physical education in the public schools, probably for the first time. He also was the first to offer teacher-training courses in physical education. **Adolph Spiess** (1810–1858) successfully introduced gymnastics into the German school systems for both boys and girls. Later, **Carl Diem**

(1882–1962) greatly contributed to the establishment of what we know today as "physical education" in the schools of Germany.

In Sweden, another type of gymnastic system was developed under the guidance of its founder, **Per Henrik Ling** (1776–1839). Using much persuasion, Ling convinced the Swedish government to establish the Royal Central Institute of Gymnastics at Stockholm in 1813 and appoint him as its director. At the institute, Ling developed and practiced what has been termed **medical gymnastics**. Medical gymnastics was designed for the relief of certain physical disabilities and consisted of exercises by the patient alone or with the aid of a trainer. Ling believed that it was not necessary to center physical training around pieces of equipment and apparatuses

Table 2.1
Other Realist Philosophers with Views on Health and Physical Activity

- **Richard Mulcaster** (1531–1611) advanced several ideas about how physical education should be taught and described the type of environment it should be taught in.
- **Francis Bacon** (1561–1626) held strong views about the importance of exercise in preventing disease (Spedding, Heath, and Ellis, 1901).
- **John Comenius** (1593–1611) emphasized that play and physical activity were important in the proper development of children (Quick, 1904).
- **John Locke** (1632–1704) supported the early Greek view that a strong body and mind were the key to optimal health.

such as in the German system, and he designed free exercises that allowed for exercise of specific muscle groups without equipment.

The movement of physical education in the schools spread from Germany and Sweden and throughout Europe in the 19th century. Both the Ling Swedish system and Jahn German system eventually were introduced in both England and the United States, where they were modified. In England, **John Maclaren** (1820–1884) from Scotland developed gymnastic exercises and wrote influential books on the subject that were used by the military.

Another important development for the promotion of physical activity in the 19th century was the resurgence of organized sports in Europe. **Pierre de Coubertin** (1863–1937) was able to unify hundreds of French athletic clubs and created sports competitions between France and England. Successful at this enterprise, de Coubertin went on to organize the revival of the Olympic games in 1896 in Greece.

THE PROMOTION OF PHYSICAL ACTIVITY AND EXERCISE IN THE UNITED STATES.

The Beginning of Physical Education in the Schools

During the late 18th and the 19th centuries, a number of individuals in the United States developed and promoted a variety of concepts, methods, and techniques related to the participation in physical activity. **Benjamin Franklin** (1706–1790) was perhaps the first American to advocate physical training in the schools in a manuscript published in 1749 (Van Dalen and Bennett, 1971). He purportedly lifted weights to improve physical strength! (Smyth, 1907; Van Doren, 1938).

Catherine Beecher (1800–1878) developed free exercises for both boys and girls and strongly advocated physical training in the schools and in the home (Beecher, 1856). Borrowing some of Beecher's ideas, **Dio Lewis** (1823–1888) compiled a number of light exercises that could be accompanied with music and performed by children and adults. Lewis went on to found the Boston Normal Institute for Physical Education and published the first American journal in physical education (Weston, 1962).

Dr. Edward Hitchcock (1828–1911) started the first major college program in physical education at Amherst College in 1861, and taught at the college for over 50 years. His program, largely based on available scientific principles of the time, served as a model for other educational programs. Hitchcock also developed the early techniques of anthropometrics, the study of human body measurement.

Dudley Sargent (1849–1924) became the director of the Hemenway Gymnasium at Harvard University in 1879. He incorporated medical and anthropometric testing (body measurements) in relation to prescription of certain exercises for the improvement of physical fitness and health. To Sargent, the goal of physical training was not necessarily the development of the physique, but rather the improvement of function. A number of future successful leaders of the physical education movement graduated from Sargent's teacher training school in Cambridge.

Various systems for the promotion of physical activity evolved over the next several decades in the United States, largely influenced by the various European systems. The German gymnastic system emphasized strength development. Its beginnings in the United States were influenced by a German Turner, Charles Beck, who taught Turner exercises at the Round Hill School in Northampton, Massachusetts. Sports were promoted through the English system. The sports of rugby, cricket, rounders, soccer, tennis and table tennis, water polo, and field hockey were all developed in England and brought to the United States. The Swedish gymnastic system, brought to the United States by Hartvig Nissen in 1883, emphasized the precise execution of movement, drills, specific exercises for particular pieces of exercise equipment, and posture-correcting movements. A system developed by American Edward Hitchcock focused on hygienic practice using exercises on light apparatus equipment. The system advocated by American Dudley Sargent emphasized individualized exercises and anthropometric measures.

The debate over which system should be used led to what some term **"The Battle of the Systems"** (Lumpkin, 2011). This debate led to an important conference in 1889 called the Boston Conference on Physical Training financed by Mary Hemenway and directed by Amy Morris Homans. During this conference, the advantages and disadvantages of the German, Swedish, and American systems were discussed. The conference helped to increase exposure of the various systems to the conference participants and eventually to the American public.

The various systems that were promoted had a large impact on increasing public attitude on the importance of physical activity. Normal, or teacher training schools, developed that were designed to train teachers in the various systems. Six such schools were started in the 1880s, such

Figure 2.8

Example of the exercises used in the Turner system.

Figure 2.9
George Wells Fitz (1860–1934). Head of the Department of Anatomy, Physiology, and Physical Training at the Lawrence Scientific School, Harvard, from 1891 to 1899 (from Kroll, 1971, Figure 6.1. Original photograph from the Harvard (Class) Album, 1898).

as the Sargent Normal School in Boston and the first professional physical education program in 1885 at Oberlin College in Ohio directed by Dr. Delphine Hanna. Eventually, state legislatures throughout the United States began passing laws requiring a certain amount of physical training in the public and private schools. The first major law was passed in the state of Ohio in 1892 (although California had passed a similar law in 1866, which was later revoked).

Physical Education as a Profession

One of the first advocates for physical education as a unique profession was **Luther Gulick** (1865–1918), a New York University Medical School graduate and former student of Delphine Hanna (Park, 1989). Gulick believed that physical education was one of the few fields that could provide profound knowledge of man through physiology, anatomy, psychology, history and philosophy. One of the major goals of physical education, according to Gulick, was the establishment and maintenance of the physiological, psychological, social, and moral development of the individual. Gulick also organized the Academy of Physical Education to bring together scholars who were doing research on physical training.

Central to the advancement of physical education as a scholarly field of study was the necessity that it be based on strong scientific grounds. Two prominent names advocating such a necessity

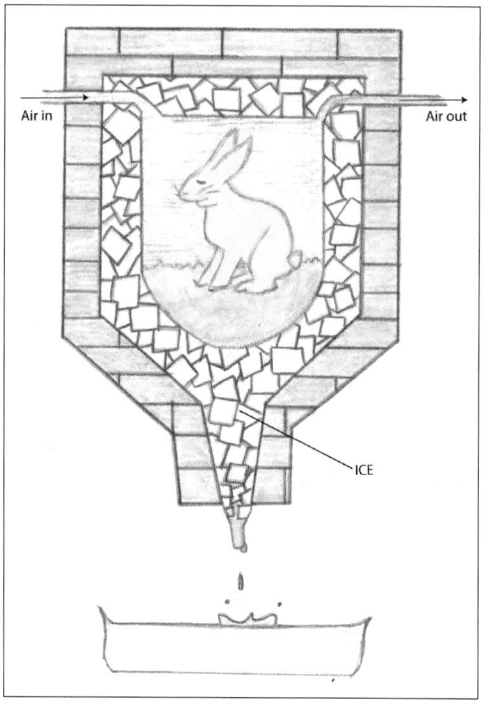

Figure 2.10
Lavoisier's calorimeter of 1780. The animal's body heat melts the ice. Measuring the amount of water formed allows estimation of the heat produced, knowing that 80 kcal of heat melts 1000 g of ice. The ice water surrounding the calorimeter provides a perfect insulation because it is at the same temperature as the ice in the inner jacket around the animal's chamber. Therefore, the insulation will neither add heat to nor take the heat from the calorimeter, (adapted from Brooks and Fahey, 1984, Figure 3-4, based on original sources and M. Kleiber, 1961).

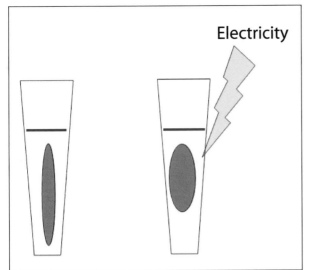

Figure 2.11
Swammerdam's experiments on the nature of muscular contraction using isolated frog muscle. On left, the muscle is suspended in saline solution. Following electrical stimulation, on right, the muscle shortens and fattens, but the level of the saline solution remains the same because the volume of the muscle did not change, contrary to animal spirit theory advocated by Descarte.

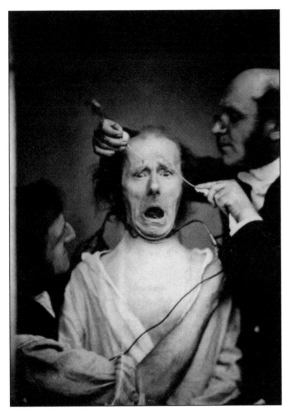

Figure 2.12
Duchenne investigating the effect of electrical stimulation of the left frontalis muscle (from Wikipedia, copyright expired).

during this time period were **Thomas Wood** (1864–1951), another Delphine Hanna student, and **George W. Fitz** (1860–1934). Wood believed that the various systems of physical training mentioned above should be replaced by a *science* of physical activity. Following the lead of Luther Gulick, Wood would later expand the aims of physical education beyond just physical development.

Similar arguments for a strong scientific basis were made by another important figure in the development of physical education as an academic field of study. A young physician named George W. Fitz became president of the department of physical education within the National Education Association. Following his medical training at Harvard, Fitz also began the first research laboratory in physical education at Harvard in 1892. Evidence of this fact comes from the many research articles he published on a variety of topics such as reaction time, muscular cramps, and measures of posture. Under his direction, the Department of Anatomy, Physiology, and Physical Training at the Lawrence Scientific School of Harvard University began the first four-year academic degree in physical education in 1891 (Kroll, 1971). He is considered by some as the *father of exercise physiology in the United States.*

The formal birth of physical education in the United States can be equated with the formation of the **Association for the Advancement of Physical Education,** later to become the *American Alliance for Health, Physical Education, Recreation and Dance.* Over 60 individuals with varied backgrounds were present at this meeting held on November 27, 1885 at Adelphi Academy, organized by **William G. Anderson** (1860–1947). The organization's first president was Edward Hitchcock with Dr. Dudley Sargent as vice president.

Other major promotions of physical activity in the United States during this time period came from the Young Men's Christian Association (YMCA). Founded earlier in Britain, the YMCA advocated a blend of spiritual, intellectual, and physical development.

There was also a tremendous increase in interest in sports, particularly at the colleges and universities in the United States. The first intercollegiate competition occurred in 1852 between Harvard and Yale universities in crew racing. By the beginning of the 20th century, most universities and colleges had athletic associations or departments that housed a variety of sports. Early on, sports were entirely a student enterprise, but due to certain ethical abuses and the formation of conferences (e.g., Ivy League, Big Ten, etc.), faculties eventually took control of athletic enterprises on the campuses.

Scientific contributions to the understanding of human movement in the 18th and 19th centuries. During this period, research and scholarly work on the study of the human body and the study of human motion began using scientific procedures. **Lavoisier** (1743–1794), a brilliant French nobleman, helped develop the foundations of modern chemistry. As illustrated in Figure 2.10, among his many accomplishments was the use of a calorimeter that estimated the amount of heat absorbed or released by a body. He greatly contributed to our understanding of oxygen and how it is used by the human body.

In the late 17th century and 18th century, **Jan Swammerdam** and a group of European scientists performed experiments that dispelled the concept of "animal spirits," leading to modern concepts of how human movement is produced.

Guillaume-Benjamin Duchenne (1806–1875) conducted many studies using electrical muscle stimulation. His work led to the development of modern neurology and better understanding of the conductivity of neural pathways.

In the next 200 years, scientists developed measuring techniques that analyzed both physiological and psychological functions. Conceptual breakthroughs in the understanding of the human engaged in physical activity also occurred that affected the development of the fields of anatomy, physiology, biology, chemistry, psychology, sociology, physical education, and eventually, kinesiology. Table 2.2 summarizes some of the major accomplishments of scientists and scholars during the period.

Table 2.2

Contributions to the Field of Kinesiology by Some of the Major Scientists and Scholars in the 17th, 18th, and 19th Centuries

Contributions	Contributor(s)
Experiments dispelling "animal spirits" theory	Swammerdam (1637–1680)
Calorimetry—measures body metabolism	Antoine Lavoisier (1743–1794)
Muscle anatomy and physiology	John Hunter (1728–1793)

Electrical changes during muscle contraction	Luigi Galvani (1737–1798)
Muscle function	Guillaume Duchenne (1806–1875)
Identification of sensory and motor pathways	François Magendie (1783–1855) Charles Bell (1774–1842)
Reflex arc	Marshall Hall (1790–1857)
Human locomotion Center of gravity concept Concept of bones as levers and many other biomechanical principles	Ernst Heinrich (1795–1878) Wilhelm Eduard (1804–1891) Eduard Friedrick Wilhelm Weber (1806–1871)
Muscle function Isometric and isotonic contractions	Adolf Fick (1829–1901)
Muscle physiology Concept of muscle hypertrophy	Wilhelm Roux (1850–1924)
Neural transmission speed	Louis Ranvier (1835–1922)
Contribution of the brain to voluntary movement Brain disorders and movement control	John Hughlings Jackson (1835–1911)
Theory of evolution Understanding relationship between animal species	Charles Darwin (1809–1882)
Psychology of behavior Psychophysiology Perception of sensation	Wilhelm Wundt (1832–1920)
Speed of mental processes	F. C. Donders (1818–1889)
Psychology and behavior Ideomotor concept	William James (1842–1910)

Speed and accuracy of movement, visual control of movement	Woodworth (1869–1962) Hollingworth (?)
Plateaus in motor learning	Bryan and Harter (1897; 1899)
Social facilitation in bicycle racers The first experiments in sport and exercise psychology	Triplett (1861–1931)
Photographic techniques in animal and human motion	Edward J. Muybridge (1821–1904) Etienne-Jules Marey (1830–1904)
Development of conflict theory Struggles between racial, ethnic and socioeconomic groups	Karl Marx (1818–1920)
Inventor of kymogragh used for measuring blood pressure and other physiological processes	Charles Verdin (invented in 1892)
Inventor of kymograph used for measuring movement	E. Zimmermann (1907)

Recap – Advocates in the US

SUMMARY

- The importance of physical activity for the masses was recognized by educational leaders in Europe.
- Several "systems" of gymnastics developed in Germany and Sweden.
- The Olympic games were revived in 1896.
- Physical education in the schools developed in the United States.
- The YMCA and the Association for the Advancement of Physical Education greatly promoted the importance of physical activity for the developing child.
- Scientific research in the study of the human body and human motion began.

THE 20TH CENTURY

The early 20th century saw the rapid rise of the United States as an international power, both economically and politically. The successful outcome of the earlier Spanish-American war resulted in territorial acquisitions in the Caribbean as well as in the Pacific. New foreign economic markets also opened up, creating greater investment opportunities for American businesses. The economic boom that resulted also allowed for greater expenditures in education and, particularly during the 1920s, an incredible number of schools were built and college enrollments dramatically increased throughout the United States. A tremendous interest in education ensued, largely brought about by the likes of a number educational psychologists such as William James, Stanley Hall, Edward Thorndike, and John Dewey. The views of these men, particularly that of John Dewey, helped shape the general educational philosophy of the United States for the entire century.

The viewpoints of James, Hall, Thorndike, and Dewey also influenced the promotion of physical activity in the schools and played a great role in the further development of physical education as a genuine field of study. The next section explores a number of factors, including Dewey's viewpoints, that affected the promotion of physical activity in the United States during the first half of the 20th century. Scientific contributions to our understanding of human movement during this time period will also be summarized.

1900 to 1930 and the Rise of Physical Education. The educational philosophy of **John Dewey** (1859–1952), expressed in his book *The School and Society* (1899), dominated the early part of the 20th century. Dewey felt that the school should be a microcosm of life itself, and that it should provide the kinds of experiences the child is apt to find in society. Thus, the school's curriculum, rather than being rigid, should be flexible in order to adapt to changes in society. Dewey's philosophy was ambitious, placing as much emphasis on the child's personal and social skills as it did on the teaching of traditional subject matter. The impact of Dewey's thinking was immediate. By 1918 health and safety, mastery of tools, techniques, spirit for learning, worthy home membership, vocational and economic effectiveness, citizenship, worthy use of leisure, and ethical character were considered viable general aims of education (Van Dalen and Bennett, 1971)

In which subjects in the school's curriculum should these aims be achieved? Many educators at the time, such as Thomas Wood and his student, **Clark Hetherington** (1870–1942) argued that through proper instruction in certain physical activities, many social as well as physical skills could be learned in physical education classes provided by the schools. These ambitious aims marked the beginning of a new philosophy of physical education, dubbed the **New Physical Education** (Wood and Cassidy, 1927). These aims were to be achieved using proper teaching methods, taking into consideration the teacher's knowledge of physiology, educational psychology, and sociology. As opposed to the regimented activities of the German and Swedish gymnastic systems, which focused primarily on physical development, the New Physical Education concentrated on the learning of games, dance, and sports, where a number of desirable cognitive and social attributes could theoretically be acquired such as obedience, subordination, self-sacrifice, cooperation, friendliness, a spirit of fair play and sportsmanship, self-confidence and self-control, mental and moral poise, good spirits, alertness, resourcefulness, courage, aggressiveness, and initiative. The learning of proper health and nutritional practices, the development of neuromuscular skill, and lifelong exercise practices were additional aims of this new and broadly expanded concept of physical education.

By 1930 nearly 40 states had passed laws requiring physical education in the public schools. With this demand, the need for quality physical education teachers increased and, as a result, the number of physical education teacher training programs reached 150 by 1930 throughout the United States. Many of these programs became affiliated with colleges and universities leading to undergraduate degrees in physical education. The first master's degree in physical education was granted in 1910 at Teachers College of Columbia University, New York, through the program started earlier by Thomas Wood. The first doctoral programs in physical education began at Teachers College and New York University later in 1927. By 1933, nearly 30 colleges throughout the United States offered graduate programs in physical education (Van Dalen and Bennett, 1971). It is also important to note that the term "physical education" was beginning to replace physical training in the context of all of these laws, signifying evidence of its growing acceptance in school curriculum (Aller, 1935).

A number of points must be made regarding the development and general acceptance of the New Physical Education concept in the early 20th century, particularly in the United States:

- The importance of physical activity in the schools became generally accepted by the populace.
- Physical education as a field of study became legitimized.
- The teaching of physical education would require considerable knowledge of physiological, cognitive, and social processes.

New Physical Education

• The field of physical education became aligned with education.

This last point was particularly important in the future direction of physical education as a field of study. Because many of the field's early advocates were from medicine, physical education might have aligned itself with the medical profession. Such an alignment was not to be. In 1891 the National Education Association first recognized physical education as a legitimate field of study, and from then on education, and not medicine, was considered the parent discipline of physical education. It is interesting to speculate how physical education would have evolved if it would have aligned itself with the field of medicine.

OTHER PROMOTERS OF PHYSICAL ACTIVITY

Sports and Athletics

The evolution of physical education as a field of study and as a legitimate subject in the school curriculum was perhaps the major promoter of physical activity in the early 20th century. However, there were other important developments as well. The interest in athletics, particularly in the universities, continued to rise. Throughout the early 20th century until World War I, the number of sports incorporated into collegiate athletics grew. An incredible number of facilities for basketball, baseball, and football were built during this time period. Football as a spectator sport became so popular that huge stadiums with seating capacities over 50,000 were commonplace, particularly in the Midwest. College football attendance skyrocketed to well over 10 million in the year 1930.

Not only was this type of physical activity popular, but it also became "big business," as universities began struggling with how to manage this new source of income. Worries about professionalism, academic standards, injuries and unethical sporting behavior (particularly in football) eventually led college administrators and concerned educators to organize a meeting in 1905 in New York. Delegates from thirteen colleges attended and decided to allow the continuation of college football, but also to develop an organization for college sports. A few weeks later, another meeting with representatives from 62 colleges was organized and subsequent discussions led to the development of the **National Collegiate Athletic Association (NCAA) in 1910.** The NCAA and the American Athletic Union (AAU) became instrumental in the regulation of amateur sports in the United States.

Intramurals

Another major contributor to the promotion of physical activity was the development of **intramural** sports. The general goals of intramurals were to provide several sport and recreational activities to the greater college student body and to encourage the pursuit of physical activity. The first intramural programs started at Ohio State University and the University of Michigan in 1913. Eventually the concept of intramurals spread to the junior and senior high schools throughout the United States.

Organizations

Other continuing contributors to the promotion of physical activity were the YMCA, the Young Women's Christian Association (YWCA), and the Boy Scouts (1910) and Girl Scouts (1912) of America.

Recreation

The play and recreational movements in America originally focused on activities for children. The Playground Association of America started in 1906 with Dr. Luther Gulick as president and President Theodore Roosevelt as honorary president. Gradually, an emphasis of recreation for adults was included. Many important physical educators were early supporters of the new recreation movement and later physical education and recreation would effectively join hands in the same national organization. As leisure time among working adults increased, city after city, starting with Chicago, began building parks for recreational and sporting activities. By 1930, nearly 1,000 cities throughout the United States had established recreational programs for their citizens.

Dance

The modern dance movement in the United States was started in the early 20th century by **Isadora Duncan** (1878–1927). This type of dance focused more on natural movement and as such provided a sharp contrast to the movements performed in the ballet. The teachings of **Gertrude Colby** at Columbia University also emphasized natural expression. The scientific study of dance was provided at Barnard College by **Bird Larson**, who advocated the importance of anatomy and biomechanics. **Margaret N. H'Doubler** developed the first dance major program at the University of Wisconsin in 1926.

Physical Therapy

The profession of **physical therapy** came into existence after the first World War, stimulated by the vast number of returning soldiers with physical handicaps and neuromuscular injuries. The American Physiotherapy Association was formed in 1921 and had its first annual meeting in 1922 at the School of Physical Education in Boston. **Mary McMillan** (1880–1959), who received her physical therapy training in England, served as the first president of the association (Hazenhyer, 1946). It is important to note that many of the early leaders of this new profession were physical educators.

SUMMARY

Recap-Other Promoters

- The early part of the 20th century saw a rapid increase in intercollegiate sports in the United States.
- Physical education became incorporated in school and college curriculums.
- The growth of intramurals and recreation and other agencies helped to raise the consciousness of the individual citizen regarding the importance of physical activity.
- The modern dance movement was started in United States by Isadora Duncan.
- The profession of physical therapy was born in the United States in 1921.

SCIENTIFIC CONTRIBUTIONS TO THE UNDERSTANDING OF MOVEMENT AND PHYSICAL ACTIVITY (1900–1930)

Physical Education Research

With the rise of physical education in the schools came a demand for tools and techniques to evaluate the effects of physical education on the physical and cognitive development of the child. One of the first tests for strength, speed, and endurance was constructed by Dudley Sargent in 1902. Additionally, in 1910 **James McCurdy** developed standards for evaluating blood pressure and heart rate. A large number of sports skill tests were also developed in this time period. Perhaps the first research laboratory in sports psychology, called the Athletic Research Laboratory, was started in 1925 by **Coleman Griffith** at the University of Illinois. Griffith examined various psychological traits of athletes. The laboratory closed down in 1932, probably as a result of lack of funds and students due to the Great Depression (Kroll, 1971). Biomechanics (then typically called *kinesiology*) laboratories within departments of physical education began to be developed. Notable leaders in this development were **Ruth Glassow** at the University of Wisconsin and **Thomas Cureton** at Springfield College. Both of these individuals argued that the laws of physics could be rightly applied to the understanding of sports skills and published many research articles and books on the topic.

Many early researchers in physical education were convinced of the existence of a general motor ability or a relatively fixed trait, which determined the capacity of any given individual to perform in a variety of motor skills. This type of trait was thought to be analogous to the intelligence quotient (better known as I.Q.), a notion based on earlier work by Alfred Binet in 1905, thought to be the general cognitive trait subserving all intellectual behavior. Many researchers in physical education thought that the higher the level of general motor ability possessed by the individual, the greater the performance in a variety of motor tasks. Researchers developed tests in an attempt to measure this type of trait. The **GMA** (General Motor Ability) **hypothesis,** as it came to be called, was the subject of intense research and debate for several decades to come.

Movement Research in Other Disciplines

Most of the early research in the new field of physical education concentrated on the evaluation of physical abilities and health related qualities of the individual. Research also continued in the other fields that greatly contributed to our understanding of human movement. Incredible advances were made in the physiology of movement, which laid the early foundations of exercise physiology. Our understanding of the nervous system also advanced with the important work of Pavlov, Bernstein, and Sherrington, to name a few. The motor learning subfield saw its early beginnings with the works of psychologists John Watson, who examined reinforcement and its effects on behavior, and E. L. Thorndike, who studied the effects of augmented feedback on motor learning. Many of these advances are summarized in Table 2.3.

Table 2.3
Contributions to the Field of Kinesiology by Some of the Major Scientists and Scholars in the Late 19th and Early 20th Centuries

Contributions	**Contributor(s)**
Reflexes Reinforcement and behavioral change	I.P. Pavlov (1849–1936)
Structure of the nervous system	Santiago Ramon y Cajal (1852–1934)
Reflexes Concept of reciprocal inhibition	Charles Sherrington (1857–1952)

Oxygen metabolism Biomechanics of locomotion	A.V. Hill (1886–1977)
Metabolism	August Krogh (1874–1949)
Lactic acid metabolism	Otto Meyerhof (1884–1951)
Concept of the synergy	N. A. Bernstein (1897–1966)
Respiration	J. S. Haldane (1860–1936)
Respiration Development of air bag for collecting expired gas	C. G. Douglas (?)
Director of Harvard Fatigue Lab (1927–1947) Altitude exercise physiology	D. B. Dill (1891–1986)
Electromyography (EMG)	Wedenski (?) E.D. Adrian (1889–1977)
Principles of behaviorism	John Watson (1878–1958)
The law of effect Augmented feedback and learning	E. L. Thorndike (1874–1949)

SUMMARY

Recap – Contributions

- Spurred on by the educational philosophy of John Dewey, physical education in schools continued to grow.
- Master's and doctoral degrees in physical education were developed in the United States.
- The field of physical education became aligned with education and not medicine.
- Collegiate and intramural sports became extremely popular.
- Recreational and dance movements contributed to the promotion of physical activity.
- The profession of physical therapy developed after World War I.
- Early research in the new field of physical education focused on the evaluation of physical education programs and their effects on the developing child.
- Research laboratories in sports psychology and exercise physiology were developed in the United States.
- Important research by the European physiologists Hill, Krogh, and Meyerhof resulted in the first Nobel prizes on muscle function and exercise.
- Research in psychology laid the early foundations for the subfields of sports psychology and motor learning.

1930 TO 1950

Continued Development of Physical Education

These two decades were marked by major world calamities—the Great Depression in the first decade, followed by World War II in the second. Both events provided major challenges to the growth of physical education. The depression severely affected the budgets of all public schools, and with it came the development of what can be termed "the 3 Rs" philosophy. Simply put, when funding for the school curriculum becomes limited, school administrators will tend to focus resources on so-called primary school subjects such as reading, writing, and arithmetic and divert funds from what some consider luxury or expendable subjects such as art, music, and vocational skills. In the minds of most school administrators at the time, physical education was also considered a secondary subject.

World War II had an effect on the curriculum of physical education. Physical education was thought to provide an excellent environment for the physical training necessary for the military. Emphasis in physical education in the public schools was on physical fitness during the war years. Following the war, there was a shift in emphasis to development of sport skills. It was thought that many of the goals of the New Physical Education could best be realized by participation in sports.

The two most prominent advocates of the New Physical Education during this period were **Jesse Williams** (1886–1966) of Columbia University and **Jay Nash** (1886–1965) of New York University. Williams's (1930) well-known statement that physical education should be considered education *through* the physical and not education *of* the physical exemplified the spirit of the New Physical Education movement and helped define the goals of physical education in the public schools. Following World War II, physical education in the public schools continued to grow. By the end of the decade, nearly all public schools in the United States had required physical education classes in the curricula for an average of approximately one and a half hours per week (Van Dalen and Bennett, 1971), resulting in a great demand for physical education teachers. A large increase in the number of physical education programs occurred in colleges and universities. By 1950 over 400 universities and colleges offered undergraduate degrees in physical education. Master's degrees became desirable for teaching at larger public secondary schools. Also, the leaders in the physical education field no longer were trained in medicine but rather held advanced degrees in physical education. All of these developments indicated the continued growth of physical education as a legitimate professional field.

- There were also important academic developments in physical education during this time period:
- Many colleges and universities began to develop research laboratories in the subfields of biomechanics, exercise physiology, sport psychology, and motor learning.
- *Research Quarterly,* a scientific journal in health and physical education ,was first published in 1930, and the *Journal of Health, Physical Education and Recreation,* a professionally oriented publication, also began early in the decade. Both of these publications grew out of the earlier *American Physical Education Review.*
- The major national physical education association in the United States changed its name to the American Association of Health, Physical Education and Recreation and became housed within the National Education Association. This move marked the continued affiliation of physical education with education as a whole and away from the medical field.

PROMOTION OF PHYSICAL ACTIVITY

Sports

In spite of the Great Depression and World War II, sports became increasingly popular around the world and certainly in North America. In the United States, the NCAA extended national tournaments into many sports and these served in the selection of athletes to the Olympic games. Intramural sports continued to grow and this period also marked the beginnings of interest in women's athletic competitions. However, only about a fourth of the colleges in the United States supported women's athletic teams before 1945 (Scott, 1945).

A dramatic occasion occurred in 1947 when Jackie Robinson took to the baseball field wearing a Brooklyn Dodgers baseball uniform. While professional baseball for African-Americans had already been established in the Negro League, Robinson became the first African-American in Major League baseball in the United States. This historic event set a precedent, allowing other African-American players to enter professional sports to become role models for aspiring African-American children in the years to come.

Following their cancellation during World War II, the Olympic games resumed in 1948 and they continue to be a premier sporting competition at the international level.

Dance

Modern dance continued to grow as a popular physical activity in the universities. By 1950 nearly half of the colleges offered at least one modern dance class; however, ballet became much less popular. With the promotional help of automobile giant Henry Ford, square dancing also became popular in the United States. Square dancing clubs sprang up all over the country (Van Dalen and Bennett, 1971).

Physical Therapy and Adapted Physical Education

World War II resulted in millions of deaths all around the world. Hundreds of thousands of people in the United States alone were left with physical disabilities, such as paralysis and loss of limbs, that restricted their mobility, their ability to function effectively in the workplace, and their participation in sport and other recreational activities. The physical therapy profession expanded after the war and many universities and colleges began to offer adapted physical education classes, which provided physical activities for persons with disabilities.

Scientific contributions to the understanding of movement and physical activity (1930–1950). Because of the proliferation of research in physical education, physiology, chemistry, and psychology, it is impossible to summarize all the major contributions to our understanding of movement and physical activity over these two decades and up to the present. Only some major contributions will be highlighted here and in the next section.

Physical Education Research

In physical education, research continued on the development of achievement tests to measure various physical abilities such as strength, endurance, jumping and throwing abilities. Efforts were made to determine norms for males and females at different ages. These tests not only provided useful developmental information about the growing child, but also could be used to classify children on the basis of physical abilities for physical education instructional purposes. Tests were also constructed to determine skill level in various sports. For example, a test on tennis ability could include separate measures of serving, volleying, and footwork. A composite score based on performance on each of the individual components was supposed to reveal a student's overall tennis ability. Much research attempted to determine the individual components within each sport that could be used to predict overall ability.

Movement Research in other Disciplines

World War II had a significant impact on research related to human performance. Many research programs were initiated at several laboratories such as the U.S. Air Force Human Resources Research Center to investigate performance capabilities and limitations of pilots, soldiers, and other personnel (Schmidt and Lee, 1999). One particular research area of interest to the military was the learning of verbal and motor skills. Perhaps the most significant theoretical work in this area came from **C. L. Hull** (1943), who attempted to explain fatigue-like effects accompanying the acquisition of these types of skills over long practice periods. While Hull's theory later proved to be incorrect, it inspired much research in both psychology and physical education for many years.

Another important conceptual advance in our understanding of the nervous system was made by **K. J. W. Craik** (1947), who argued that humans perceive environmental events and produce movement in an intermittent as opposed to a continuous manner. His ideas, as well as others during this period (Wiener, 1948), had much to do with the creation of the analogy that the human brain can be likened to a computer. He also sparked the shift of interest in psychology from the behaviorist viewpoint to the so-called "information processing approach." Unlike the behaviorist black box viewpoint, the brain was viewed as an *active* processor of information. This information had to pass through several processing stages such as stimulus identification, response selection, and response execution. The information processing approach is still very popular today in the fields of psychology, computer science, and kinesiology.

Using electrical stimulation techniques, the Canadian neurosurgeon **W. Penfield** (1891–1976), determined that movements associated with various body parts are represented in different areas of the motor cortex in the brain, called **somatotopic representation** (Penfield and Rasmussen, 1950). Also, movements related to fine precision and function are more highly represented in the motor cortex. For example, the fingers and lips take up a much larger portion of the motor cortex than, say, the legs. Penfield's work had a great impact on our understanding of the brain's function in the control of movement. The work of Penfield represented the rapid developing field of neurophysiology. One major focus of neurophysiology is on the contribution of the nervous system to the control of movement.

The German physiologist **Erich von Holst** (1908–1962) performed many experiments on the coordination of movements in animals and humans (von Holst, 1939). His work was largely unknown in the western hemisphere until it was translated in 1973. However, von Holst's biomechanical techniques and conceptualization about the nervous system's role in the control of movements were far ahead of his time.

Flying aircraft, driving tanks and detecting the presence of enemies on radar screens involves the interaction or *interface* between the human and the machine or device. Building machines to fit human capabilities became the central focus of the field of **human factors** or ergonomics, which developed soon after World War II. Initially this area focused on military related tasks, but it eventually widened to the study of human interaction with all types of machines and equipment such as automobiles, computers, and typewriters.

SUMMARY

- World War II had an impact on the curriculum of physical education, with an emphasis on physical fitness and military training.
- By 1950 virtually every major university offered degrees in physical education.
- The popularity of collegiate and intramural sports continued to rise.
- Research in physical education focused on the development of tests to evaluate sport skills.
- The psychological work of Hull and Craik contributed to our theoretical understanding of human performance.
- The field of neurophysiology emerged with the work of Sherrington, Penfield, and many others during this period.
- Human factors research began during the war, as interests developed in the "interface" between the human and the machine.

1950 TO THE PRESENT—THE EMERGENCE OF KINESIOLOGY AS A FIELD OF STUDY

In this section, we continue to explore the various promotions of physical activity and the scientific contributions to understanding of the structure and functions of the human body engaged in physical activity. In addition, the beginnings of kinesiology as a field of study are documented. The development of the field up to the present and speculations as to where the field might be headed in the future also are discussed.

The promotion of physical activity. A major promotion of physical activity occurred as organized sports "exploded" during the second half of the 20th century. The three major sports in the United States, football, baseball, and basketball grew enormously at the professional level in that the number of professional teams in each sport significantly increased after World War II. Professional tennis and golf also expanded. An important factor in this expansion was the use of television in bringing sports competitions into homes across the nation. Sports figures such as Jim Brown in football, Mickey Mantle in baseball, Wilt Chamberlain in basketball, Billy Jean King in tennis, and Arnold Palmer in golf became heroes and heroines to millions of aspiring young athletes. The viewing of professional sports on television had, without doubt, a large effect on the development of youth sports. Age-group clubs in other sports such as in swimming, soccer, gymnastics and hockey also sprang up all over the country.

The findings from a relatively obscure research paper in the *Journal of Health, Physical Education and Recreation* published in 1953 were brought to the attention of then president of the United States, Dwight D. Eisenhower. The purpose of the study by Hans Kraus, a physician, and Ruth Hirschland was to compare the muscular strength and flexibility of American youths with a European sample of children from Italy and Austria (Kraus and Hirschland, 1953). The results of the study were staggering. They showed that 60 percent of the American youths failed the test, compared with only about 9 percent of the European children. These results were quickly picked up by the newspapers and media and were turned into the makings of a national health crisis. Eisenhower, an avid golfer, was so alarmed that in 1955 he assembled some prominent sports figures to examine the youth fitness issue. As a result, a Youth Fitness Conference was scheduled to be held in Denver, Colorado, in the fall of 1955. Ironically, President Eisenhower suffered from a heart attack and the conference was postponed. One of Eisenhower's personal physicians, Dr. Paul D. White, prescribed an exercise program to facilitate Eisenhower's recovery. Following the conference on Physical Fitness of American Youth in 1956, Eisenhower established the President's Council on Youth Fitness. From this time on, physical activity (or the lack of it!) became a nationally recognized issue and concern.

There were other important developments as well that contributed to the general awareness of the importance of physical activity and exercise:

- Autopsies on U.S. soldiers killed in the Korean War in the early 1950s indicated the presence of coronary artery disease (Enos, Holmes, and Beyer, 1953). These reports were most alarming given that the average age of the soldiers was 22.1 years!
- In 1972 Title IX of the Educational Amendments Act was passed by Congress, stating that women have equal access to sport, fitness, and physical education opportunities (Siedentop, 1990). Title IX contributed to the rapid expansion of women's sports at all levels of competition.
- A very popular book was published by Dr. Kenneth Cooper (1968) entitled *Aerobics*. Cooper argued that the key to good health was through aerobic conditioning. Aerobic conditioning improves the body's ability to deliver oxygen to the muscles, heart, and other organs. In the book, Cooper provided a number of exercise programs for males and females of different ages and fitness levels.

THE FITNESS BOOM (OR BUST?)

All of these events contributed to what has been called the **Fitness Boom,** which began in the 1970s and continues today. This phenomenon represented a significant change in attitudes and awareness toward exercise and diet habits as well as a heightened awareness of the detrimental effects of smoking and excessive alcohol consumption. The increasing popularity of marathon races, 10K runs, and triathlons involving elite athletes as well as the average citizen has been taken as some evidence for this boom. The last 10 years has also seen the rapid development of aerobic classes throughout the United States. Health and fitness clubs have sprung up in nearly every city in the United States, and their popularity is rising in other parts of the world. Communities all around the nation are building bicycle and jogging trails and recreational centers that house swimming pools, running tracks, and weight rooms.

Yet, in spite of this marked increase in participation in physical activity by the populace and a heightened awareness of the importance of exercise and proper diet, a large segment of our population still does not exercise regularly and/or maintain a healthy, nutritious diet. While it is very difficult to obtain accurate estimates from both children and adults of the type and frequency of physical activity, most available evidence is indeed discouraging. Recent surveys indicate that only approximately 15–20 percent of adults engage in sufficient physical activity to maintain an adequate fitness level (Caspersen, Christenson, and Pollard, 1986; U.S. Department of Health and Human Services, 1996). This lack of adequate exercise contributes to many health problems such as heart attacks, hypertension, and low back pain. The public cost associated to these health problems in the form of Workmen's Compensation, lost work hours, and higher insurance rates is enormous, and amounts to billions of dollars each year (Villeneuve, Weeks, and Schweid, 1983).

The data on children are equally discouraging. A summary of findings from the National Children and Youth Fitness Study II was published in the *Journal of Physical Education, Recreation and Dance* (see Ross and Pate, 1987). The study tested nearly 5,000 children all over the United States and showed that 6- and 9-year-olds were not only fatter than a comparable group of children 20 years earlier, they performed poorly on cardiopulmonary endurance (one half and one mile runs) and upper body strength tests. Other recent studies on children of similar age indicate poor eating habits (Public Voice for Food and Health Policy, 1991). In every socioeconomic group, children eat too much fat and salt, and their diets do not contain enough fiber. Recent data show that one out of every five teenagers in the United States smokes cigarettes, a habit that eventually contributes to a large number of diseases (U.S. Department of Health and Human Services, 1991). These and other similar studies paint a rather bleak picture of the general health and fitness level of a nation supposedly in the midst of a so-called Fitness Boom. We must conclude from all of these studies that the Fitness Boom has, on the one hand, increased society's *appreciation* for the value of proper physical activity, while on the other hand, it has affected only a small percentage of people's lifestyles.

The evolution of kinesiology as a field of study. While physical education has been largely considered a professional field going back to its early days, there was always a recognition that it contained a formidable academic content and was worthy of serious scientific study. As I have already mentioned, Thomas Wood and George Fitz were among some of the first strong advocates of physical education as a legitimate academic enterprise. Wood was skeptical of the physiological benefits of the various gymnastic systems (e.g., German, Swedish) imported from Europe. He felt that there was limited scientific evidence to support their views. Fitz was particularly interested in physiological changes due to exercise and in the performance capabilities of the athlete. But in the early days of physical education, the views of Wood and Fitz were in the minority. The primary emphasis and purpose of physical education until the early 1960s was on the development of physical education teachers.

There was one particular and seemingly unrelated event that had a significant effect on turning attention in physical education toward the academic content of the field: the **space race**. The launching of the space satellite

Sputnik in 1957 by the former Soviet Union instigated a competitive spirit between the Soviet Union and the United States that caused many scientific disciplines to reexamine their academic content. The academic contents of most disciplines were scrutinized as to their scientific merits and contributions to society, and physical education did not escape this scrutiny. The academic content of physical education was severely criticized by James Conant (1963) and James Koerner (1963), with Conant even calling for abolishment of graduate programs in physical education.

Another very important event causing a shift of attention to the academic content of physical education was the passage of the **Fischer Bill** in the state of California in 1961. The Fischer Bill came about because of the prevailing belief that the academic content of teacher preparation programs was weak in many areas. The public was outraged when it learned that a large percentage of teachers who had earned college degrees in physical education not only were school administrators but were also mathematics, physics, and foreign language teachers in many schools. The Fischer Bill contained criteria for academic subjects as well as certification requirements for elementary, secondary, and junior college teaching. It stated that a teacher must first graduate with a degree in an academic subject before taking professionally oriented coursework leading to teacher certification. Along with many other fields of study such as athletics, art, home economics, music, and vocational education, physical education was defined as a *nonacademic subject*. Nonacademic subjects could no longer qualify as legitimate undergraduate degrees (a similar bill was passed in Texas in 1987). After over a half century of growth, popularity, and respect in educational circles, physical education was now considered a secondary, nonessential subject.

In response to these criticisms came a highly influential paper published by **Franklin Henry** (1904–1993) from the University of California at Berkeley (Henry, 1964). In this paper, Henry eloquently argued the case for physical education as an academic discipline with unique subfields of study. Henry identified these areas as "kinesiology and body mechanics; the physiology of exercise, training and environment; neuromotor coordination, the kinesthetic senses, motor learning and transfer; emotional and personality factors in physical performance; and the relation of all of these to human development, the functional status of the individual, and his ability to engage in motor activity. They also include the role of athletics, dance, and other physical activities in the culture and in primitive as well as 'advanced' societies" (p. 33). His plea was for the field to concentrate on the development of these subfields of knowledge through scientific research.

Henry's argument formed the basis for what was to be called the **discipline movement** within physical education, and served as an inspiration to those who were already actively engaged in research in the various subfields he described. Henry's now classic paper also provided a framework for considerable debate and discussion on issues such as:

1. What are the appropriate academic subfields in physical education?
2. Should the focus of research in physical education be on the effectiveness of teaching methods, on sport and athletic performance, or on the psychological, physiological, and biomechanical aspects of human movement and physical activity in a more generic sense? and
3. If emphasis is placed on academic content, is the term "physical education" the most appropriate label for the field?

The subsequent debates on these and other related issues by physical educators throughout North American were extensive and continue today. Let us briefly examine each of these issues.

The Subfield Issue. By the 1960s, the most developed academic area of study within physical education was probably exercise physiology. By this time, most departments of physical education at major universities and colleges in North America had developed exercise physiology laboratories. Laboratories in the areas of biomechanics, motor learning, motor development, and sport psychology (social-psychological factors in performance) were also in place. Research activity in these and other areas of physical education increased rapidly in the next three decades, and scientific journals were created to publish this work. A number of scientific societies were formed to provide a format for scientists to present their work at annual meetings. Table 2.4 summarizes a significant sampling of this information. These scientific societies and scholarly journals have provided scientists interested in human movement a format to present the results of their research. It is clear from Table 2.4 that the number of societies and journals is extensive, and this provides strong support to the notion that the various subfields have become legitimate academic enterprises.

The Major Focus of the Field of Study. What should be the focus of inquiry in the science of physical activity? I believe the following three proposals have proven to be the most popular:

Teacher Training—This is considered to be the traditional view of physical education. The argument for this view states that the primary purpose of the physical education degree is to train physical education teachers. Students would study the learning of sport skills and the acquisition of teaching skills to be used in the elementary and secondary schools (e.g., Siedentop, 1990).

Sports—This proposal states that the focus of inquiry of physical education should be on sports. Students would take courses in exercise physiology, biomechanics, sports psychology, sociology, and so forth in order to learn about athletic performance as well as the role of sport in society (e.g., Whited, 1970).

Kinesiology (also Exercise Science, Human Movement Science)—Advocates of this viewpoint state that the primary focus of inquiry is the study of human movement in a variety of settings and the structure and function of human body while engaged in physical activity (e.g., Newell, 1990).

The name change issue. Many leaders in the field argued that the name "physical education" was no longer appropriate, given the shift in focus away from teacher preparation to academic content. They believed that a new name should be created that reflected this academic content. Also, because physical education was considered a nonessential subject, the field developed a poor reputation and lost respect in the eyes of many people both inside and outside of the educational establishment. Terms like "jock" were used to represent students who were majoring in physical education at universities and colleges. It was for these two reasons that many leaders in physical education thought the time had come to change the name of the field. Unfortunately, there was no agreement as to what this new name should be. Names such as sport and exercise science, human movement studies, human performance, and of course, kinesiology were used to represent departments that shifted focus away from the teaching preparation orientation and toward the scholarly or discipline-oriented approach.

Finally, clinical kinesiology is the scientific investigation of the human movement related problems in education, recreation, and rehabilitation settings. The well-established fields of physical education, recreational service, and corrective or physical therapy can be conceptualized as part of clinical kinesiology.

Most departments of physical education in universities and colleges have had to struggle with the above issues. Partly because of economic reasons, departments of physical education have had to choose between supporting either a high-quality teacher training program or a research-oriented program. With limited resources, it is very difficult for departments to support both types of programs. There has also been increased pressure on departments of physical education to become more research oriented. As such, many former departments of physical education have eliminated (or reduced the scope of) their teacher preparation program emphasis and changed their name to kinesiology (or some related term). The first departments of kinesiology in Canada were at the University of Waterloo in Ontario (1967) and at Simon Fraser University in British Columbia in 1972. Strongly influenced by Franklin Henry's disciplinary focus for physical education, the former Department of Physical Education became the Department of Kinesiology and Physical Education at California State University at Hayward in 1972. The first department of kinesiology in the United States with little emphasis in physical education was at UCLA in 1973. Since that time, the number of departments focusing on the scholarly understanding of human physical activity has increased all over North America and the world.

Over the last 30 years, there has been a rapid increase in scholarly activity in the various subfields of kinesiology. This has necessitated the formation of a number of research societies and scholarly journals associated with the subfields (see Table 2.4). The creation of research societies and scholarly journals is an important indication of the development of kinesiology as an academically oriented field of study.

While the metamorphosis from physical education to kinesiology was occurring, it became clear that other career paths (in addition to physical education and coaching) were possible for kinesiology graduates. As pointed out in Chapter 1, a popular career option for qualified kinesiology graduates has been physical therapy. Other medical and health related professions have become good options for kinesiology graduates. A degree in kinesiology has become generally recognized as providing excellent preparation for fields related to health and wellness.

In 1926 influential leaders in physical education developed the Academy of Physical Education, which later evolved into the Kinesiology Council and finally into the American Academy of Kinesiology and Physical Education (AKPE). The AKPE was formed in the 1960s within the American Association for Health, Physical Education, Recreation and Dance to help prevent any further splintering of discipline advocates from those supporting teacher preparation. The purpose of the academy was to: promote the development of kinesiology as a science and as an academic discipline, provide a professional outlet for those people specializing in one or more of the subfields in kinesiology, and promote the advancement of knowledge in all aspects of kinesiology. The academy consisted of influential scholars in kinesiology and physical education and eventually expanded to cover all the major subdivisions of kinesiology. Progress was made over the next decade through the creation of some scholarly publications and academic meetings.

Figure 2.13
A photograph taken of Franklin Henry during an interview session at his apartment in Oakland, California, on October 13, 1990.

There was growing sentiment that a national organization, separate from AAHPERD, was needed to represent the academic discipline of kinesiology. In the late 1980s, several advocates of the discipline approach began a series of meetings that ultimately resulted in the formation of the American Kinesiology Society (AKS) in 1991. The purpose of AKS was "to promote, protect, and advance the interests of the study of human movement and/or physical activity in all related professional, disciplinary and performance contexts" (*The American Kinesiologist*, 1991). Unfortunately, the permanent establishment of AKS did not materialize. However, in 2007, the American Kinesiology Association (AKA) was founded and its membership consists of departments of kinesiology in universities throughout the United States. At this writing, approximately 115 departments are members of the organization. It is certainly hopeful that this organization will become an effective promoter for the field of Kinesiology on a national level.

In summary, the focus on academic content and scholarly research related to the understanding of human movement and physical activity led to the creation of departments of kinesiology. Career paths in medical related professions, health and wellness, and others are now viable options for kinesiology majors. The formation of several academic societies and scholarly journals associated with the subfields of kinesiology has occurred over the last 30 years.

Table 2.4

Scientific Societies and Scholarly Journals in Supporting Areas and Kinesiology Subfields

	Scientific Societies	Scholarly Journals
Anatomy	American Association of Anatomists	Journal of Anatomy
	Canadian Association of Anatomists	Anatomical Record
	International Congress of Anatomists	American Journal of Anatomy
		Acta Anatomica
Physiology	Biophysical Society	European Journal of Applied Psychology
	Physiological Society	American Journal of Physiology
	Society for Neuroscience	Journal of Applied Physiology
	Federation of Societies in Experimental Biology	Canadian Journal of Physiology
	Canadian Physiological Society	Canadian Journal of Physicology and Pharmacology
	Scandinavian Physiological Society	Muscle and Nerve
Biomechanics	International Society of Biomechanics	Journal of Applied Biomechanics
	American Society of Biomechanics	Research Quarterly for Exercise and Sport
	Canadian Society of Biomechanics	Canadian Journal of Sport Sciences
		Medicine and Science in Sport Exercise
Exercise Physiology	American College of Sports Medicine	Medicine and Science in Sports Exercise
	American Association of Cardiovascular and Pulmonary Rehabilitation	International Journal of Sports Medicine
	International Group on Biochemistry of Exercise	Journal of Cardiopulmonary Rehabilitation
		Research Quarterly for Exercise and Sport
		(others in Physiology listed above)
Motor Control, Motor Learning and Motor Development	North American Society for the Psychology of Sport and Physical Activity (NASPSPA)	Journal of Motor Behavior / Motor Control
	Canadian Society for Psychomotor Learning and Sport Psychology (CSPLSP)	Perceptual and Motor Skills
		Human Movement Science
		Journal of Human Movement Studies

Sport and Exercise Psychology	NASPSPA	Journal of Sport and Exercise Psychology
	CSPLSP	The Sport Psychologist
	Association for Applied Sport Psychology	Journal of Personality
	American Psychological Association (Division 47, Exercise and Sport Psychology)	Journal of Applied Sport Psychology
		International Journal of Sport Psychology
		Research Quaterly for Exercise and Sport
		(other Psychology journals)

HIGHLIGHT

One of the strongest advocates for "kinesiology" as a field of study was **Jerry Barham** (1963; 1966) from Colorado State College at Greeley (now called the University of Northern Colorado). Barham believed that the central focus of the field was the study of human movement, which could be broken down into three major subdivisions: *general* kinesiology, *applied* kinesiology, and *clinical* kinesiology.

The study of general kinesiology entails prerequisite coursework in physics, anatomy, physiology, and psychology. Following this coursework, the student would then take classes in the various specialized areas of kinesiology:

—structural kinesiology (anatomy);
—mechanical kinesiology (biomechanics);
—physiological kinesiology (exercise physiology);
—psychological kinesiology (motor learning and motor control; social-psychological factors; and
—maturational kinesiology (motor development).

Applied kinesiology has to do with the various performance areas related to human movement such as dance, sports, and various types of exercises. These areas can be scientifically examined using theories and facts generated from general kinesiology.

ISSUES IN THE FUTURE OF KINESIOLOGY

In this section, two important issues and challenges facing kinesiology as a field of study will be mentioned: the name change, and a parent organization for the field of kinesiology.

Name Change—As departments of physical education continue to change their names reflecting the renewed focus on academic content, many scholars believe it is important that an appropriate name is chosen that adequately represents this content. There has been much debate on what this content should be and what name should be used to represent it (e.g., Kretchmar, 1989; Newell, 1990 for differing views). In 1989, the American Academy of Kinesiology and Physical Education passed a resolution that the term "kinesiology" be used to represent the subject matter, the academic departments and the degrees related to the study of movement. While this resolution cannot be enforced, many scholars believe the academy, composed of some of the most prominent scholars in the field, has sent a message that kinesiology is the preferred term.

Parent Organization—There are also some scholars who believe that a national organization representing the field of kinesiology is important to establish. A national organization can promote the field and its interests on a national scale, much like other national organizations in different disciplines. It can also serve to stimulate research by allowing the membership to: 1) present their research at annual meetings; 2) interact with their colleagues at different universities; and 3) compete for grant monies provided by the national organization for research projects. It is hopeful that the AKA will fill this need.

CHAPTER SUMMARY

- The promotion, interest, and participation in physical activity has waxed and waned throughout history. Some civilizations and cultures have actively promoted participation in physical activity while others have not. We have seen that early humans engaged in dance to mimic the movements of animals whom they worshiped and respected. By all estimations, humans were quite physically active as they hunted and fished for food.

- Dance as an art form probably developed to a high degree in ancient Egypt. From there, other dance forms spread to a number of ancient civilizations. The Greeks promoted physical activity through athletic competitions like the Olympic Games and through dance. The importance of physical exercise for military training was realized by the Spartans and later by the Romans. However, to the Romans, athletic competitions and dance were considered to be entertainment and physical activity (other than required for physical labor) was not advocated for the masses or for nobility.

- During the Middle Ages, the Christian church in western Europe largely scorned dance and physical activity related to recreation. However, the social and moral code of chivalry endorsed a variety of forms of sports and athletic competitions.

- In the 18th and 19th centuries, a number of educational leaders in Europe saw the importance of physical activity for the masses. A variety of gymnastic systems developed in Germany, Sweden, and later, England. These systems spread to the United States, where they were further modified. Physical education in the public schools began in the United States during the 19th and 20th centuries. As the need for qualified physical education teachers grew, master's and doctoral degrees in physical education were developed at most major universities.

- A number of important events in the latter half of the 20th century led to the so-called Fitness Boom, which increased public awareness of the importance of physical activity and proper nutrition. But as we have seen, participation in regular exercise and proper diet is still limited to a small segment of the population.

- Competition with the former Soviet Union led to an increase in interest in the scholarly content of academic subjects in the school curriculum. This interest forced many physical education departments to emphasize the scholarly understanding of human movement and physical activity. Eventually, many departments of physical education changed their name and their focus to kinesiology (or some related term).

- A major goal for the field of kinesiology is to continue to promote the participation in physical activity and healthy lifestyles to a much larger segment of the population across all socioeconomic levels and age groups, and to scientifically contribute to our understanding of the structure and function of the human body engaged in physical activity.

IMPORTANT TERMS

Adolph Spiess (1810–1858) - successfully introduced gymnastics into the German school systems for both boys and girls

Alfonso Borelli (1608-1679) - considered to be the father of modern biomechanics of the locomotor system

American Kinesiology Association – founded in 2007 is considered the parent organization of the field of Kinesiology

Ancient Greece – a civilization that produced many philosophers who embraced the importance of physical activity, the connection between the body and mind, and the birthplace of the Olympic Games

Ancient Rome – a militaristic civilization that dominated the entire European and Mediteranean area for a hundreds of years.

Andreas Vesalius (1514–1564) – made significant contributions to our understanding of human anatomy

Animal spirits – a concept developed by Erasistratus that spirits called pneuma circulated through the nerves

Archimedes (287–212 B.C.) - developed the first work on hydrostatic principles that govern floating bodies and the motions used in swimming

Aristotle (384–322 B.C.) – an ancient Greek philosopher who provided several treatises on the analysis of human and animal motion

Asceticism - a lifestyle promoted by the early Christian church characterized by abstinence from various sorts of worldly pleasures often with the aim of pursuing religious and spiritual goals

Association for the Advancement of Physical Education – the first national organization in the United States that promoted the field of physical education

Ballet – a dance form originating during the Italian Renaissance period

Battle of the systems – a debate among physical educators in the late 19th century over the advantages and disadvantages of the various gymnastic systems

Benjamin Franklin (1706–1790) – one of the Founding Fathers of the United States who, among many accomplishments, also promoted the importance of different types of physical activity

Bird Larson – early leader in physical education

C. L. Hull – early researcher who investigated fatigue-like effects in motor learning

Carl Diem (1882–1962) – early German leader in physical education

Carlo Blasis (1787–1878) – early Italian who helped develop the ballet dance form

Catherine Beecher - (1800–1878) developed free exercises for both boys and girls and strongly advocated physical training in the schools and in the home

China – ancient China is credited with the development of many sports

Chivalry – a moral and social code practiced by males during the Middle Ages of Europe

Chorus – groups of performers in ancient Greek dances

Clark Hetherington (1870–1942) - early leader in physical education

Coleman Griffith (1893 – 1966) – considered to be the Father of American Sport Psychology

Court dances – dances developed during the Middle Ages

Dance of Death – a popular dance form during the Middle Ages

Dio Lewis - (1823–1888) compiled a number of light exercises that could be accompanied with music and performed by children and adults.

Discipline movement – influenced by Frankline Henry, it emphasize the scientific study of physical activity

Dudley Sargent - (1849–1924) an early leader in the develop of physical training

Edward Hitchcock (1828-1911) – developed many types of physical exercises

Egypt – ancient Egypt developed many dance forms and sports

Erasistratus (304–250 B.C.) - developed the theory of pneumatism, which suggested the phenomenon of life was associated with pneuma, or spirits, that resided in the body and caused movement

Erich von Holst - (1908–1962) performed many experiments on the coordination of movements in animals and humans

Fitness boom – is associated with an increase in awareness of the importance of physical activity in the late 1960's and early 1970's

Franklin Henry (1904–1993) – founder of the disciplinary movement in physical education that helped in the development of kinesiology

Franz Nachtegall (1777–1847) - who introduced compulsory physical education in the European public schools, probably for the first time.

Friedrich Jahn (1778–1852) - was instrumental in the development of the Turnplatz (play or exercise ground).

Galen (129–199 A.D.) - made several contributions to our understanding of human anatomy.

Galileo Galilei (1564–1642) - laid the groundwork for the development of classical (Newtonian) mechanics and thus to modern-day biomechanics.

General motor ability hypothesis – this idea, later disproven, was that success in all motor skills was dependent on only one ability, called general motor ability.

George Fitz (1860–1934) – considered by many as the father of exercise physiology

Gertrude Colby - one of the early developers of modern dance who developed a method of dance that allowed a dancer to move freely through space, creating as many shapes, using different body parts, and different movements

Human factors - involves the study of all aspects of the way humans relate to the world around them, with the aim of improving operational performance, safety, through life costs and/or adoption through improvement in the experience of the end user.

Humanism - was an attempt during the Renaissance period to resurrect the classical literature of the Greeks and Romans.

India – this ancient civilization was dominated by the religions of Hinduism and Buddhism

Intramurals - the general goals of intramurals were to provide several sport and recreational activities to the greater college student body and to encourage the pursuit of physical activity.

Isaac Newton (1642–1727) - the co-inventor of the calculus, Newton formulated his famous three laws of motion in the Principia, that formed the foundation of physics and modern biomechanics

Isadora Duncan (1878–1927) – a major contributor to modern dance

James McCurdy – an early leader in physical education

Jay Nash (1886–1965) – an early leader in physical education

Jean-Georges Noverre (1810–1840) - one of the early founders of the ballet dance form

Jean-Jacques Rousseau (1712–1778) – an early French philosopher who contributed to our understanding of human development

Jerry Barham – conceptualized kinesiology as a legitimate field of study

Jesse Williams (1886–1966) - of Columbia University, an early leader in physical education known for his statement that physical education should be considered education through the physical and not education of the physical

Johann Guts Muths (1759–1839) - from Germany, who is considered the grandfather of modern physical education

John Dewey (1859–1952) – an early philosophical leader in education

John Maclaren (1820–1884) – an early Scottish developer of gymnastic exercises

Jujitsu – a major type of martial arts

Jules Perrot (1810–1892) – an early French developer of the ballet dance form

K. J. W. Craik (1914-1945) – an influential psychologist who was one of the founders of cognitive science and the information processing approach to the study of brain function

Kinesiology Council – an early group of leaders in the field of kinesiology

Kung fu – a Chinese martial art form

Leonardo da Vinci (1452 –1519) – an Italian Renaissance figure who made several contributions to our understanding of human anatomy

Luther Gulick (1865–1918) – an early and influential advocate of physical education

Mapheus Vegius (1405–1458) a Humanist philosopher who advocated the importance of moderate exercise

Margaret N. H'Doubler (1889-1982) developed the first dance major program at the University of Wisconsin in 1926.

Martin Luther (1483 –1546) - a German monk and the leader of the Protestant Reformation, saw certain types of physical activity as important in counteracting the negative effects of gambling and drinking as well as promoting good health

Mary McMillan (1880-1959) - served as the first president of the American Physiotherapy Association

Massage – a manipulation technique that has its origins in many civilizaitons

Medical gymnastics – developed by Per Henrik Ling, was designed for the relief of certain physical disabilities and consisted of exercises by the patient alone or with the aid of trainer

Moralism - a type of individualistic spirit that ran against traditional Christian philosophy in western Europe during the Renaissance period

National Collegiate Athletic Association – founded in 1910, became instrumental in the regulation of amateur sports in the United States

New Physical Education – advocated by Jesse Williams and Jay Nash, emphasized education through the physical not of the physical

Niels Stensen (1638-1686) – made significant contributions to our understanding of human anatomy including the cardiovascular and muscular systems

Olympic festivals – ancient Greek athletic games starting in 776 B.C.

Palaestra - a type of physical education of ancient Roman male youths

Pantomime – a type of non-spoken entertainment common in ancient Roman times

Per Henrik Ling (1776-1839) – developed a type of exercise called medical gymnastics in Sweden

Petrus Paulus Vergerius (1349-1420) – a humanist who promoted games and exercise as a preparation for the military.

Physical therapy – a type of health care profession involved in the promotion of optimal health using scientific principles to prevent, assess and correct movement dysfunction

Pierre de Coubertin (1863-1937) – helped revive the Olympic games in 1896

Plato (428-347 B.C.) – an ancient Greek philosopher

Realism – a viewpoint in the Renaissance period that led to the development of modern scientific inquiry

René Descartes (1596-1650), a French philosopher, mathematician and scientist

Ruth Glassow a biomechanics pioneer in physical education at the University of Wisconsin-Madison

Scholasticism – primarily led by the monks of the early Christian Church, this view emphasized the importance of facts as the basis for knowledge, truth, and wisdom

Socrates (471–399 B.C.) – a philosopher in ancient Greece who emphasized the importance of dance

Somatotopic representation – the representation of parts of the body in various areas of the brain

Space race – this was a historical competition between the United Sates and Soviet Union that resulted in the emphasis of science in school curriculums

Spectacula – games of entertainment in ancient Rome

Thomas Cureton (1941-1969) – from the University of Illinois, published many papers in the areas of fitness and exercise

Thomas Wood (1864–1951) – an early advocate for the science of physical education

Vittorino de Feltre (1378–1446) – a Humanist who believed in the importance of moderate exercise

W. Penfield (1891–1976) – a Canadian neurosurgeon who discovered somatotopic representation in the brain

William Anderson (1860–1947) – an early leader in physical education

William Harvey (1578–1657) – provided the first accurate account of the blood circulation

Yoga - a physical, mental, and spiritual discipline, originating in ancient India

Young Men's Christian Association – this organization was an early promoter of physical activity around the world

INTEGRATING KINESIOLOGY: PUTTING IT ALL TOGETHER

1. Based upon what you have read about ancient civilizations, the Middle Ages, the Renaissance, and the development of physical education and kinesiology in more modern times, what are some factors that contribute to the physical activity patterns of people in a society?

2. Compare and contrast the athletic contests of the ancient Greeks and Romans.

3. Are there sporting events today that remind you of the ancient Roman athletic events called spectacula?

4. During the Middle Ages, the western Christian church placed many restrictions on certain physical activities. What are today's attitudes of various religions toward different types of physical activities?

5. What should be taught in physical education (e.g., sports skills, fitness, values, etc.) has been the subject of debate for over a century. Based on your understanding of various "systems" brought to the United States from Europe and of the "content" within the New Physical Education proposed earlier this century, what do you think should be taught today in physical education?

6. How do you think physical education would have evolved if it had aligned itself with the field of medicine instead of education?

7. What were some of the commonalities between Franklin Henry's view of physical education and Jerry Barham's view of kinesiology?

KINESIOLOGY ON THE WEB

http://www.aakpe.org/This is the Web site for the Academy of Kinesiology and Physical Education. It contains information on the definition and history of kinesiology. Also at this site, you can listen to interesting interviews of prominent leaders in kinesiology and physical education, including the late Dr. Franklin Henry. (http://www.aakpe.org/ShowArticle.cfm?id=84)

REFERENCES

Adrian, E. D. (1925). Interpretation of the electromyogram. *Lancet*, 2:1229–33, 2:1283–86.

Aller, A. S. (1935). The rise of state provisions for physical education with the public schools of the United States. Doctoral Dissertation. University of California, p. 163.

The American Kinesiologist (1991). *The American Kinesiology Society*, Vol. 1, No. 1.

Barham, J. (1963). Organizational structure of kinesiology. *The Physical Educator*, Vol. 20. No. 3, 120–21.

Barham, J. (1966). Kinesiology: Toward a science and discipline of human movement. *Journal of Health, Physical Education and Recreation*. October, pp. 65, 67, and 68.

Beecher, C. E. (1856). *Physiology and Calisthenics for Schools and Families*. New York: Harper and Row.

Brooks, G. A., and Fahey, T. D. (1984). *Exercise Physiology: Human Bioenergetics and Its Applications*. New York: Macmillan.

Bryan, W. L., and Harter, N. (1889). Studies on the telegraphic language. The acquisition of hierarchy of habits. *Psychological Review, 6*, 345–75.

Bryan, W. L., and Harter, N. (1897). Studies in the physiology and psychology of telegraphic language. *Psychological Review, 4*, 27–53.

Caspersen, C. J., Christenson, G.M., and Pollard, R. A. (1986). Status of the 1990 physical fitness objectives: evidence from NHIS 1985. Public Health Report; 101: 587–92.

Chryssafis, J. (1930). Aristotle on kinesiology. *Journal of Health and Physical Education, 1*, No. 7.

Conant, J. B. (1963). *The Education of American Teachers*. New York: McGraw-Hill.

Cooper, K. H. (1968). *Aerobics: With a Foreword by Richard L. Bohannon and a Preface by William Proxmire*. New York: M. Evans; distributed in association with Lippincott, Philadelphia.

Coulton, G. G. (1923). *Five Centuries of Religion*. Cambridge, England: Cambridge University Press.

Craik, K. J. W. (1948). The theory of the human operator in control systems. II. Man as an element in a control system. *British Journal of Psychology, 38*, 142–48.

Dewey, J. (1899). *The School and Society*. Chicago: University of Chicago Press.

Dodson, D. W. (1954). The integration of Negroes in baseball. *Journal of Educational Sociology, 28*, Oct., 73–75.

Donders, F. C. (1869). On the speed of mental processes. 1868–1869. In W. G. Koster (ed.). *Attention and Performance II: Acta Psychologica*, 1969, 30, 412–31.

Ellis, H. (1923). *The Dance of Life*. Boston: Houghton-Mifflin. P. 54.

Enoka, R. M. (1988). *Neuromechanical Basis of Kinesiology*. Champaign, IL: Human Kinetics.

Enos, W. F., Holmes, R.H., and Beyer, J. (1953). Coronary disease among United States soldiers killed in action in Korea. *Journal of the American Medical Association, 152*, 1090–93.

Gardiner, E. N. (1930). *Athletics of the Ancient World*. Chicago: Ares.

Gardiner, E. N. (1955). *Athletics of the Ancient World*, Oxford: Clarendon Press.

Harmon, A. M. (1936). *Lucian with an English Translation*. Cambridge, MA: Harvard University Press.

Hazenhyer, I. M. (1946). A history of the American Physiotherapy Association. *The Physiotherapy Review, 26*, No. 1. 174–84.

Henry, F. M. (1964). Physical education: An academic discipline. *Journal of Health, Physical Education and Recreation*. September, pp. 32, 33, 69.

Hill, A. V. (1927). *Muscular Movement in Man*. New York: McGraw-Hill.

Hollingworth, H. L. (1909). The inaccuracy of movement: With special reference to constant errors. *Archives of Psychology*. No. 13.

Hull, C. L. (1943). *Principles of Behavior*. New York: Appleton-Century-Crofts.

Inglis, B. (1965). *A History of Medicine*. Cleveland: World.

Jeannerod, M. (1985). *The Brain Machine*. Cambridge, MA: Harvard University Press.

Kaelber, L. (1998). *Schools of Asceticism: Ideology and Organization in Medieval Religious Communities*. University Park: Pennsylvania State University Press, 1998.

Kleiber, M. (1961). *The Fire of Life: An Introduction to Animal Energetics*. New York: Wiley.

Koerner, J. D. (1963). *The Miseducation of American Teachers*. Boston: Houghton-Mifflin.

Kraus, H., and Hirschland, R. P. (1953). Muscular fitness, and health. *Journal of Health, Physical Education and Recreation, 24*, 17–19.

Kretchmar, R. S. (1989). Exercise and sport science. *Big 10 Leadership Conference Report*. Champaign, IL: Human Kinetics.

Kroll, W. P. (1971). *Perspectives in Physical Education*. New York: Academic.

Lawler, L. B. (1964). *The Dance in Ancient Greece*. Middleton, CT: Wesleyan University Press.

Lonsdale, S. (1981). *Animals and the Origins of Dance*. New York: Thames and Hudson.

Lumpkin, A. (2011). *Introduction to Physical Education, Exercise Science, and Sport Studies*. New York: McGraw-Hill, 8th Edition.

Meerloo, J. A. M. (1960). The Dance: From Ritual to Rock and Roll—Ballet to Ballroom. Philadelphia: Chilton.

Miller, D. M. (1985). Images of ancient Greek sportswomen in the novels of Mary Renault. *Arete: The Journal of Sport Literature. 3:1*, 11–16.

Newell, K. M. (1990). Physical activity, knowledge types, and degree programs. *Quest, 42*, 243–68.

Park, R. J. (1989). The second 100 years: Or, can physical education become the renaissance field of the 21st century? *Quest, 41*, 1–27.

Penfield, W., and Rasmussen, T. (1950). *The Cerebral Cortex of Man: A Clinical Study of Localization of Function*. New York: Macmillan.

Poliakoff, M. (1987). *Combat Sports in the Ancient Greek World*. New Haven, CT: Yale University Press.

Pleket, H. W. (1975). Games, prizes, athletes and ideology. *Arena, 1*, 49–89.

Public Voice for Food and Health Policy (1991). Cited in *Boulder Daily Camera*, Knight-Ridder Tribune News, Sept. 12.

Quick, R. H. (1904). *Essays on Educational Reformers*. New York: D. Appleton.

Rasch, P. J., and Burke, R. K. (1963). *Kinesiology and Applied Anatomy: The Science of Human Movement*. Philadelphia: Lea and Febinger.

Sarton, G. (1954). *Galen of Pergamon*. Lawrence: University of Kansas Press.

Scanlan, T. F. (1984). Olympic dust, the Delphic laurel, and the Isthmian toil: Horace and Greek Athletics. *Arete, 1:2*, 163–75.

Schmidt, R. A. (1988). *Motor Control and Learning: A Behavioral Emphasis*. Champaign, IL: Human Kinetics.

Scott, M. G. (1945). Competition for Women in American Colleges and Universities. *Research Quarterly, 16*, 70–71.

Shawn, T. (1946). *Dance We Must*. London: Dennis Dobson.

Siedentop, D. (1990). *Introduction to Physical Education, Fitness, and Sport*. Mountain View, CA: Mayfield.

Singer, C., and Underwood, E. A. (1962). *A Short History of Medicine*. New York: Oxford University Press.

Slowikowski, S. S. (1989). The symbolic *Hellanodikai*. *Aethlon: The Journal of Sport Literature, 7:1*, 133–41.

Smyth, A. H. (1907). *The Writings of Benjamin Franklin*. Volume 5: New York: Macmillan.

Spedding, J. L., Heath, D. D., and Ellis, R. L. (1901). *The Works of Francis Bacon*. London: Longman.

Steindler, A. (1935). *Mechanics of Normal and Pathological Locomotion in Man*. Springfield: Charles C. Thomas.

Thorndike, E. L. (1927). The law of effect. *American Journal of Psychology, 39*, 212–22.

Todd, T. (1985). The myth of the muscle-bound lifter. *National Strength and Conditioning Association Journal, Vol. 7*, No. 3. 37–41.

Twietmeyer, G. (2008). A theology of inferiority: Is Christianity the source of kinesiology's second-class status in the academy? *Quest, 60*, 452–66.

U.S. Department of Health and Human Services (1991). Cited in *Boulder Daily Camera*, Sept. 14, 1991, Associated Press.

U.S. Department of Health and Human Services (1996). *Physical Activity and Health: A Report of the Surgeon General*. Atlanta, GA: U.S. Department of Health and Human Services, Centers for Disease Control and Prevention, National Center for Chronic Disease Prevention and Health Promotion.

Van Buskirk, L. (1928). Measuring the results of physical education. *Journal of Educational Method, 7*, 221–29.

Van Doren, C. (1930). *Benjamin Franklin*. New York: Viking.

Villeneuve, K., Weeks, D., and Schweid, M. (1983). Employee fitness: The bottom line. *Journal of Physical Education, Recreation and Dance. 54*, 35–36.

Von Holst, E. (1973). Relative coordination as a phenomenon and as a method of analysis of central nervous function. In R. Martin (ed. and trans). The collected papers of Erich von Holst: Vol. 1. *The Behavioral Physiology of Animals and Man.* (pp. 33–135). Coral

Gables: University of Miami Press. (Original work published in 1939).

Wuest, D. A., and Bucher, C.A. (1995). *Foundations of Physical Education and Sport.* St. Louis: Mosby.

Weston, A. (1962). *The Making of American Physical Education.* New York: Appleton-Century-Crofts.

Whited, C. V. (1970). *Sport Science, the Modern Disciplinary Concept of Physical Education.* In the Proceedings of NC-PEAM. Pp. 223–30.

Williams, J. F. (1930). Education through the physical. *Journal of Higher Education, 1,* 279–82.

Wood, T. D., and Cassidy, R. (1927). *The New Physical Education.* New York: Macmillan.

Woodworth, R. S. (1899). The accuracy of voluntary movement. *Psychological Review* 3, (2, Whole no. 13).

Chapter Three
Anatomical and Physiological Systems

Methods of Describing the Human Body

The Anatomical Position

The Anatomical Planes and Axes

The Skeletal System

Composition of Bones

Types of Bones

Function of Bones

Joints (or Articulations) and Connective Tissue

Scientific Terminology of Joint Movement

The Muscular System

Types of Muscular Tissue

Characteristics or Qualities of Muscle Tissue

The Structural Classification of Skeletal Muscles

Contraction of Skeletal Muscle

Motor Units

Types of Muscle Fibers

Principles of Motor Unit Recruitment

Types of Skeletal Muscle Contractions

CHAPTER THREE

Anatomical and Physiological Systems

"I shall examine that true book of ours, the human body of man himself."

— Vesalius (1651), the Father of Modern Anatomy

"He is a composite of chemicals-in-tissues within a framework of bones, a series of joints, a system of muscles, and network of blood vessels and nerves. He is designed for motion."

— Benjamin Ricci (1967)

STUDENT OBJECTIVES

1. To describe the major structural components of the skeletal, muscular, nervous, cardiovascular, and respiratory systems in the human body.
2. To identify the major functions of these systems in human movement and physical activity.

In this chapter, we will investigate some of the systems within the human body that influence movement and physical activity such as the skeletal, muscular, nervous, respiratory, and cardiovascular systems. While not all the physiological systems of the body are included in this chapter, the reader should gain an appreciation of the importance of each system that is discussed to the production of human movement.

It is convenient and quite traditional to study each system independently. We must recognize, however, that these systems are structurally and functionally interconnected, and much more research is needed to more fully understand the interactions among these systems during movement and exercise. We begin our exploration of the human body by first examining methods of describing human anatomy.

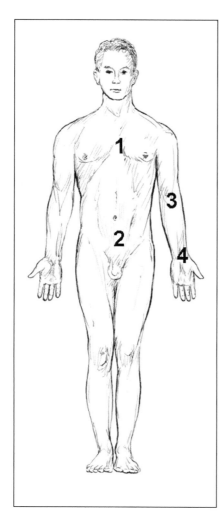

Figure 3.1

The anatomical position

METHODS OF DESCRIBING THE HUMAN BODY

The Anatomical Position

To avoid confusion, scientists have developed terms to describe the relative location of body parts based on the **anatomical** position (see Figure 3.1).

From this position it is possible to locate body parts relative to one another using the specific directional terms in Table 3.1. The anatomical position is the position of the human body standing erect with the arms by its sides and the palms of the hands facing forward.

Let's try using some of this terminology to describe the positions of the body marked with the number 1, 2, 3, and 4 in Figure 3.1. The following statements are true:

1 is superior to 2.

2 is inferior to 1.

2 is medial to 3.

3 is lateral to 2.

3 is proximal to 4.

4 is distal to 3.

These anatomical terms can be used to locate parts of the body relative to one another, whether internally or on the surface of the body.

TABLE 3.1
Anatomical Direction Terms and Their Descriptions

Anatomical Direction	Description
Superior	higher or above (also, toward the head)
Inferior	lower or below (also, away from the head
Lateral	away from the midline of the body
Medial	toward the midline of the body
Proximal	closer to the trunk
Distal	farther from the trunk
Anterior (ventral)	toward or at the front of the body
Posterior (dorsal)	toward or at the back of the body
Superficial	at or near the surface of the body
Deep	farther from the surface of the body

Figure 3.2

Three major (cardinal) anatomical planes

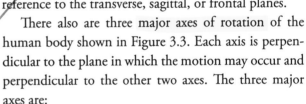

The Anatomical Planes and Axes

The structure of the human body can be further described with reference to the imaginary anatomical planes and axes. Figure 3.2 illustrates the three major (or cardinal) anatomical planes that correspond to the three dimensions of space. As one can see, the cardinal planes are perpendicular to one another and their common point of intersection is the center of the body (i.e., center of mass). The cardinal planes also are referred to as the midplanes because they travel through the center of the body and separate the body into equal halves by mass. The three cardinal planes are defined as follows:

1. The *cardinal transverse* plane divides the upper and lower halves of the body equally;
2. The *cardinal sagittal* plane divides the right and left halves of the body equally;
3. The *cardinal frontal* (or *coronal*) plane divides the anterior and posterior halves of the body equally.

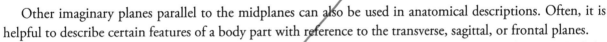

Other imaginary planes parallel to the midplanes can also be used in anatomical descriptions. Often, it is helpful to describe certain features of a body part with reference to the transverse, sagittal, or frontal planes.

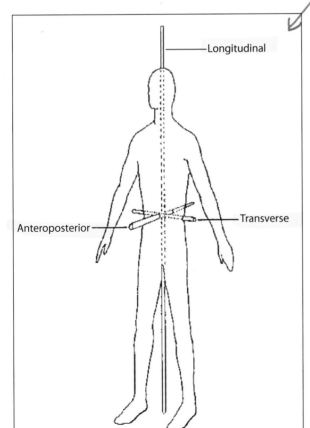

There also are three major axes of rotation of the human body shown in Figure 3.3. Each axis is perpendicular to the plane in which the motion may occur and perpendicular to the other two axes. The three major axes are:

1. The *transverse axis* passes through the body from side to side and is associated with movement in the sagittal plane;
2. The *anteroposterior axis* passes through the body from front to back and is associated with movement in the frontal plane;
3. The *longitudinal axis* passes through the body from top to bottom and is associated with movement in the transverse plane.

Figure 3.3

Three major axes of rotation of the human body

Figure 3.4

Anterior and posterior views of the human skeleton. From Wingerd, B.(2007).
The human body: Essentials of anatomy and physiology.
University Readers, San Diego. pg. 115.

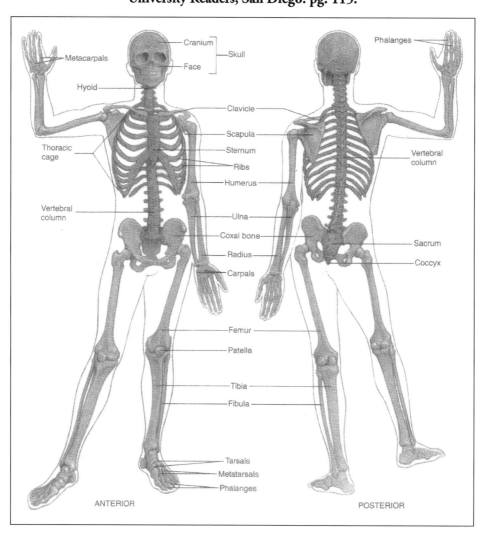

THE SKELETAL SYSTEM

A glance at the anterior and posterior views of the adult human skeleton shows not only the large number of bones, but also the variety of bones within the human body (see Figure 3.4). As shown in Table 3.2, there are a total of 206 bones in the skeletal system, and these are divided into two sections: the **axial skeleton**, made up of the skull, vertebral column, and thorax, and the **appendicular skeleton**, consisting of the upper and lower extremities.

Composition of Bones

What are bones made of? The outside layer of bone or the *compact tissue* is very hard due to inorganic mineral salts such as calcium carbonate and calcium phosphate, which make up about two-thirds of the dry weight of the bone. The remaining one-third of the bone is composed of organic material (20 percent of this being water). The interior spongy material or *cancellous* tissue is common in large flat bones and the knobby ends of long

TABLE 3.2

The Two Sections of Human Skeletal System

Section	Location	Number
Axial	cranium	8
	face	14
	ear	6
	hyoid	1
	vertebral column	26
	ribs	24
	sternum	<u>1</u>
		80 total
Appendicular	clavicles	2
	scapulae	2
	humerus bones	2
	radii	2
	ulnae	2
	carpals in the hands	16
	metacarpals in the hands	10
	phalanges in the fingers	28
	pelvic bones	2
	femurs	2
	patellas	2
	fibulas	2
	tibias	2
	tarsals in the feet	14
	metatarsals in the feet	10
	phalanges in the toes	<u>28</u>
		126 total

bones. With age or disuse, the amount of inorganic materials in bone decreases, making the bones more brittle and more susceptible to fracture or breakage.

Types of Bones

There is a variety of shapes and sizes of bones. In general, the shapes and sizes of bones can be categorized into four types: long, short, flat, and irregular. Figure 3.5 illustrates some examples of these four types of bones. *Long* bones like the femur are cylindrical with large knobby ends where ligaments and tendons attach near the joint (see below). These bones have evolved to endure great stresses like those created during locomotion. The

Figure 3.5
Examples of the four types of bones

knobby end of a long bone is called an *epiphysis* and the long shaft is termed a *diaphysis*. *Short* bones like the metatarsals of the foot are small and are composed of spongy material on the inside and a thin but hard exterior. *Flat* bones like the sternum also contain a spongy interior and hard surface, and generally have evolved to protect the vital organs of the body. *Irregular* bones like the vertebra are oddly shaped for a variety of purposes. All living and healthy bones are covered by a layered fibrous membrane called the *periosteum*, which contains both blood vessels and elastic fibers.

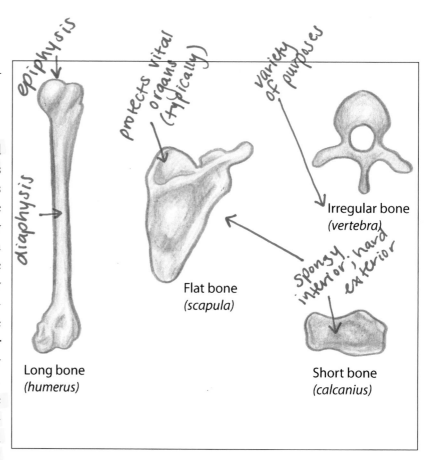

epiphysis

diaphysis

protects vital organs (typically)

variety of purposes

Irregular bone
(vertebra)

Flat bone
(scapula)

spongy interior, hard exterior

Long bone
(humerus)

Short bone
(calcanius)

Function of Bones

Because bones are hard, the uninformed person may think that the only purpose of bones is to provide structure and form for the body, much like the wooden frame of a house. Nothing could be further from the truth! There are actually five major functions of bones in the skeletal system:

5 Major Functions

1. Bones provide form and structure to the human body;
2. They provide protection for the internal organs (e.g., the skull protects the brain from severe external forces);
3. The bones are analogous to levers in a pulley system by providing attachments for muscle;
4. Bones (particularly the long and flat types) produce red and white blood cells as well as blood platelets essential for clotting;
5. Bones also store minerals such as calcium and phosphorus.

Joints (or Articulations) and Connective Tissue

The point in the skeletal system where two or more bones meet (or articulate) is called a **joint**. Depending on the type of joint, the bones may or may not actually be attached to one another by connective tissue called **ligaments**. Ligaments serve to prohibit motion in an undesirable plane and they limit the range or extent of normal movement. Bones are connected to the surrounding muscles by white fibrous tissue called **tendons** (see Figure 3.6). Joints are classified according to the amount of movement they allow. In this regard, there are three types of joints (see Figure 3.7):

1. *Immovable* (or Synarthroses) *Joints* involve the articulation of two or more bones that have been fused together. These types of joints do not permit movement by muscular force, but they do provide protection. For example, the immovable joints in the cranium, called the sutures, protect the brain by providing a "cushion" if an external force is applied. In addition, syndesmosis joints consist of dense fibrous tissue that bind bones together, like the radius and ulna.
2. *Slightly Movable* (or Amphiarthroses) *Joints* have a restricted range of motion because of the particular structure of the bones and the connective tissue. Types of these are symphysis joints located between the two pubic bones. Other types are synchrondrosis joints that separate the epiphyses from the diaphyses within long bones.
3. *Freely Movable* (or Diarthroses) *Joints* permit a wide range of movement. These types of joints are located in the upper and lower extremities such as the shoulder, elbow, wrist, hip, knee, and ankle joints. There are six types of diarthrodial joints:

3 types of Joints

① **Gliding Joints**—only permit gliding movements across adjacent bones. The joints of the carpal bones of the wrist and tarsal bones of the foot are good examples.

② **Hinge Joints**—allow for movement in only one plane of motion. There are hinge joints at the elbow and knee.

③ **Condyloid Joints**—characterized by an oval-shaped head of a bone (called a condyle) that fits into a shallow cavity of another bone. The joints between the carpals and metacarpals are good examples. Condyloid joints allow for motion in two planes.

④ **Pivot Joints**—allow for only one kind of movement, namely, rotation around the longitudinal axis of the bone. Turning of the head is permitted by the rotation of the atlas (first cervical vertebra supporting the skull) around the odontoid process of the axis (the second cervical vertebra).

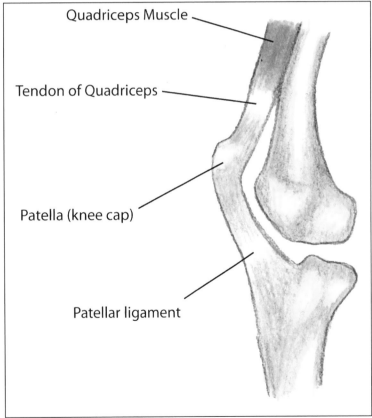

Figure 3.6
Example of a ligament and a tendon (in the knee joint)

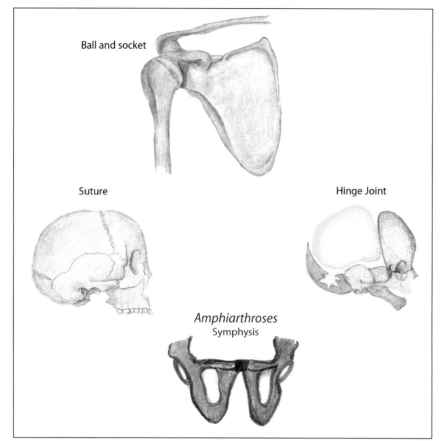

Ball and socket

Suture

Hinge Joint

Amphiarthroses
Symphysis

Figure 3.7
Type of joints

⑤ **Saddle Joints**—are shaped like a saddle and permit movement in two planes of motion. The only true saddle joint is the carpometacarpal joint of the thumb.

⑥ **Ball and Socket Joints**—permit the greatest amount of movement. Examples of these joints are the shoulder and hip joints.

Joints serve to limit or constrain the type of motion around them. In this way, each joint possesses a certain number of **degrees of freedom** (or planes of motion) in which the joint can move. For example, the hinge joint at the elbow and pivot joint of the atlas and axis allow only one degree of freedom while the ball and socket joint at the shoulder has three degrees of freedom. While the structural constraints of joints limit their degrees of freedom, the human body possesses an incredible total number of degrees of freedom, allowing much flexibility in the way it can move in the environment. The evolution of our arms and legs has greatly contributed to this increase in flexibility. As a challenge, how many degrees of freedom are there in the joints of the index finger in your hand? (Answer: 4. Remember that the metacarpal phalangeal joint is a condyloid joint allowing for *two* degrees of motion, unlike the other two joints that allow for only one degree of freedom in each.)

Scientific Terminology of Joint Movement

Beginning kinesiology students must be aware of some of the scientific terms used to describe joint motion and movement. **Flexion** and **extension** of a joint refers to a decrease and an increase, respectively, of the angle formed by the articulating bones. **Abduction** refers to a movement away from the midline of the body or body part, while **adduction** refers to a movement toward the midline of the body or body part. **Rotation** refers to the turning of a bone or segment around its vertical axis. **Elevation** means moving toward the superior position, while **depression** is moving toward the inferior position. **Supination** is the lateral (outward) rotation of the forearm and **pronation** is the medial (inward) rotation of the forearm. **Inversion** is the lifting of the *medial* border of the foot and **eversion** is the lifting of the *lateral* border of the foot. **Dorsiflexion** is the moving of the top of the foot toward the shin, while **plantar flexion** is the moving of the foot downward (i.e., pointing the toes). Finally, **circumduction** is a circular or cone-like movement of a body segment, such as the motion required to draw a circle using only the wrist joint. How many degrees of freedom (planes of motion) are required to produce circumduction: one, two or three? (Answer: 2).

· Key Terms ·

THE MUSCULAR SYSTEM

All types of movement, from the complex dance routines of the ballerina to a simple act of blinking one's eye, are made possible by the muscular system of our bodies. Our ability to move and function depends on over 600 muscles in our bodies, which comprise nearly 50 percent of total body mass. The muscles throughout the body vary considerably in size and shape.

Types of Muscular Tissue

There are three major types of muscular tissue: smooth, cardiac, and skeletal. Smooth muscle tissue is predominantly located in the walls of internal (visceral) organs such as the intestines, stomach, the pupil of the eye, and the blood vessels. Smooth muscle tissue, the fibers of which are smaller than the other two types, is controlled by the autonomic nervous system. Contraction of the smooth muscle fibers in the blood vessels, for example, impedes blood flow and causes blood pressure to rise. Cardiac muscle tissue comprises the walls and various partitions of the heart. Interestingly, the cardiac tissues in the heart can contract rhythmically without receiving impulses from the nervous system. Fortunately, we do not have to "will" our cardiac muscle tissues to contract! In addition, cardiac muscle tissue fibers run together to form a continuous branching network. Skeletal muscle tissue (also called "striated" or striped because of its appearance) is the most predominant type in the human body. There are over 430 skeletal muscles throughout the body that are mostly voluntarily controlled. It is these skeletal muscles that specifically contribute to our ability to perform a wide variety of skilled movements. Figure 3.8 illustrates the superficial skeletal muscles from both an anterior and posterior view of the anatomical position.

Characteristics or Qualities of Muscle Tissue

All muscle tissue, regardless of type, has four distinct characteristics: extensibility, elasticity, excitability, and contractility. Extensibility means that the muscle tissue can be stretched beyond its normal resting length, something like a rubber band. Also like a rubber band, muscle tissue has elasticity in that it will return to its normal resting length following stretch. Excitability refers to the fact that muscle tissue is able to receive and respond to a wide variety of stimulation. Finally, contractility means that muscle tissue can shorten in length following stimulation. The contractions of smooth and cardiac muscle tissue are slow relative to skeletal muscle tissue.

The Structural Classification of Skeletal Muscles

While there is a wide variety of skeletal muscle shapes, they can be generally classified into four categories: longitudinal, radiate, fusiform, and penniform (diagonal). Figure 3.9 illustrates these four types of classifications, with the latter type further divided into sub-categories. Longitudinal muscles have fibers that run parallel with the long axis of the muscle. Radiate muscles have fibers that fan outward from a single attachment. Fusiform muscles are composed of muscle fibers that are in the form of a spindle. Fusiform muscles are designed for speed of movement and allow for a greater range of movement. However, fusiform muscles do not produce as great a muscular force compared to penniform muscles. Penniform muscles have fibers that are arranged somewhat like a feather. The fibers of penniform muscles emanate diagonally away from a tendon. In general, the larger the muscle, the more force it can produce. This is because the larger muscle generally contains more individual muscle fibers that contract to produce muscular force and the fibers are usually thicker, which implies more contractile tissue. Now let us examine the interior of the skeletal muscle and clarify what we mean by the term *muscle contraction*.

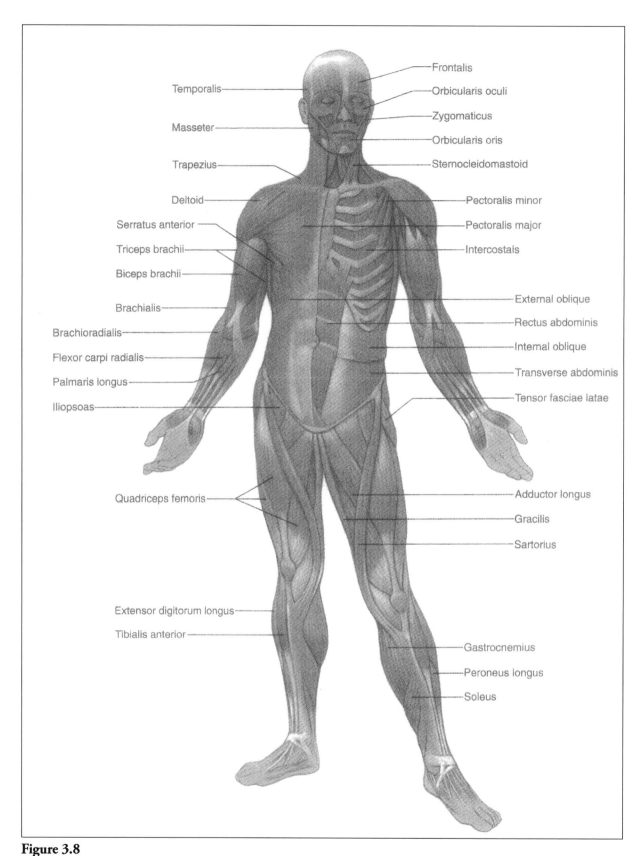

Figure 3.8

Anterior (left) and posterior (right) views of the skeletal muscles. From Wingerd, B. (2007). *The human body: Essentials of anatomy and physiology.* **University Readers, San Diego. Figure 7.12 and Figure 7.13, pgs. 160–161.**

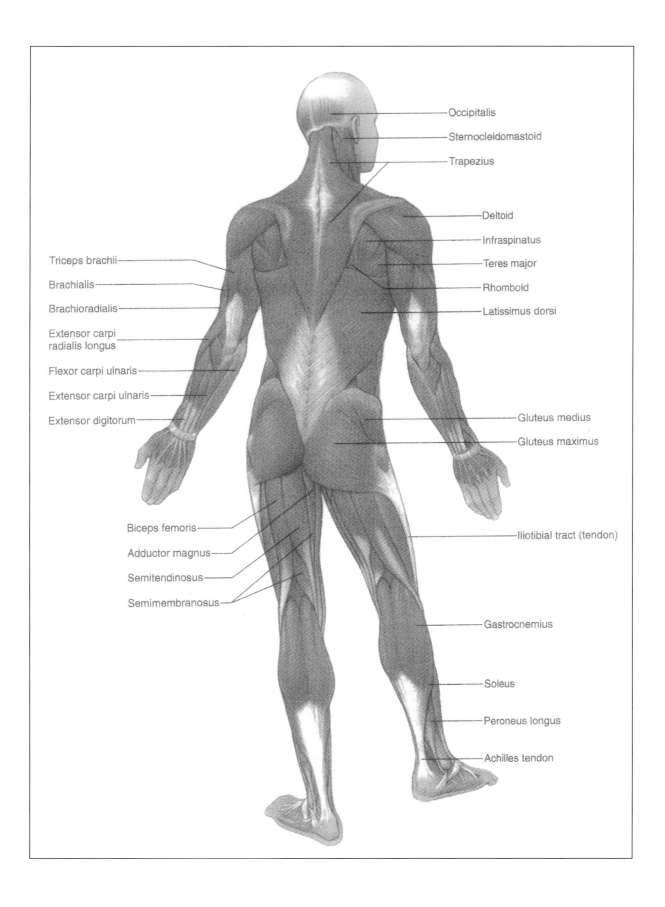

Occipitalis

Sternocleidomastoid

Trapezius

Deltoid

Infraspinatus

Teres major

Rhomboid

Latissimus dorsi

Triceps brachii

Brachialis

Brachioradialis

Extensor carpi radialis longus

Flexor carpi ulnaris

Extensor carpi ulnaris

Extensor digitorum

Gluteus medius

Gluteus maximus

Biceps femoris

Adductor magnus

Semitendinosus

Semimembranosus

Iliotibial tract (tendon)

Gastrocnemius

Soleus

Peroneus longus

Achilles tendon

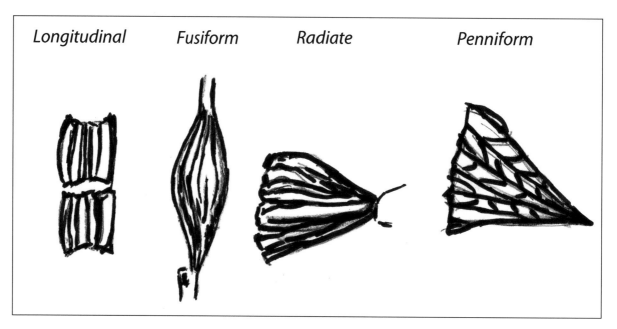

Figure 3.9
Four categories of muscle

Contraction of Skeletal Muscle

When a muscle is stimulated by a nerve impulse, its overall length becomes shorter and the muscle appears fatter (see Figure 3.10). To Galen, Descartes, and countless scientists up until the 18th century, these changes were due to the muscle filling up with "animal spirits," a mysterious substance supposedly produced by the brain. This hypothesized muscular response is similar to blowing air into a balloon, which would imply an increase in muscle volume. But based on experiments by Swammerdam and others in the 18th century, we now know this is *not* what happens when a muscle is stimulated. An understanding of muscle anatomy can tell us why.

Figure 3.11 shows progressively magnified sections of a fusiform skeletal muscle in the gastrocnemius. In **a**, you can see the belly of the muscle and its attachments to tendons on both ends. In **b**, a small area of the muscle is magnified showing individual muscle fibers. As you can see, muscles consist of a great number of muscle fibers (cells). In **c**, a small section of the muscle fiber is magnified showing that it is composed of many **myofibrils**, which appear striped or striated with light and dark bands. A closer look shows that there are actually several areas or zones of differing patterns, namely the Z lines, I bands, A band, and H zone. Finally, in **d** we can see that between each Z line (an area called the **sarcomere**), the light and dark zones are caused by an interesting

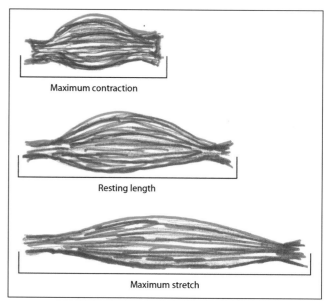

Maximum contraction

Resting length

Maximum stretch

Figure 3.10
The shape of a skeletal muscle depends on its overall length.

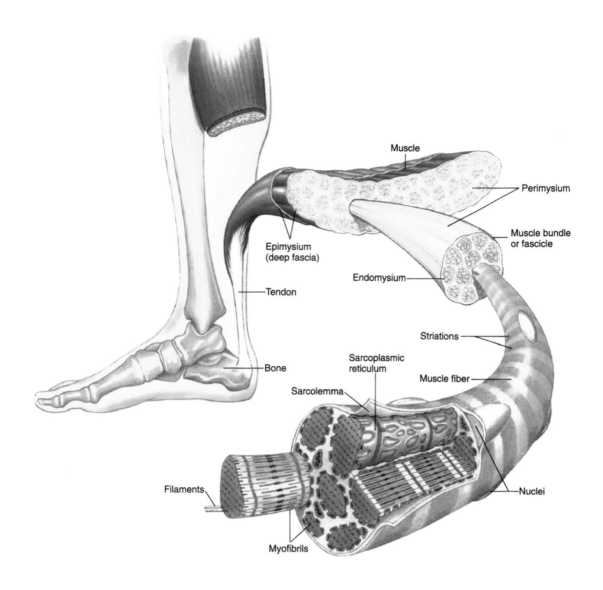

Figure 3.11
Progressively magnified sections of a skeletal muscle. From Wingerd, B. (2007). *The human body:*
Essentials of anatomy and physiology. **University Readers, San Diego. Figure 7.2, pg. 146.**

microfilaments

arrangement of thick and thin protein filaments. The **actin molecule** is thinner than the **myosin molecule**, and the two types of molecules are interlaced with one another. According to the "sliding filament" theory (Huxley, 1958), when the muscle is stimulated, biochemical changes occur, which cause the actin molecules to slide past the myosin molecules such that the sarcomere becomes shorter or contracts. When this contraction process is multiplied throughout the muscle, the *total* length of the muscle becomes shorter and the muscle, as a whole, becomes fatter due to increasing overlay of the two filaments. Thus, the shortening of the muscle occurs because of myofibrillar contraction, not because of a change in muscle volume, as Galen and Descartes thought. In Figure 3.12, we can see the relative positions of the actin and myosin molecules depending on whether the muscle is stretched, relaxed, or contracted.

 Motor Units. There are motor neurons (e.g., *alpha* motor neurons) whose cell bodies originate in the spinal cord and whose axons carry nerve impulses to a set of muscle fibers in a muscle. When a nerve impulse travels

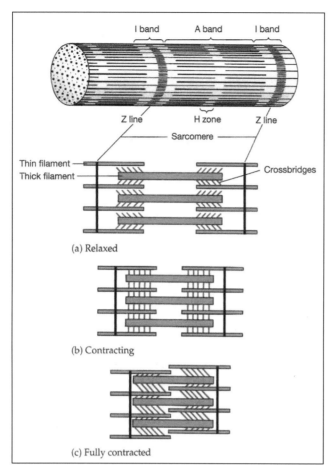

Figure 3.12

The relative positions of actin and myosin filaments in relaxed (a), contracting (b) and fully contracting (c) muscle. From Wingerd, B. (2007). *The human body: Essentials of anatomy and physiology.* University Readers, San Diego. Figure 7.8 pg. 153.

down the motor neuron, it will cause a contraction of the associated set of motor fibers. A motor neuron and its set of associated muscle fibers in the muscle is called a **motor unit**. The ratio between the motor neuron and the number of muscle fibers it stimulates varies considerably throughout each muscle and across muscles. Smaller ratios are associated with finer control of overall muscular contraction. The motor units in muscles of the eye, fingers, lips, and tongue have low ratios (between 1:10 to 1:100). The motor units of the leg and trunk muscles, for example, have very high ratios (up to 1:2000).

Types of Muscle Fibers. There are three major types of muscle fibers, partly distinguished by the manner in which they respond to stimulation from the motor nerve. **Fast twitch** fibers contract rapidly and produce greater forces, but also fatigue rather quickly. **Slow twitch** fibers contract slowly, produce smaller forces, and are more resistant to fatigue. **Fast oxidative-glycolytic** (FOG) fibers contain characteristics of both fast and slow twitch fibers, and have force- and fatigue-resistant characteristics intermediate to the two other fiber types. Fast twitch fibers are white in color while slow twitch and FOG fibers are redder. The distribution of the three fibers varies across the muscles and differs considerably across individuals. In general, individuals who possess a greater proportion of fast twitch fibers tend to be more powerful while individuals with more slow twitch fibers tend to have more muscular endurance. Research has suggested that elite athletes in different sports tend to have different distributions of slow and fast twitch fibers. Figure 3.13 illustrates the percentage of slow and fast twitch fibers in marathon runners, sprinters and untrained individuals. Sprinters tend to have more fast twitch fibers and marathoners tend to have greater numbers of slow twitch fibers while untrained people have about the same distribution of fast and slow twitch fibers. So, while having more fast twitch or slow twitch muscle fibers does not guarantee that one will become an Olympic sprinter or marathon runner, respectively, the distribution of muscles fibers is a contributing factor to success in certain sports or physical activities.

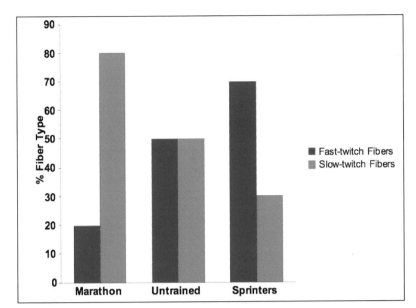

Figure 3.13
Percentage of fast and slow twitch muscle fibers in different types of athletes, (adapted from Hay, J..G., and Reid, J. G. *Anatomy, Mechanics, and Human Motion.* Prentice Hall. 1988)

Principles of Motor Unit Recruitment. As stated above, when the muscle fibers in a given motor unit are stimulated by a motor nerve, all the fibers in the unit contract together. This is called the **all-or-none principle**. This principle does not apply to the whole muscle, however. Governing motor unit control over the *whole* muscle is the size principle and gradation of muscular contraction principle. The **size principle** states that motor units containing smaller muscle fibers are recruited before larger muscle fibers during a muscular contraction. The **gradation of muscular contraction principle** states that the number of motor units recruited during a muscular contraction depends on the intensity of the task. A smaller number of motor units is required for weak contractions and a larger number of motor units is required for strong contractions. These latter two principles allow for efficient use of motor units during a variety of muscular contractions.

Rate coding is another type of principle of motor unit recruitment. Rate coding refers to the fact that individual motor units may fire at different frequencies, typically between 3–50 impulses per second. Rate coding varies with fiber type and type of movement (Hamill and Knutzen, 1995).

Types of Skeletal Muscle Actions. When a skeletal muscle contracts and its overall length changes, this is termed an **isotonic** action. There are two types of isotonic action: concentric and eccentric action. **Concentric action** results in an overall shortening of the muscle such that the proximal and distal ends of the muscle move toward the center of muscle. An example of a concentric action is the shortening of the biceps muscle when we flex our elbow. **Eccentric action** results in an overall lengthening of muscle such that the two ends of the muscle move away from the center of the muscle. Suppose you are arm wrestling with someone and are "losing the battle!" As your arm is about to be pinned, your opponent's arm muscles are undergoing concentric (or shortening) action but your arm muscles are undergoing eccentric (or lengthening) action (Piscopo and Baley, 1981, p. 26). Another major type of action is an **isometric action** in which the muscle fibers contract, but the limb does not move. For example, suppose that while arm wrestling, you and your opponent are at a standstill. The relative amount of force produced by each of you is the same. Consequently, there is no significant change in the positions of your arms. Both you and your opponent are undergoing isometric action.

Muscle Spindles. Lying in parallel to skeletal muscle fibers are muscle spindles, which are receptors designed to signal changes in the length of the muscle. There is also evidence that muscle spindles contribute to the sensation of movement along with other movement receptors located in the joints and tendons. Muscle spindles are the only receptors in the body that receive efferent or motor input. Figure 3.14 illustrates the anatomy of the muscle spindle along with its various motor and sensory pathways.

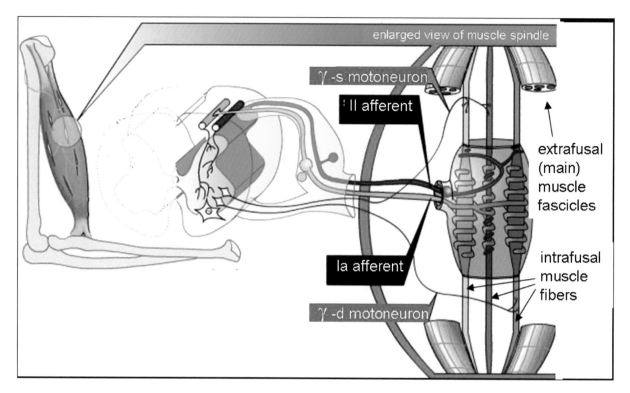

Figure 3.14

Mammalian muscle spindle showing typical position in a muscle (left), neuronal connections in spinal cord (middle) and expanded schematic (right). The spindle is a stretch receptor with its own motor supply consisting of several intrafusal muscle fibers. The sensory endings of a primary (group Ia) afferent and a secondary (group II) afferent coil around the non-contractile central portions of the intrafusal fibers. Gamma motoneurons activate the intrafusal muscle fibers, changing the resting firing rate and stretch-sensitivity of the afferents (Source: http://commons.wikimedia.org/wiki/File:Muscle_spindle_model.jpg. Copyright in the Public Domain.)

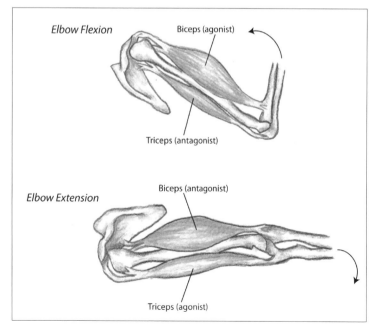

Figure 3.15

Example of agonist and antagonist muscles. In elbow flexion (above), the biceps is an agonist muscle. In elbow extension (below), the triceps is an agonist muscle.

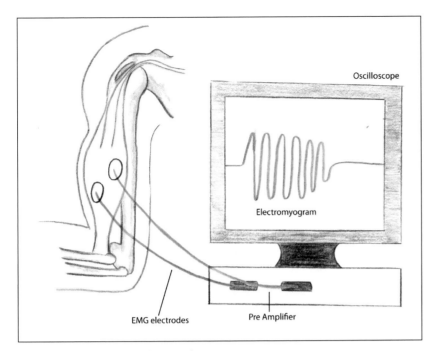

Figure 3.16
Electromyography (EMG).
EMG electrodes placed on the surface of the skin above the biceps muscle pick up the minute voltage changes as a result of a muscle contraction. These voltages are amplified and then displayed on an oscilloscope.

Oscilloscope

Electromyogram

EMG electrodes   Pre Amplifier

Functions of Skeletal Muscles

Individual muscles can only pull in one direction, they cannot push. To assist in joint movement, it is necessary for more than one muscle to be attached around the joint. Surrounding most joints are *sets* of muscles that pull on tendons that are attached to the bones of a given joint. An individual skeletal muscle may serve several major uses or functions during movement: as an agonist, antagonist, stabilizer, neutralizer, or synergist.

Agonist and Antagonist Muscles. When a given muscle contracts and causes the joint to move in a desired direction, it is called an **agonist muscle**. For example, during elbow flexion, the biceps brachii muscle is the agonist muscle for this movement (the brachioradialis and the brachialis also are elbow flexors). The agonist muscle may be a *prime mover* if it is the major muscle involved in the initiation of the movement. A muscle is an *assistant mover* if it plays a minor role in agonist activity. On the opposite side of the joint is the triceps muscle, which for elbow flexion is the **antagonist muscle**. The antagonist muscle often relaxes while the agonist contracts to allow for smooth joint movement. The antagonist muscle also helps stop or brake the movement, particularly in fast movements (Lestienne, 1979). It should be obvious that whether a muscle is an agonist or an antagonist depends on the direction of the movement. If I *extend* my elbow joint, the triceps is now the agonist muscle and the biceps is the antagonist (see Figure 3.15).

Stabilizers. A muscle can be a stabilizer when it helps to prevent the movement of one end of a prime mover, thereby allowing another muscle to work effectively. To produce pure elbow flexion, it is necessary to stabilize the shoulder joint so that the elbow flexors can contract effectively and pull the lower arm toward the upper arm.

Neutralizers. An individual muscle may also be a neutralizer when it neutralizes the effects of a movement caused by another muscle. For example, the left and right external oblique muscles contribute to left and right trunk rotation, respectively. However, during a sit-up exercise, for example, these two muscles also cause trunk flexion when contracted together because their individual contributions to trunk rotation are canceled out or neutralized.

Synergists. Muscles acting as synergists assist in various ways to allow a prime mover or another muscle to work effectively. A synergist muscle can act as an assistant mover, a stabilizer, or a neutralizer. Any of these three functions can therefore be considered synergistic.

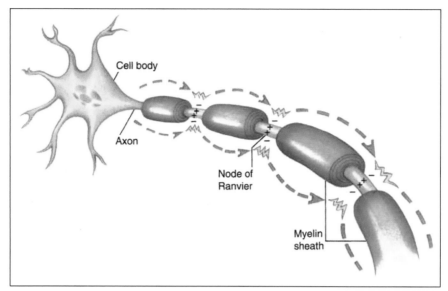

Figure 3.17

The neuron is the fundamental component of the nervous system.
Figure 8.9, *The Human Body: Essentials of Anatomy and Physiology*, ed. Bruce Wingerd, pp. 189. Copyright © 2009 by Cognella, Inc.

Methods for Studying Skeletal Muscle Function

There are several ways to study the function of muscle as it relates to movement. The five major methods are dissection, biopsy, palpation, electrical stimulation, and electromyography. Let us briefly examine each method.

Dissection. This procedure, along with palpation, was the first procedure used to understand muscle function. As you will recall, Galen and Vesalius were among some of the early prominent scientists to use dissection in understanding muscle anatomy. Dissection is a surgical procedure used on cadavers to expose the muscle (and other internal organs) for observation.

Biopsy. The muscle biopsy procedure was developed in the late 1960s by Swedish exercise physiologists in order to understand how the muscle metabolizes various nutrients. The technique involves inserting a hollow needle into the muscle through the skin. A stylet is inserted into the needle and small piece of muscle tissue is removed, quickly frozen and then later analyzed biochemically or examined under a microscope (Brooks, Fahey, and White, 1996). This technique is extensively used today by exercise physiologists all over the world.

Palpation. This is a technique of examining skeletal muscle by touch with the hands. It is particularly useful for determining areas of muscle soreness and tenseness, but it is also used to determine whether certain muscles are active during movement and exercise. The technique is used by medical doctors, physical therapists, athletic trainers, and researchers. **Electrical Stimulation**. In the last chapter we learned that Galvani discovered that muscle tissue is sensitive to electrical stimulation. Since that time, electrical stimulation has been used for therapeutic and research purposes.

Electromyography. Muscles are not only sensitive to electricity, but they also produce it during contraction. Minute voltages of electricity can be picked up by special electrodes. The signal can be amplified and either displayed on an oscilloscope or stored in a computer for further analysis (see Figure 3.16). Using electromyography (EMG), electrical activity from individual motor units or from the whole muscle can be detected. Details about this technique can be examined in an excellent book on the subject by Basmajian and De Luca (1985).

THE NERVOUS SYSTEM

Thought, emotions, intentions, actions, sensations, perceptions, metabolic functions—all of these activities would be impossible without our nervous system. The primary functional unit in the nervous system is the nerve cell or **neuron**. The late astrophysicist Carl Sagan was well known for his remark that in the universe there are "billions and billions" of stars. Equally impressive is the fact that there are well over 100 *billion* neurons in the nervous system and the number of interconnections among them is almost countless! Because of this complexity, one might say that the brain, and the nervous system as a whole, are like a computer. However, even the most advanced computer is not as complex as the human nervous system. In spite of this complexity, much is known about the anatomy of the nervous system and the role the nervous system plays in the control of movement and physical activity. Let us begin with a description of the neuron. Following this description, we will explore the anatomical and functional divisions of the nervous system.

The Neuron

The basic or fundamental component of the nervous system is the neuron (see Figure 3.17). The neuron has two basic intrinsic properties: irritability and conductivity. **Irritability** (or excitability) refers to the ability of the neuron to respond to stimulation. **Conductivity** refers to the ability of the neuron to send information to other structures. In fact, the neuron is designed to send or propagate impulses to muscles, glands, or other neurons.

There are three major parts to a neuron: the cell body, dendrites, and axon. The nucleus of the neuron is contained in the **cell body** (or soma). **Dendrites** carry nerve impulses *toward* the cell body and **axons** carry nerve impulses to other neurons and structures. Axons may vary from under one millimeter to over one meter in length! Figure 3.18 illustrates the three major types of neurons: sensory (afferent), motor (efferent), and interneurons. Sensory neurons contain receptors that are sensitive to specific types of stimulation such as temperature, pain, sound, light, and movement. Motor neurons carry impulses that eventually stimulate the muscles and glands. Finally, interneurons lie entirely within the central nervous system. Nearly 99 percent of all neurons in the body are interneurons. Interneurons have two main functions. First, they serve as connectors to

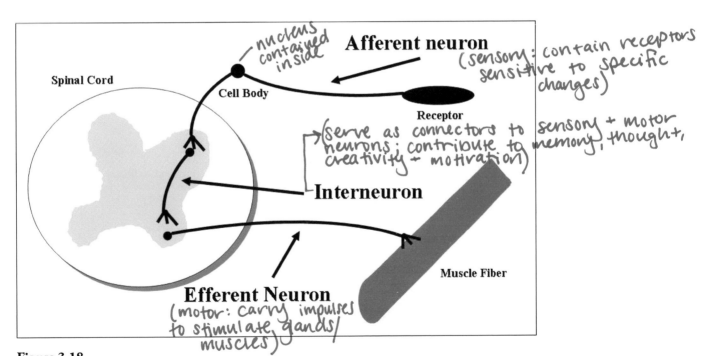

Figure 3.18

Three types of neurons (afferent, interneuron, and efferent).

sensory and motor neurons. Second, interneurons are with a variety of mental phenomena including memory, thoughts, creativity, and motivation.

The irritability of a neuron depends on its own intrinsic or internal condition and on surrounding influences. Some neurons may receive up to 100,000 inputs! These inputs can be facilitating in that they tend to cause the neuron to send an impulse, or they can be inhibitory, which reduce the likelihood of the neuron sending an impulse. Each neuron has its own **threshold** of activation. The threshold is like the act of shooting a gun (Sherwood, 1991). Squeezing the trigger increases the likelihood that the gun will fire. However the gun will not fire until the necessary force is reached. Once this force is attained, the gun fires. In an analogous fashion, once the threshold of the neuron is reached, the neuron will send an impulse down its axon in an "all-or-none" fashion.

The speed of the nerve impulse down an axon depends on whether the axon is surrounded by myelin fibers that serve as a coating or insulator, just like rubber around an electric wire. **Myelin** is primarily a fatty substance that insulates the impulse from surrounding, and potentially interfering, electrical activity. The speed of myelinated axons is up to 50 times *faster* than unmyelinated axons! It is known that the disease **multiple sclerosis** causes *de*myelination in various neurons throughout the nervous system, which slows the speed of nerve impulse transmission and increases the amount of interference.

An impulse traveling down an axon of a neuron will eventually reach what is called a **synapse**, or the junction between adjacent neurons. When the impulse reaches the synapse, some type of neurotransmitter stored at the end of the axon is released. **Neurotransmitters** alter the receptibility of the so-called *post*-synaptic neuron to the impulse. Some synapses are excitatory in that they facilitate the transmission of the impulse, whereas others are inhibitory in that they reduce the likelihood that the impulse will be carried on. Some diseases (e.g., Parkinson's disease) and drugs (e.g., alcohol) can alter the effectiveness of neurotransmitters.

[margin handwritten note: excitatory vs inhibitory synapses]

Anatomical Divisions of the Nervous System

The nervous system can be structurally divided into the **central nervous system** (CNS) and the **peripheral nervous system** (PNS). The CNS includes the brain and the spinal cord. The brain is surrounded by the cranium, and the spinal cord is contained within the bones of the vertebral column. The PNS is made up of all nervous tissue outside of the CNS that carries nerve impulses to and from the skin, muscles, glands, and visceral tissue.

Central Nervous System. The CNS is composed of the brain and spinal cord and is responsible for receiving and interpreting information picked up by various sense organs about the external (outside the body) and internal (within the body) environment. The CNS is also responsible for sending motor signals to glands and muscles to promote secretions and movement, respectively.

The Brain

Shown in Figure 3.19 is what some have termed "the most complicated, mysterious, awesome organ on earth"—the human brain (Sherwood, 1991, p. 85). Weighing approximately 3 lb. (1.36 kg), the adult brain stores 100 trillion bits of information over a 70-year period. This incredible amount of information is equal to 500,000 sets of Encyclopaedia Britannica, which, when stacked, would reach almost 442 miles (713km) high!

The exterior of the brain is wrinkled like a walnut, and these wrinkles are called **convolutions**. From an evolutionary point of view, the brain may have grown faster than the cranium. To find more space within the cranium, the brain may have folded in on itself causing the convolutions. The thin outer covering of the brain is called the **cerebral cortex**, responsible for a variety of cognitive functions. The brain is divided

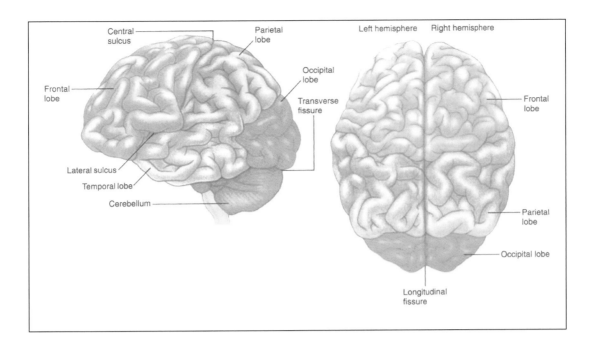

Figure 3.19

The human brain showing the four major regions or lobes, and various brain structures discussed in the text. The basal ganglia, thalamus, and hypothalamus are located deep within the brain. From Wingerd, B. (2007). *The human body: Essentials of anatomy and physiology.* University Readers, San Diego. Figure 8.18a, pg. 199.

into the right and left hemispheres, which are connected by a thick band of 300 million neuronal axons, called the **corpus callosum**. There are four major regions of the brain called cortical lobes: the frontal, temporal, parietal, and occipital lobes.

The brain is thought to function as an integrated whole. However, various regions of the brain have been shown to contribute to certain functions:

BRAIN DIVISIONS

Brain Stem (Midbrain, Pons, Medulla)

1. Responsible for regulation of certain muscular reflexes associated with posture.
2. Regulation of cardiovascular, respiratory, and digestive systems.
3. Helps control arousal of cerebral cortex.
4. Acts as a sleep center.

Cerebellum

1. Assists in the maintenance of balance and muscle tone.
2. Contributes to coordination of voluntary movement.

Hypothalamus

1. Assists in the maintenance of body temperature.
2. Helps control thirst and food intake.
3. Helps regulate urine output.
4. Contributes to the control of emotional behavior.

Thalamus

1. Serves as a major "relay station" for all synaptic input.
2. Contributes to some crude awareness of sensation.

Basal Ganglia

1. Contribute to the initiation of voluntary movement.
2. Assist in the coordination of slow voluntary movement.
3. Act as an inhibitor of muscle tone.

Cerebral Cortex

1. Responsible for complex mental functions such as thinking, memory, creativity, and decision making and language.
2. Interprets conscious sensory information.
3. Responsible for the voluntary control of movement.

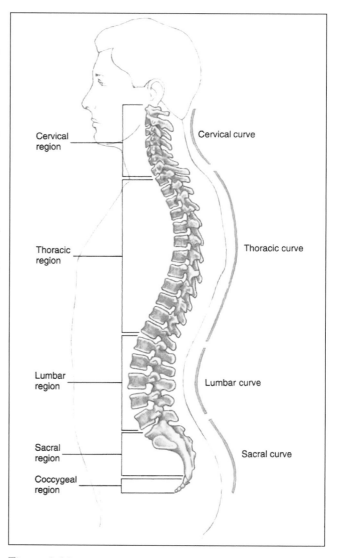

Figure 3.20

Four major regions of the spinal cord. From Wingerd, B. (2007). *The human body: Essentials of anatomy and physiology.* University Readers, San Diego. Figure 6.11a, pg. 122.

The Spinal Cord

The traditional view of the spinal cord was that it only served as a relay station between the brain and more distal organs. Today we know the spinal cord is involved in much more sophisticated activities. It is true that the spinal cord serves as a relay station in that it contains the motor and sensory pathways that link the brain to the muscles, glands, sensory receptors, and other vital organs. Research has also shown that the spinal cord is a major integration center of descending motor signals and sensory information. The spinal cord also plays a major role in the control of many reflexes and in the coordination of locomotion.

The spinal cord extends from the base of brain through the vertebral canal and connects to the spinal nerves (which are part of the peripheral nervous system). Figure 3.20 illustrates the four major sections or regions of the spinal cord: cervical, thoracic, lumbar, and sacral. It contains 31 pairs of spinal nerves, with each pair of nerves

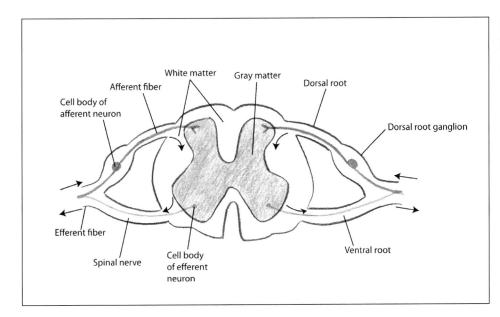

Figure 3.21
Cross sections of the spinal cord.

Labels in figure: White matter, Gray matter, Dorsal root, Afferent fiber, Cell body of afferent neuron, Dorsal root ganglion, Efferent fiber, Spinal nerve, Cell body of efferent neuron, Ventral root

supplying efferent and afferent neurons to a specific region of the body called a **dermatome**. Figure 3.21 shows a cross section of the spinal cord and the associated spinal nerves.

The interior H-shaped region of the spinal cord is gray matter composed of unmyelinated motor neurons and interneurons. The cell bodies of the sensory neurons lie outside of the spinal cord. The surrounding white matter of the spinal cord is made up of myelinated sensory and motor neurons within the major pathways to and from the brain, respectively. In general, the motor pathways are contained in the front or anterior portion of the cord, and the sensory pathways lie in the back or posterior part of the cord.

The Peripheral Nervous System. The peripheral nervous system is that part of the nervous system anatomically lying outside the central nervous system. The peripheral nervous system includes the 31 pairs of spinal nerves containing both sensory and motor nerves as well as sensory receptors for vision, touch, taste, hearing, smell, and proprioception. There are also 12 pairs of cranial nerves emerging from the brain that can carry sensory, motor, or both types of information.

Functional Divisions of the Nervous System

Functionally, the nervous system can be divided into the somatic nervous system and the autonomic nervous system. The somatic nervous system is involved in the voluntary control of skeletal muscles. The autonomic nervous system is responsible for controlling smooth muscle, cardiac muscle, and the glands. This system is largely involuntary and subconscious—that is, it is not necessary for you to willfully control such processes as circulation, digestion, sweating, breathing, and the dilation of the pupils in the eyes. The autonomic nervous system can be further subdivided into the sympathetic and parasympathetic nervous systems. The sympathetic nervous system is largely facilatory, while the parasympathetic nervous system is largely inhibitory. Thus, these nervous systems tend to act in opposition to one another. For example, the effects of sympathetic stimulation causes an increase in heart rate, while parasympathetic stimulation causes a decreased heart rate. The sympathetic nervous system typically dominates in life-threatening situations, while the parasympathetic dominates in quiet, relaxing situations.

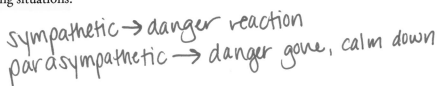

sympathetic → danger reaction
parasympathetic → danger gone, calm down

THE CARDIOVASCULAR SYSTEM

The cardiovascular system is composed of the heart, blood vessels, and the blood. The <u>heart</u>, a muscular organ, acts as a <u>pump</u> by forcing the blood through the myriad of blood vessels in the body. The <u>blood carries essential nutrients</u>, oxygen, carbon dioxide, and other important substances to and from all parts of the body. Figure 3.22 illustrates the path of circulation of the blood in relation to the heart. **Pulmonary circulation** carries blood to and from the lungs, where oxygen and carbon dioxide are exchanged during respiration. **Systemic circulation** carries blood between the heart and the various organ systems. Let us begin with a brief overview of the anatomy of the heart.

Anatomy of the heart. The heart is located in the center of the chest cavity (see Figure 3.23). It is composed primarily of cardiac muscle tissue and is the approximate size of a large, clenched fist. The <u>right and left sides of the heart function as two separate pumps</u>. Each side contains two separate chambers separated by valves that only allow blood to flow in one direction. The upper chambers are called the **atria** (atrium is singular), and receive blood returning to the heart. The lower chambers are called **ventricles**, and receive the blood from the atria and pump blood away from the heart.

The Blood Vessels. The **arteries** are blood vessels that carry blood away from the heart, and the **veins** are blood vessels that carry blood to the heart. The **capillaries** are very small blood vessels that serve as the sites for exchange of materials (e.g., oxygen, carbon dioxide, nutrients) between the blood and tissues. The average diameter of a capillary is 7 micrometers, where 1 micrometer is equal to one millionth of a meter! Between the arteries and capillaries are the **arterioles**, which help distribute blood flow throughout the tissue. Finally, the **venules** connect the capillaries to the veins.

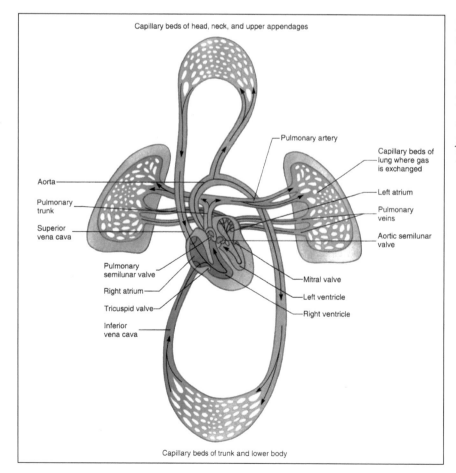

Figure 3.22

The path of circulation of the blood in relation to the heart. From Wingerd, B.(2007). *The human body: Essentials of anatomy and physiology.* **University Readers, San Diego. Figure 12.8, pg. 291.**

Figure 3.23

The human heart (Source: Figure 12.3, The Human Body: Essentials of Anatomy and Physiology, pp. 285. Copyright © 2009 by Cognella, Inc.)

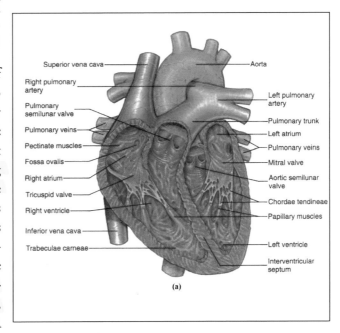

(a)

Electrical Activity of the Heart. Nearly all of the cardiac muscle cells are contractile cells that do the mechanical work of pumping blood. However a small number of what are called **autorhythmic cells** have a peculiar but nonetheless important property—namely, initiating and transmitting the nerve impulses that cause the cardiac muscle to contract. Interestingly, the autorhythmic cells are capable of sending rhythmic nerve impulses by themselves. Cardiac cells responsible for autorhythmic behavior are located in four areas of the heart: the **sinoatrial node** (**SA node**), the **atrioventricular node** (**AV node**), **the bundle of His**, and the **Purkinje fibers** (see Figure 3.24). The SA node is considered the "pacemaker" of the heart, because its resting rhythm is about 70 impulses per minute (the normal resting heart rate), and it dominates the slower rhythms of the other autorhythmic cells. The coordinated interaction among these cells is responsible for contraction of the heart chamber. However, lack of sleep, anxiety, and certain drugs like caffeine, alcohol, and nicotine can affect the coordination of the SA node with the other autorhythmic cells.

Heart Rate, Stroke Volume, and Cardiac Output. There are several mechanical measures of heart function. **Heart rate** is the beats per minute of the heart, and as mentioned above, is approximately 70 beats per minute at rest, established by the dominance of the SA node. **Stroke volume** is the volume of blood pumped per beat (or stroke). **Cardiac output** is the volume of blood pumped by each ventricle per minute and is the product of heart rate and stroke volume. Therefore,

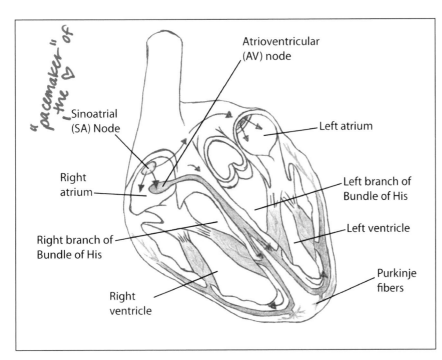

Figure 3.24

The autorhythmic cells of the heart. From Wingerd, B.(2007). *The human body: Essentials of anatomy and physiology.* **University Readers, San Diego. Figure 12.10, pg. 293.**

$$\text{Cardiac Output (CO)} = \text{Heart Rate x Stroke Volume} \qquad (3.1)$$

Suppose that your resting heart rate is 70 beats per minute and that your stroke volume is 65 milliliters per beat. An easy calculation shows that your total cardiac output is 4,550 milliliters per minute, or 4.55 liters per minute. A "not so easy" calculation shows that over a course of a year your heart pumps over 2 million liters of blood per year! But this only represents the blood pumped at rest. During exercise, your cardiac output can climb to between 20–40 liters per minute! Clearly, your heart can be considered the "workhorse of the body."

Blood Supply to the Heart. At rest, the heart only demands about 3 percent of the total cardiac output to the body. However, as does all living tissue in the human body, the heart requires a steady supply of blood that contains the necessary nutrients and oxygen for metabolism. Branching from the aorta, the right and left **coronary arteries** are responsible for delivering blood to the heart (see Figure 3.25). Blood supply to the heart depends on changes in oxygen demands. During exercise, the oxygen demands are very high and the blood supply to the heart is increased primarily through an increase in the dilation of the coronary blood vessels. Coronary artery disease however can impede the blood supply to the heart. An inadequate blood supply to the heart may lead to a number of complications such as cardiac pain (called angina pectoris), or actual damage to the cardiac tissues.

1. The pumping of blood by the heart;
2. The contraction of the skeletal muscles causing the blood to squeeze through one-way valves in the veins back to the heart;
3. Gravity tends to attract the flow of blood in the direction of the body parts nearest the ground (i.e., the legs during upright stance, the head when the body is inverted).

Blood flow throughout the circulatory system must be maintained at desirable rates at all times so that oxygen and essential nutrients are delivered to the tissues and waste products are removed. The flow of blood (F) through a vessel depends on two major variables: the blood vessel's **pressure gradient** (D P), which represents the difference in pressure from one end of the vessel to the other, and its **resistance** (R). The following expression represents this relationship in a quantifiable manner:

$$\mathbf{F = D\ P/R} \qquad (3.2)$$

Blood flow always moves from areas of high pressure to low pressure, much like the flow of water through a garden hose after opening the faucet. Following the contraction of the heart, pressure is high at the beginning of a blood

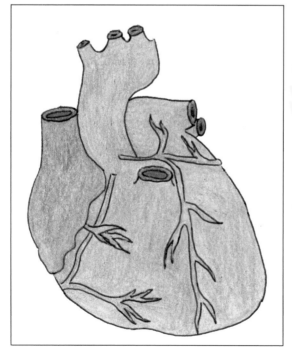

Figure 3.25

The right and left coronary arteries. Blood Flow. The flow of blood through the circulatory system is due to the following three factors:

vessel and the pressure is low farther away from the heart. Because of this difference (or gradient) in blood pressure, the blood flows through the blood vessel from the region of high pressure (nearest the heart) to the region of low pressure (farthest from the heart). We can say that blood flow in a vessel is *directly* proportional to the pressure gradient because as the DP increases, F also increases. However, the above expression also indicates that blood flow in a vessel is *inversely* proportional to increases in the resistance of that vessel. So, the higher the resistance or hindrance to blood flow, the lower the flow of blood through that vessel.

Factors Affecting Resistance to Blood Flow. The three factors affecting resistance to blood flow through the vessel are: 1) blood viscosity; 2) vessel length; and 3) vessel radius. Viscosity is the resistance to flow of a fluid. Blood viscosity is therefore related to the "thickness" of the blood. If the blood viscosity increases, resistance to blood flow through the vessel increases. Excessive increases in the number of red blood cells result in higher blood viscosity. Vessel length and its radius also affect resistance to blood flow. As blood flows, it makes contact with the inner walls of the vessel. The greater the surface area of contact, the more the resistance to blood flow. Therefore, the longer the vessel the greater the contact area and thus the greater the resistance. In addition, the smaller the radius of the vessel, the greater the resistance, because for a given amount of blood, there is more contact with the inner walls of the vessel.

Hypertension and Hypotension. The term **hypertension** means an unusually high blood pressure (over 150/90 mm Hg) and **hypotension** refers to low blood pressure (below 100/60 mm Hg). Chronic hypertension puts undue stress on the heart and the blood vessels. Patients with this disease may suffer from hemorrhages of small blood vessels leading to bloody noses, retinal damage, or even severe stroke. Hypotension may result from heart failure or loss of blood volume, to name just a couple of the possible causes. Essentially, severe and sustained hypotension results in inadequate blood supply to various tissues and organs and leads to a number of serious complications, or even death.

HIGHLIGHT

Measuring Blood Pressure

We all have had the experience of having our blood pressure measured by a physician or nurse. Recall that an inflatable cuff, called a **sphygmomanometer** (invented in 1896 by S. Riva-Rocci), is applied to your upper arm and tightly inflated so that the pressure produced by the device exceeds that of the arterial blood pressure, momentarily halting blood flow through the brachial artery. This device indicates the height to which a column of mercury (Hg) is pushed by pressure in the inflatable cuff. The higher the height of the mercury, the greater the pressure.

Immediately following contraction of the left ventricle during the cardiac cycle, blood is pumped into the aorta. The aorta swells with blood, much like the filling of a balloon, before releasing the blood farther downstream to the other arteries such as the brachial artery in your arm. The arterial pressure during **systole** (contraction of the heart) is highest at this point, amounting to 120 mm Hg in an average healthy adult. During **diastole** (relaxation and filling of the heart), the arterial pressure is lowest, around 80 mm Hg, again in an average healthy adult. Therefore, arterial blood pressure oscillates between systolic and diastolic pressure during each cardiac cycle.

How does the physician use the sphygmomanometer to measure systolic and diastolic blood pressure? By placing the end of a listening device, called a **stethoscope**, just distal to the cuff over the brachial artery, the physician or nurse can actually hear different patterns of sounds caused by the blood flow. When the cuff pressure exceeds the arterial pressure at the beginning of the procedure, no sound is heard because of the lack of blood flow. When the cuff pressure is gradually reduced to just below the arterial pressure, the blood flow actually becomes turbulent as it squeezes through the artery past the cuff at some critical point. The turbulent blood flow can be heard, and the physician notes the pressure of the cuff that is the high point of systolic blood pressure, around 120 mm Hg. The physician continues to release the pressure of the cuff until no sounds can be heard through the stethoscope, when blood flow returns to normal. This point represents the lowest diastolic blood pressure, around 80 mm Hg. It turns out that the average systolic over the diastolic blood pressure is around 120/80 mm Hg.

The Respiratory System

Our very existence as living beings depends on the ability to extract oxygen (O_2) from the air and to release carbon dioxide (CO_2) back into the air. The importance of oxygen to our survival can be illustrated when one tries to hold one's breath for as long as possible. Practical experience indicates that this is possible for only a short period of time before the craving of O_2 by our bodies becomes too great (and too painful!). Ironically, it is not the deprivation of O_2 but the related increase in CO_2 that serves as the major stimulus for the rate of breathing. But it is the O_2 we breathe that is eventually used in combination with food (which is broken down by the digestive system) to produce energy necessary for all life's body functions. Many simple organisms, such as worms, are able to absorb O_2 directly into their skin. However,

most animals, including humans, breathe in air (called **inhalation**) to extract O_2, and breathe out air (called **exhalation**) to remove the CO_2 using specialized organs. The removal of CO_2 is important to maintain the proper acid/base balance in the blood.

The mechanical act of breathing is called **ventilation**. Fortunately, ventilation is another body function that we do not have to think about, although at times, we can voluntarily increase or decrease its rate. Thus, ventilation is primarily under the control of the autonomic nervous system and is only one aspect of a more complex process called respiration.

The actual exchange of O_2 and CO_2 between the body and the external environment is called **respiration**. Respiration is a complex process in that it involves a number of systems within the body including the muscular, circulatory, and nervous systems. These systems contribute to the delivery of O_2 from the air into the bloodstream and then to the cells in exchange for end products of cellular metabolism—primarily CO_2 and water. Technically, respiration can be subdivided into external respiration

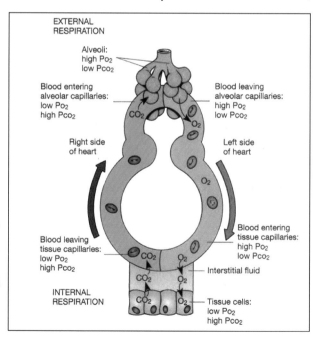

Figure 3.26

External and internal respiration. From Wingerd, B.(2007). *The human body: Essentials of anatomy and physiology*. University Readers, San Diego. Figure 14.13, pg. 352.

and internal (or cellular) respiration. **External respiration** refers to all the activities involved in the exchange of O_2 and CO_2 between the atmosphere and the lungs, between the lungs and the blood, and between the blood and tissues. **Internal respiration** involves the exchange of O_2 and CO_2 between the various cells throughout the body and the bloodstream. The activities of external and internal respiration are illustrated in Figure 3.26.

In this section, we will briefly explore four major aspects of respiration: the anatomical structures and mechanics of respiration, gas exchange during respiration, gas transport during respiration, and the neural control of respiration.

Anatomical Structures and Mechanics of Respiration.

The anatomy of the respiratory system is illustrated in Figure 3.27. Air enters the body either through the mouth or the nasal passages. From there, it travels into the **pharynx** (throat), which serves as a common passage for both the respiratory and the digestive systems. Air travels from the pharynx to the **trachea** (windpipe), then into the left or right bronchi. The **bronchi** further subdivide into smaller **bronchial tubes** that connect with the alveoli. The **alveoli** are small, thin-walled, inflatable sacs surrounded by pulmonary capillaries, where the O_2 and CO_2 are exchanged. There are roughly 300 million alveoli in the lungs, with a total surface area of nearly 75 square meters, nearly 25 times the surface area of the skin (Asimov, 1963) or about the size of a tennis court (McArdle, Katch, and Katch, 1996)!

Air moves in and out of the lungs due to changes in its pressure gradient inside and outside of the body. Inhaling actually causes the air pressure within the body to become much lower than the atmospheric pressure, causing the air to move from a region of high pressure (the atmosphere) to low pressure (the lungs). Inhalation is caused by the contraction of several muscle groups—the diaphragm and the external intercostal muscles. The **diaphragm** is a large, dome-shaped muscle located below the lungs; when it contracts, it descends downward, increasing the volume of the thoracic cavity (see Figure 3.28). Contraction of the **external intercostal muscles**, whose fibers lie between the adjacent ribs, further help in enlarging the thoracic cavity to facilitate the inhalation of air. Exhalation, for the most part, is accomplished passively without any significant muscle contraction. However, one can actively force air out of the lungs through the contraction of the **abdominal and internal intercostal muscles**.

Measures of Ventilation. The measures of ventilation relate to the different categories of the changes in lung volumes and capacities during respiration. The average total volume of air the lungs can hold, nearly 6 liters for men and 4.2 liters for women, is called the **total lung capacity (TLC)**. The volume of inhaled and exhaled air during one normal breath at rest, called the **resting tidal volume (TV)**, is much less, only 500 ml. The **vital capacity (VC)** is the maximum volume of air

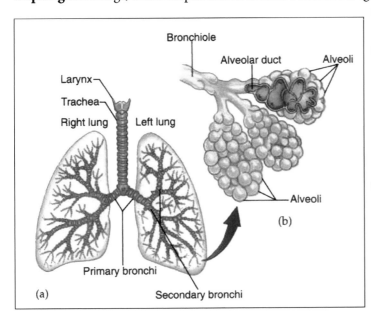

Figure 3.27
Anatomy of the respiratory system. From Wingerd, B.(2007). *The human body: Essentials of anatomy and physiology*. University Readers, San Diego. Figure 14.7a and b, pg. 344.

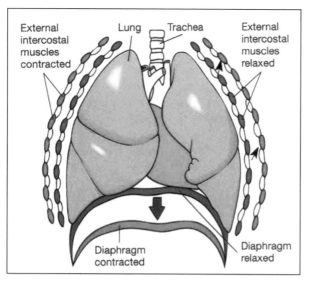

External intercostal muscles contracted

Lung   Trachea

External intercostal muscles relaxed

Diaphragm contracted

Diaphragm relaxed

Figure 3.28

The diaphragm, intercostal, and extracostal muscles. From Wingerd, B. (2007). *The human body: Essentials of anatomy and physiology.* University Readers, San Diego. Figure 14.11a, pg. 348.

exhaled following a maximal volume of air inhaled. The **residual volume (RV)** is the volume of gas remaining after maximum exhalation. These and other volumes can be measured using a device called a **spirometer**.

Air flow in and out of the lungs is directly proportional to the pressure gradient or the difference between atmospheric and intra-alveolar pressure and inversely proportional to the resistance of the airways (nasal passages, pharynx, trachea bronchi, bronchioles, alveoli) . The greater the pressure gradient, the greater the air flow. However, if the airways reduce in diameter for whatever reason, resistance goes up and air flow goes down as shown in the following expression:

$$F = \frac{\Delta P}{R} \qquad (3.3)$$

where F = air flow rate

ΔP = air pressure gradient between the atmosphere and the alveoli

R = airway resistance

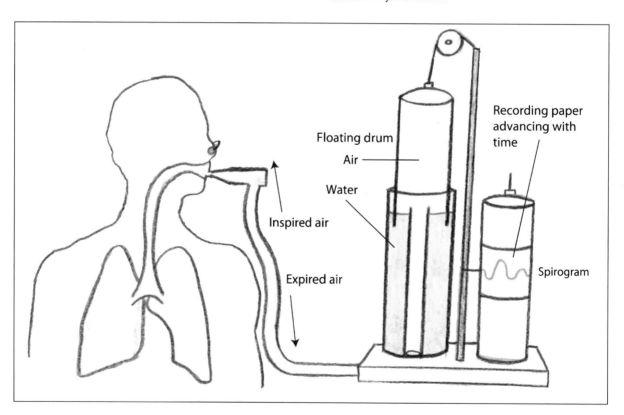

Floating drum

Air

Water

Inspired air

Expired air

Recording paper advancing with time

Spirogram

Figure 3.29

Measures of ventilation.

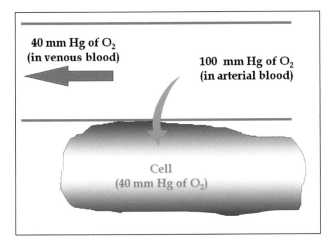

Figure 3.30

Diffusion of O2 and CO2 through the alveoli walls and tissue cells.

40 mm Hg of O₂
(in venous blood)

100 mm Hg of O₂
(in arterial blood)

Cell
(40 mm Hg of O₂)

Many abnormal conditions, from minor stuffy noses to a variety of pulmonary diseases (e.g., asthma, bronchitis, and emphysema), can reduce the size of the airways, and thus increase the resistance to air flow.

Gas Exchange During Respiration.

Once inhaled air reaches the alveoli, O2 and CO2 must be exchanged. **Partial pressure** of a gas refers to the individual pressure of a gas exerted within a mixture of gases (like the atmosphere). The total atmospheric pressure is roughly 760 mm Hg at sea level and O2 exerts about 160 mm Hg or nearly 21 percent of total atmospheric pressure. The pressure exerted by CO2 in the atmosphere is negligible at .3 mm Hg. It is important to note that gases dissolved in a liquid, like the blood, also exert a partial pressure.

O2 Exchange. The O2 partial pressure in the alveoli at the time of inhalation, approximately 100 mm Hg, is less than atmospheric pressure. The blood returning from the body tissues carries an O2 partial pressure of 40 mm Hg. Therefore, due to the pressure gradient, O2 diffuses through the alveoli walls into the blood in the pulmonary capillaries (see Figure 3.30).

CO2 Exchange. The partial pressure of CO2 in the alveoli is 40 mm Hg, but in the blood, within the pulmonary capillaries, it is 46 mm Hg. Therefore, CO2 diffuses from the blood to within the alveoli.

In sum, the exchange of O2 and CO2 in the lungs is accomplished due to the partial pressure gradients of these two gases in the inhaled air (within the alveoli) and the blood (in the pulmonary capillaries surrounding the alveoli). Following this exchange, the blood, now rich in O2, circulates to the left side of the heart, where it is pumped by the left ventricle to the body tissues.

Gas Transport During Respiration.

The transportation of both O2 and CO2 in the blood is accomplished by a molecule called **hemoglobin** (Hb). Hemoglobin is made up of two combined molecules: **heme** and **globin**. Only about 1.5 percent of the diffused O2 is dissolved into the blood. The remaining 98.5 percent becomes chemically bound to the *heme* part of Hb within the red blood cells (the red blood cells are called **erythrocytes**). The erythrocytes represent around 43 percent of the volume of blood. This portion of the blood volume is called the **hematocrit** (see Figure 3.31). Once the erythrocytes reach the capillaries throughout the body, the O2 molecules are released by Hb, and then diffuse into the cells due to the pressure gradient of O2 in the blood and in the cells.

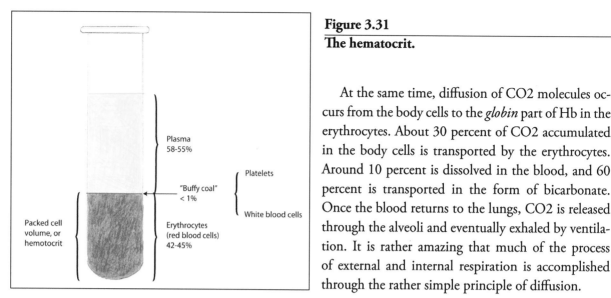

Figure 3.31
The hematocrit.

At the same time, diffusion of CO2 molecules occurs from the body cells to the *globin* part of Hb in the erythrocytes. About 30 percent of CO2 accumulated in the body cells is transported by the erythrocytes. Around 10 percent is dissolved in the blood, and 60 percent is transported in the form of bicarbonate. Once the blood returns to the lungs, CO2 is released through the alveoli and eventually exhaled by ventilation. It is rather amazing that much of the process of external and internal respiration is accomplished through the rather simple principle of diffusion.

Neural Control of Ventilation.

The Rhythm of Ventilation. The control of ventilation is primarily the responsibility of the autonomic nervous system. The rhythmic cycle of inhalation and exhalation is controlled by a complex set of *pacemaker* neurons in the respiratory center within the brain stem. Impulses are continuously sent to the diaphragm and external intercostal muscles, causing them to alternately contract and relax (see Figure 3.32).

The Magnitude of Ventilation. The respiratory center in the brain stem receives information about the content of O2 in the blood by **peripheral chemoreceptors** located in the aorta and carotid arteries. If and when the partial pressure of O2 becomes dangerously low, the chemoreceptors alter the input to the brain stem

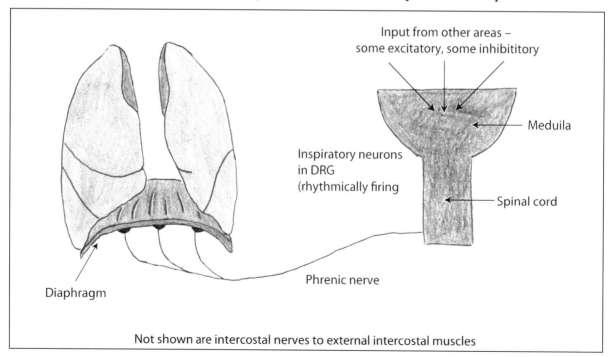

Figure 3.32
Neural control of the respiratory muscles.

so that the *magnitude* of ventilation will increase. **Central chemoreceptors** located near the respiratory center help monitor the buildup of CO2 in the blood. If the CO2 content becomes too high, these chemoreceptors inform the respiratory center to increase the magnitude of ventilation. Thus, the control of the magnitude of ventilation is regulated by chemoreceptors that help monitor both the O2 and the CO2 content in the blood.

CHAPTER SUMMARY

- The structural "foundation" of the human body is provided by the skeletal system, comprised of over 200 bones of various shapes and sizes. The bones of the skeletal system serve a number of functions, including the production of the red and white blood cells. The wide variety of joints provides for great dexterity in terms of how we produce movements.

- The muscular system provides a means for the human body to move in the environment through the process of the contraction of muscle fibers. Muscles only pull, they cannot push. Therefore, sets of muscles around each joint allow for the production of force in many directions of movement. The three major types of muscles are skeletal, cardiac, and smooth; they control voluntary movement, contractions of the heart, and control of subconscious muscle activity of the visceral organs, respectively.

- The nervous system allows for communication among the various organs throughout the body. The major anatomical divisions of the nervous system are the central and peripheral nervous systems. The central nervous system is composed of the brain and the spinal cord. The peripheral nervous system is all the nervous tissue lying outside of the spinal cord, which carries both efferent and afferent information to and from the various organs of the body. The functional divisions of the nervous system are the somatic and the autonomic nervous systems. The somatic nervous system is responsible for the voluntary control of the skeletal muscles. The autonomic nervous system controls involuntary actions.

- The cardiovascular system is composed of the heart, blood vessels, and the blood. The major function of the cardiovascular system is the transportation of oxygen, carbon dioxide, and other nutrients to and from the cells throughout the body. The heart is a muscular organ that acts as a pump by forcing blood through the various blood vessels. Contraction of the heart is initiated by autorhythmic cells capable of sending nerve impulses by themselves. There are a number of measures of heart function such as heart rate, stroke volume, cardiac output, and blood pressure that can be used to assess the performance of the heart and the cardiovascular system in general.

- The respiratory system is designed to facilitate the exchange of O2 and CO2 between the body and the atmosphere. External respiration refers to all the physiological activities that occur between the atmosphere and the lungs, between the lungs and the blood, and between the blood and tissues in the lungs. Internal respiration refers to the processes responsible for the exchange of O2 and CO2 between the blood and cells of the body. There are many specialized organs involved in the mechanics of respiration. The exchange of O2 and CO2 between the body and the atmosphere, as well as between the blood and the cells, is accomplished by diffusion due to partial pressure gradients. Both O2 and CO2 are carried in the bloodstream, either dissolved or chemically bound to hemoglobin molecules. The rhythm of ventilation that results in inhalation and exhalation is controlled by pacemaker cells in the brain stem. The magnitude of ventilation is primarily dependent on chemoreceptors located in the aorta and carotid arteries and near the respiratory center in the brain stem.

"The machinery of the body is all of a chemical or physical kind. It will all be expressed some day in physical and chemical terms."

— A.V. Hill (1927)

STUDENT OBJECTIVES

1. To identify the three major sources of energy used during movement and exercise.
2. To understand how energy is utilized during human movement and exercise.
3. To distinguish between anaerobic and aerobic metabolisms.
4. To describe the training methods used to improve the three major energy systems.
5. To understand the ventilatory and cardiovascular responses to exercise.
6. To understand the negative effects of physical inactivity on physiological function.
7. To discern the effects of exercise on the incidence of cardiovascular disease.

IMPORTANT TERMS

Abdominal and internal intercostal muscles – muscle groups assisting in ventilation

Abduction – movements away from the body

Actin and myosin molecules – molecules responsible for muscle contraction

Adduction – movements towards the body

Agonist and antagonist muscles – muscles surrounding a joint that initiate and help brake a movement, respectively

All-or-none principle – all the muscles fibers in a motor unit are activated simultaneously

Alveoli – tiny air sacs in the lungs

Anatomical planes and axes – help describe how movements are performed

Anatomical position – a reference position of the human body

Arteries – blood vessels that carry oxygenated blood away from the heart

Arterioles – blood vessels that connect the arteries to the capillaries

Atria – upper chambers of the heart

Atrioventricular node – one type of cardiac cell assisting in the rhythmical beating of the heart

Autonomic nervous system – a part of the nervous system controlling subconscious physiological events

Autorhythmic cells – help control the rhythmical beating of the heart

Axial and appendicular skeletons – the two major sections of the skeletal system

Axons – the part of a neuron that carries impulses away from the cell body

Basal ganglia – a part of the brain important in the initiation of movement

Biopsy – a procedure used to remove muscle tissue

Blood flow – affected by the pressure gradient and resistance within the blood vessels

Brain – a major part of the central nervous system

Brain stem – important area of the nervous system that help to regulate many subconscious physiological event

Bronchial tubes – small airways in the lungs

Bundle of His – an area of the heart helping to regulate the rhythmical beating of the heart

Capillaries – small blood vessels

Cardiac muscle – muscle tissue of the heart

Cardiac output - the volume of blood pumped by each ventricle per minute

Cell body - the bulbous end of a neuron containing the cell nucleus

Central and peripheral nervous systems – the central nervous system contains the brain and spinal cord and the peripheral nervous system contains the peripheral nerves

Cerebellum – an important part of the brain assisting in the coordination of movement

Cerebral cortex – the outer layer of the brain

Circumduction - a circular or conelike movement of a body segment

CO_2 partial pressure – the amount of pressure exerted by carbon dioxide

Composition of bones – bones are made of organic and inorganic material

Concentric – the shortening of muscle tissue

Conductivity - the ability of the neuron to send information to other structures

Contractility – the ability of muscle tissue to shorten

Convolutions – the 'wrinkles' or fissures on the surface of the brain

Coronary arteries – blood vessels supplying the heart

Corpos callosum - a wide, flat bundle of neural fibers connecting the left and right cerebral hemispheres

Degrees of freedom – the independent motions of a joint

Dendrites – parts of a neuron carrying impulses toward the cell body

Depression - moving towards the inferior position

Dermatome – an area of the body supplied by a spinal nerve

Diaphragm – a flat muscle assisting in ventilation

Diastole – blood pressure during the relaxation of the heart

Dissection – a surgical procedure for removing a body part

Dorsiflexion - the moving of the top of the foot toward the shin

Eccentric – activation of the muscle as it lengthens

Elasticity – a characteristic of muscle tissue to return to normal resting length following stretch

Electrical stimulation – a way to excite muscle and nervous tissue

Electromyography – a measure of the electrical activity of muscle

Elevation - moving towards the superior position

Erythrocytes – red blood cells

Eversion - the lifting of the lateral border of the foot

Excitability – human tissue that reacts to stimulation

Exhalation – expelling of air during ventilation

Extensibility – the characteristic allowing for the stretching of muscle tissue

Extension – an increase in joint angle formed by articulating bones

External intercostal muscles – assist ventilation

External respiration - all the activities involved in the exchange of O2 and CO2 between the atmosphere and the lungs, between the lungs and the blood, and between the blood and tissues

Fast oxidative-glycolytic fibers – muscle fibers containing characteristics of both fast and slow twitch fibers that are fatigue resistant

Fast twitch fibers – muscle fibers that contract rapidly and produce greater forces, but also fatigue rather quickly

Flexion – a decrease in joint angle formed by articulating bones

Function of bones – bones have a variety of functions

Fusiform muscles - muscle that are in the form of a spindle and designed for speed of movement and allow for a greater range of movement

Globin – a molecule that helps form hemoglobin

Gradation of muscular contraction principle - the number of motor units recruited during a muscular contraction depends on the intensity of the task

Heart rate – the frequency of the beating of the heart

Hematocrit - the volume percentage (%) of red blood cells in blood

Heme – a molecule that helps form hemoglobin

Hemoglobin – a molecule that transports both O2 and CO2 in the blood

Hypertension – high blood pressure

Hypotension – low blood pressure

Hypothalamus – an area of the brain that helps regulate arousal

Inhalation – the part of ventilation that brings air into the lungs

Internal respiration – also called cellular respiration

Inversion - the lifting of the medial border of the foot

Irritability - the ability of the neuron to respond to stimulation

Isometric – muscle activity without a change in length of the muscle

Isotonic – muscle activity that results in a change in length of the muscle

Joint – the joining of two or more bones

Ligaments – connective tissue between two bones

Longitudinal muscles – muscles that have fibers that run parallel with the long axis of the muscle

Motor units – a motor neuron and the muscle fibers it stimulates

Multiple sclerosis – a disease affecting the nervous system

Muscle contraction – a process that occurs following stimulation of muscle fibers

Muscle spindles – sensory receptors that lie in parallel to muscle fibers

Myelin – a fatty tissue surrounding some nerve tissue

Myofibrils – compose many muscle cells

Neuron – a nerve cell

Neurotransmitters – a substance that facilitates the transmission of an action potential from one nerve cell to another

Neutralizers – a muscle that minimizes the effects of another muscle

O2 partial pressure – the pressure exerted by oxygen

Pacemaker neurons - cells in the brain stem that help control the rhythm of ventilation

Palpation – the use of the hand in certain therapeutic techniques

Partial pressure – the amount of pressure exerted by a gas

Penniform muscles - have fibers that are arranged somewhat like a feather

Peripheral and central chemoreceptors - located in the aorta and carotid arteries, these receptors monitor the oxygen content in the blood

Pharynx – a major airway for ventilation

Plantar flexion – a movement of the foot downward (e.g., pointing the toes)

Pressure gradient – the difference in pressure from one end of a vessel to the other

Pronation - the medial (inward) rotation of the forearm, for example

Pulmonary circulation – the flow of blood to and from the heart and lungs

Purkinje fibers - cardiac cells assisting in the autorhythmic behavior of the heart

Radiate muscles - have fibers which fan outward from a single attachment

Residual volume - the volume of gas remaining after maximum exhalation

Resistance – impedance of air or blood flow

Respiration – the process of exchanging carbon dioxide and oxygen between the atmosphere and human body

Resting tidal volume - the volume of inhaled and exhaled air during one normal breath at rest

Rotation – angular motion

Sarcomere – at the molecular level, the part of the muscle containing actin and myosin

Sinoatril node – a group of pacemaker cells that help to regulate heart rate

Size principle - states that motor units containing smaller muscle fibers are recruited before larger muscle fibers during a muscular contraction

Skeletal muscle – muscles that are under voluntary control

Slow twitch fibers - contract slowly, produce smaller forces and are more resistant to fatigue

Smooth muscle – a type of muscle under subconscious control

Somatic nervous system – helps to control voluntary movement

Sphygmomanometer – a device used to measure blood pressure

Spinal cord – a part of the central nervous system

Spirometer – a device used to measure ventilation

Stabilizers - helps to prevent the movement of one end of a prime mover thereby allowing another muscle to work effectively

Stethoscope – a listening device used by physicians

Stroke volume – the amount of blood pumped per beat of the heart

Supination - is the lateral (outward) rotation of the forearm, for example

Sympathetic and parasympathetic nervous systems – the two parts of the autonomic nervous system

Synapse – the gap between two adjacent neurons

Synergists – muscles that assist the prime mover or another muscle to work effectively

Systemic circulation – the flow of blood between the heart and all parts of the body except the lungs

Systole – arterial pressure during the contraction of the heart

Tendons – connective tissue between muscles and bones

Thalamus – an important 'relay station' in the brain

Threshold – the amount of stimulation needed before a neuron is activated

Total lung capacity - the average total volume of air the lungs can hold

Trachea – an important airway for ventilation

Types of bones- there are different types of bones

Types of joints – there are different types of joints

Veins – blood vessels that carry de-oxygenated blood to the heart

Ventilation – the process of breathing air in and out of the lungs

Ventricles – the lower chambers of the heart

Venules – blood vessels that connect the capillaries to the veins

Vital capacity - the maximum volume of air exhaled following a maximal volume of air inhaled

INTEGRATING KINESIOLOGY: PUTTING IT ALL TOGETHER

1. Using anatomical terms, how would you describe the position of the following body parts?
   - The top of the head relative to the toes
   - The right thumb relative to the left shoulder
   - The right shoulder relative to the right elbow
2. Using the scientific terminology of joint movement, describe the action of the following joints:
   - The elbow joint as you reach forward to grasp a coffee cup
   - The knee joint as you stand up from sitting in a chair
   - The shoulder joint as you draw a large circle on a chalkboard
3. Why do you think it is more difficult to breathe at high altitudes? [Hint: You must take into account the partial pressure of oxygen in the atmosphere as well as in the lungs].
4. Using other references, identify diseases affecting the skeletal, muscular, cardiovascular, and nervous systems that are not mentioned in this chapter. Do these diseases affect human movement? In what way?
5. Measure your resting heart rate for five days in a row as soon as you wake up in the morning. How consistent is your heart rate across these five days? What factors might affect the variability of heart rate?

KINESIOLOGY ON THE WEB

en.wikipedia.org/wiki/Anatomical_terms_of_location—This Web site provides detailed information about the anatomical position and important anatomical terms

www.innerbody.com—This Web site contains information n the structure and function of various anatomical and physiological systems. Images and animation are also provided.

REFERENCES

Asimov, I. (1963). *The Human Body—Its Structure and Operation*. Boston: Mentor.

Basmajian, J. V., and De Luca, C. J. (1985). *Muscles Alive: Their Functions Revealed by Electromyography*. Baltimore: Williams and Wilkins.

Brooks, G. A., and Fahey, T. D. (1984). *Exercise Physiology: Human Bioenergetics and Its Applications*. New York: Macmillan.

Castiglioni, A. (1958). *A History of Medicine*. Translated from the Italian and edited by E. B. Krumbhaar. New York: A. A. Knopf.

Hamill, J., and Knutzen, K. M. (1995). *Biomechanical Basis of Human Movement*. Baltimore: Williams and Wilkens, p. 121.

Hay, J. G., and Reid, J. G. (1988). *Anatomy, Mechanics, and Human Motion.* Englewood Cliffs, NJ: Prentice Hall.

Huxley, H. E. (1958). The contraction of muscle. *Scientific American*, 199, 66–82.

Lestienne, F. (1979). Effects of inertial load and velocity on the braking process of voluntary limb movements. *Experimental Brain Research*, 1979, 35, 407–18.

McArdle, W. D., Katch, F. I., and Katch, V. L. (1996). *Exercise Physiology: Energy, Nutrition, and Human Performance.* Philadelphia: Lea and Febiger.

Piscopo, J., and Baley, J. A. (1981). *Kinesiology: The Science of Movement.* New York: John Wiley and Sons.

Ricci, B. (1967). *Physiological Basis of Human Performance.* Philadelphia: Lea and Febiger.

Sherwood, L. (1991). *Fundamentals of Physiology: A Human Perspective.* St. Paul, MN: West.

Chapter Four
Exercise Physiology Foundations

Energy Utilization During Movement and Exercise

Energy

The Laws of Thermodynamics

Metabolism

Nutrients

Measurement of Energy in Nutrients

Importance of ATP

ATP and Muscle Contraction

Sources of ATP Production

Immediate energy sources

Non-oxidative energy sources

Oxidative energy sources

Training and the Improvement of the Energy Sources

Immediate energy sources

Non-oxidative energy sources

Oxidative energy sources

Exercise Training Techniques

CHAPTER SUMMARY

Important Terms

Integrating Kinesiology: Putting It All Together

Kinesiology on the Web

References

CHAPTER FOUR

Exercise Physiology Foundations

n the last chapter, we examined a variety of physiological systems within the human body that play a major role in movement and exercise. In this chapter, we explore how the human body responds to the demands of exercise and adapts (or changes) with repeated bouts of exercise. Most of us, at one time or another in our lives, have engaged in some type of exercise program, sport, or recreational activity. Some activities demand strength or very forceful short duration movements such as swimming a 50-meter sprint, others require endurance or the ability to sustain an activity over longer periods, such as a 10K run. These type of activities require the production of energy to meet the demands placed on the body. Where does this energy come from? How is it produced? Is the process of energy production the same for different types of activities? How does physical training enhance the body's ability to produce energy? What are the ventilatory and cardiovascular responses to exercise? What role does exercise play in the incidence of cardiovascular disease? These are some of the interesting questions addressed in this chapter.

ENERGY UTILIZATION DURING MOVEMENT AND EXERCISE

The living human body requires energy to perform all activities such as breathing, eating, thinking, and moving around in the environment. Without energy, or more specifically, the conversion of energy, these activities would be impossible. What is energy? What is meant by the "conversion" of energy? How does the conversion of energy allow the human body to perform different types of activities during movement and exercise? The answer is not a trivial one and we must begin by briefly exploring the laws of thermodynamics that govern the exchange of all energy in the universe.

Energy. Energy is an abstract quantity because it has no shape or mass so it is commonly defined in terms of the potential force it may produce upon matter and the resulting displacement of that matter. Energy can be defined as the capacity to do work, and the work produced may be expressed as the following:

$$W = F \times d \hspace{4cm} (4.1)$$

where W is work, F is force and d is the displacement through which the force is applied. In nature, there are many types of energy that produce work (e.g., heat, mechanical, chemical, etc.). The energy in heat can be used to boil water. Energy in chemical reactions within a battery can produce electricity. The lifting of a box off the floor is a result of mechanical energy produced by our muscles. It is clear that work can be produced by different forms of energy.

The Laws of Thermodynamics. Thermodynamics is a branch of physics that deals with how heat can be converted into or exchanged with other forms of energy. According to the **first law of thermodynamics**, the total amount of energy in the universe remains constant and therefore energy can be neither created nor destroyed. However, energy may be *converted* from one form to another. For example, to produce the work required to walk, chemical energy stored in the body must be eventually converted to mechanical energy in order to produce body motion. But the conversion of one form of energy is never perfect. The **second law of thermodynamics** states that some amount of energy will be lost in the conversion of one form of energy to another. Often, as in the case of muscle contraction, the conversion of chemical to mechanical energy results in the loss of energy in the form of heat. In fact, it has been estimated that only about 30 percent of the energy is released to perform the work necessary for muscle contraction (particularly concentric muscle action). The other 70 percent is lost in the form of heat energy (Brooks, Fahey, and White (1996).

The total energy of a given system includes its **potential energy** and **kinetic energy**. Potential energy is the *capacity* of doing work that a body has because of its position or configuration (Luttgens and Wells, 1992). For example, the potential energy of a ball is higher at the edge of a cliff than after it lands. As soon as the ball falls from the cliff, its potential energy decreases, but its kinetic energy increases. Kinetic energy of a body is the energy it possesses due to its motion. The faster a body moves, the greater its kinetic energy. Once the body stops moving, its kinetic energy is zero. Thus, the kinetic energy of the ball is high when it is falling to the ground, and it is zero at rest on the ground.

As we will learn in the next section, the human body stores potential energy in the chemical bonds of various molecules. Kinetic energy is released when these bonds are broken. Thus, a body's potential energy can be transformed to kinetic energy to do work. The kinetic energy is subsequently used for performing the work necessary in muscle contraction, building new tissue, transmission of nerve impulses and other important physiological activities (McArdle, Katch, and Katch, 1996).

Metabolism. The total of all the chemical processes by which substances within the human body are transferred to different forms of energy is called **metabolism**. As mentioned above, the breakdown of larger molecules into smaller molecules releases energy and this specific form of metabolism is called **catabolism**. There is also a metabolic process, called **anabolism**, referring to the chemical reactions necessary to form larger molecules from smaller ones. Both catabolism and anabolism are necessary for the generation of energy needed to sustain all body activities at rest, called **basal metabolism**, and during exercise or physical activity.

Nutrients. The energy required to perform all life-sustaining activities such as respiration, circulation, nerve transmission, and muscle contraction, is extracted from the food we eat by our digestive system. Foods contain **nutrients**, which are defined as substances used in the body to promote growth, maintenance, and repair of tissues. There are six classes of nutrients: carbohydrate, protein, fat, vitamins, minerals, and water. Of these, only the first three are catabolized to release the energy needed to sustain life.

Each nutrient is a molecule constructed from the union of two or more atoms. Basic chemistry informs us that there are over 100 different types of atoms. The atoms of carbon, hydrogen, nitrogen, and oxygen are considered the building blocks from which the nutrients of carbohydrates, protein, and fat are made. When these molecules are chemically bound together, they hold potential energy. Kinetic energy necessary to produce work is released when these molecules are broken down during catabolism.

Measurement of Energy in Nutrients. The amount of potential energy in carbohydrates, fat, and protein differs, and is measured in units called **calories** (1,000 calories = 1 kilocalorie). A calorie is defined as the amount of heat necessary to raise the temperature of a gram of water by one degree centigrade. Table 4.1 indicates the number of kilocalories (kcal) in one gram of each nutrient.

Importance of ATP. The immediate source of energy for muscle contraction and the energy needed for all other body functions is derived from a high-energy molecule called **adenosine triphosphate (ATP)**. ATP consists of a large molecule called adenosine and three phosphate groups (see Figure 4.1).

When ATP is catabolized or broken down into adenosine diphosphate (ADP) and inorganic phosphate (Pi), energy is released that can be directly used to do cellular work. ATP is stored in most cells, particularly muscle cells, where it can be easily used to provide the energy necessary for muscle contraction.

ATP and Muscle Contraction. In the last chapter, we discussed that muscle contraction may involve the sliding of thin filaments past the thick filaments, according to the sliding-filament theory. The breaking down of ATP near the thin and thick molecules provides the energy required to allow this process to occur. The nerve impulse to the muscle causes a sequence of biochemical events (such as the release of the neurotransmitter acetylcholine [ACh], and the production of a muscle action potential) eventually leading to a splitting of the ATP (located on the myosin cross bridges) and a release of energy. The released energy causes the myosin cross bridges to combine with the myosin binding sites on the actin molecule (see Figure 4.2) and to move toward the center of the sarcomere in an action called the **power stroke**. Once a given power stroke is complete, the myosin cross bridge detaches from the actin molecule and combines with another myosin binding site further along the actin molecule. Then this whole cycle is repeated.

Like the cogs on a ratchet, the myosin cross bridges continue to move back and forth with each power stroke effectively pulling the actin filaments toward the center of a sarcomere. As the Z lines are drawn toward each other, the sarcomere shortens and the muscle fibers produce tension (see Figure 4.3). When muscle stimulation stops, the muscle relaxes, the myofilaments go back to their original positions, and the muscle cell resumes its resting length. To reiterate, the process of muscle contraction is not due to "animal spirits" filling up the muscle as Descartes thought, but rather is due to complex biochemical events.[1]

Table 4.1

Caloric Content of Nutrients

Conversion of grams (g) to kilocalories (kcal)

1 g carbohydrate = 4 kcal
1 g fat = 9 kcal
1 g protein = 4 kcal

Note: Per gram, fat contains more potential energy, in the form of kcal, than the other two nutrients. However, within the human body, the conversion of fat to energy takes more time. Even though carbohydrate contains less energy per gram, the body can produce energy more quickly using carbohydrate.

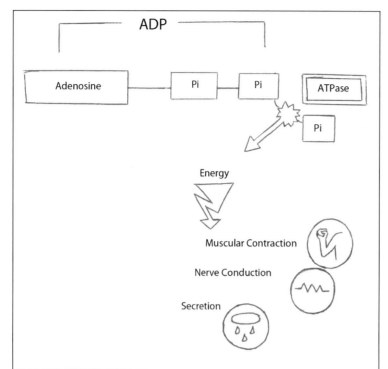

Figure 4.1

ATP consists of the large molecule adenosine and three smaller phosphate components. When ATP is broken down (to form ADP), energy is released that can be used to perform biological work.

Sources of ATP Production. The supply of ATP within the muscle cell is limited, and mechanisms are needed to replenish it. There are three sources of ATP production: immediate energy sources, non-oxidative energy sources, and oxidative energy sources. The first two sources can produce ATP without the presence of the oxygen molecule (O_2), called **anaerobic metabolism**, whereas the third energy source, called **aerobic metabolism**, requires O_2. Let us now examine each energy source and describe how ATP is produced.

Immediate energy sources. There are three sources of immediate energy within the muscle cell. The first source of immediate energy is ATP itself (mentioned above). In the presence of water (H_2O) and enzymes called ATPases, ATP is split into ADP and Pi in a process called **hydrolysis**. The second immediate energy source is from the compound creatine phosphate (CP), which provides a reserve of phosphate to make ATP. The interaction between creatine phosphate, ADP, and an enzyme called creatine kinase yields creatine (C) and ATP in a process termed **rephosphorylation**. The third immediate energy source involves the generation of ATP from two ADP molecules. The interaction of two ADP molecules with an enzyme called myokinase yields ATP and adenosine monophosphate (AMP).

These three immediate energy sources exist in the muscle cells near the contractile elements of actin and myosin. As the ATP is catabolized, energy is produced for muscle contraction. However, the amount of ATP supplied by the first immediate energy source is so limited that it cannot sustain muscle contraction for more than a few seconds. The second and third immediate energy sources can sustain muscle contraction for less than one minute.

Other non-oxidative energy sources. A second major energy source also involving anaerobic metabolism involves the breakdown of a simple sugar called glucose and a stored carbohydrate, called glycogen, made up of many glucose molecules. The breakdown of glucose and glycogen is called **glycolysis** and **glycogenolysis**[2], respectively. This second energy source is required when the first *immediate* energy sources of ATP and CP are depleted. Glycolysis and glycogenolysis provide enough energy to sustain fairly powerful muscular contractions for several seconds, up to approximately one minute.

Glucose is broken down in a series of 10 or 11 enzymatically regulated reactions and yields the production of both ATP and a substance called pyruvic acid (which can be quickly converted to lactic acid). Rapid

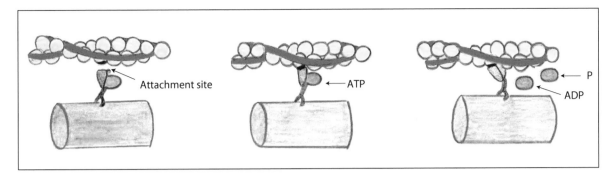

Figure 4.2

Actin-myosin interaction as proposed by the sliding filament hypothesis. During the power stroke, the myosin head is proposed to bind to the actin molecule and rotate. The rotation of the myosin head causes the actin filament to move, causing muscle contraction.

non-oxidative glycolysis results in very little ATP production. In addition, if the intensity of muscular activity continues without oxygen, lactic acid is produced. The continued accumulation of lactic acid in muscle tissue is related to an increase in **muscular fatigue**, preventing the muscle from producing the desired force. However, it is not clear that lactic acid actually causes muscle fatigue, as other factors are involved. In fact, lactic acid may actually be used as an additional energy source (Brooks, 2001).

Oxidative energy sources. The first two major sources of energy allow an individual to perform physical activities without the use of oxygen, albeit for only a short period of time. For example, it is possible to run the 100-meter dash while holding one's breath (provided that one finishes the event before the immediate energy

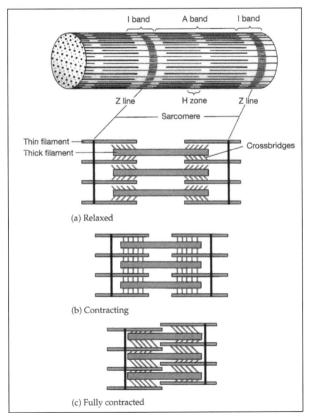

Figure 4.3

The relative positions of actin and myosin filaments in relaxed (a), contracting (b) and fully contracting (c) muscle. During contractions, the Z lines move toward one another, effectively, causing the muscle to shorten in length. From Wingerd, B.(2007). *The human body: Essentials of anatomy and physiology*. University Readers, San Diego. Figure 7.8, pg. 153.

Table 4.2

Characteristics of the Three ATP Energy Sources

Characteristic	Immediate	Non-oxidative	Oxidative
Speed of reaction	Very rapid	Rapid	Slow
Type of fuel	ATP and CP	glucose and glycogen	carbohydrate, fat, and protein
Amount of ATP produced	very limited	limited	almost unlimited
By products		lactic acid	CO2 and H2O
Oxygen required?	no	no	yes
Types of activities	sprints, weight lifting, shot put	100 to 400 m run	long endurance events
Duration of activity	0 to 10 sec	10 sec to 1 min	1 min and up

source and the nonoxidative sources are depleted!). Intense muscular activities of a minute or longer must be supported by **oxidative metabolism**. Oxidative metabolism is a complex process whereby glucose is broken down and combined with oxygen to yield ATP, plus the by-products of water (H2O) and carbon dioxide (CO2).

Oxidative metabolism is a complicated process that begins with a conversion of pyruvic acid to a substance called acetyl coenzyme A (CoA) within the mitochondria when oxygen is used. Following this conversion, acetyl CoA enters a series of reactions called the **Krebs cycle**, and finally a process called the **electron transport chain**. The Krebs cycle and the electron transport chain are the body's means to produce large amounts of ATP from a single molecule of glucose. The *oxidative* breakdown of glucose is far more extensive and time consuming than non-oxidative metabolism, but provides a substantial amount of energy required to sustain physical activity for long time periods.

Both stored fat and protein can also enter the Krebs cycle and the electron transport chain after they have first been broken down to acetyl CoA (see Figure 4.4). Thus, all three of the nutrients (carbohydrates, fat, and protein) can be catabolized to produce ATP in the presence of oxygen. The amount of potential energy produced from fat is twice that of carbohydrates and protein on a kcal per gram basis (see Table 4.1). However, carbohydrate oxidative metabolism is the most efficient. Because of this efficiency, during hard intensive exercise, the body chooses to metabolize mostly carbohydrates. However, during sustained physical activity, the body metabolizes fats that contain a large reservoir of potential energy.

Table 4.2 provides a summary of the characteristics of the three energy systems. There is an overlap of use of the three energy systems depending on the type of physical activity. Figure 4.5 shows the relative amount

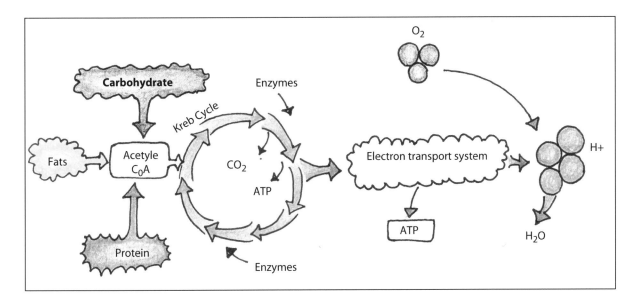

Figure 4.4

The oxidative energy source. Carbohydrates, fats and protein are catabolized to acetyl CoA that enters the Krebs cycle and the electron transport system. Complex enzymatic activity produces large amounts of ATP (in the presence of O2) and the by-products of CO2 (carbon dioxide) and H20 (water).

of energy potential of each the three energy sources depending on the time elapsed of the physical activity. As discussed above, activities less than 10 seconds in duration rely primarily on the immediate energy sources.

The use of the immediate energy source rapidly diminishes after this time period. The nonoxidative energy sources are used extensively between 10 seconds and 1 minute of the activity. After 1 minute, the oxidative energy sources are predominantly used.

Training and the Improvement of the Energy Sources. The delivery of energy from each of the three energy sources described above can be improved by certain types of exercise training techniques. Let us briefly discuss how certain exercise training techniques affect the improvement in the delivery of energy from each source.

Immediate Energy Source

The immediate energy sources are relied on during short explosive types of physical activity such as the start in a sprinting race, a shot put, or a weight lifter's attempt at lifting his/her personal best in one maximal effort. One way to improve the immediate energy source is through strength training using the progressive resistance technique (see below). Research has shown that this type of training results in an enlargement of the muscle tissue (**hypertrophy**). Hypertrophy does *not* increase the *relative* concentrations (i.e., percentage) of ATP, CP, or any of the previously mentioned enzymes (e.g., ATPase, creatine kinase, myokinase) in the muscle. The relative concentrations of these substances are the same in untrained and trained muscle. However, because the muscle is larger following progressive resistance strength training, the *absolute* concentrations of ATP, CP, and the other enzymes are greater in the immediate energy source. The net result of strength training is that muscle contraction is more forceful.

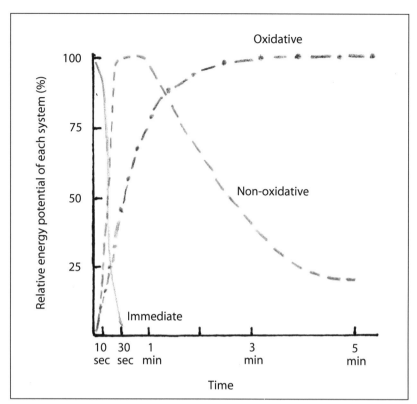

Figure 4.5

The use of the three energy sources as a function of the duration of exercise. In short duration activity, the immediate and non-oxidative energy sources are used. In activities of long durations, the oxidative energy source is primarily used.

Other Non-Oxidative Energy Sources

Other non-oxidative energy sources, such as glycolysis, are more dominant during activity durations between approximately 4 seconds and 1 minute. Events such as the 100-meter dash in running and the 50-meter sprint in swimming place great demands on the non-oxidative energy sources.

There is a substantial body of evidence that indicates the benefits of what has been called **interval training** on improvements in these types of sprinting activities (see below). As mentioned earlier, the processes of glycogenolysis and glycolysis are important in the production of non-oxidative energy. However, the research is less clear on how interval training actually affects these processes.

Oxidative Energy Sources

Oxidative energy sources provide the most energy for physical activities over a minute in duration. To improve the delivery of energy from this source requires either **interval training or endurance training** (see below). Pyruvate (a product of glycolysis), fat, and protein metabolism occur within the mitochondria. Much of ADP is phosphorylated to ATP, also within the mitochondria. The mitochondria are so important for the production of energy for activities longer than 1 minute in duration that they have been dubbed "the powerhouses of the cell." Endurance training is thought to increase the *mitochondrial mass* within the muscle perhaps by as much as 100 percent, either due to an increase in the number of mitochondria or the interconnections among them (Brooks, Fahey, and White (1996). The enhancement of mitochondrial mass by endurance training greatly facilitates the metabolism of lipids (fats) thereby releasing a large supply of potential energy required for intensive exercise over long durations. The increases in mitochondrial mass also appear to facilitate the clearance of lactic acid formed during exercise. These two factors, among several others, greatly increase the endurance capacity of the individual.

Table 4.3
LEVEL OF STRESS

Too Low	Appropriate Amount	Too High
no adaptation functional capacity	adaptation	injury

Exercise Training Techniques. Improving the immediate, anaerobic, and aerobic energy sources requires certain exercise training techniques. Training techniques, along with some other training principles, are briefly described below.

Specificity of Training—This principle states that only those systems or body parts affected by the stress of training will make adaptations. For example, a training protocol designed to improve the strength of the leg muscles should result in hypertrophy of those target muscles in the leg, However, no adaptation is expected in the arm muscles due to the stress of training the leg muscles.

Overload Principle and Progressive Resistance Training—This principle states that properly exercising the body above and beyond what it is accustomed to (stress), leads to physiological adaptations that result in a greater functional capacity. For example, proper strength training using weights provides the stress required for the body to make adaptations. As a result, the targeted skeletal muscles grow larger (hypertrophy), increasing their force production capability. However, as summarized below, too little stress results in no adaptation and too high stress results in injury. In general, the level and type of stress required to improve the three energy sources is different (see Table 4.3).

Progressive resistance training incorporates the overload principle to maximize the effects of exercise. Progressive resistance training is the gradual increase of the stress level during training over some extended time period. For example, if you have a goal to run the marathon (26.2 miles) and have never trained for this distance, your training program should consist of a gradual increase in the distance over several months. You may start out running only 2–3 miles early in your training program. As your body adapts to the stress of working out at this distance, you can then increase your mileage to, say, 4–6 miles. After several months of proper training you should be able to cover the entire 26.2 miles. The point is that your body requires a certain amount of time to adapt to the stress imposed on it by your training program before new and greater levels of stress are imposed. Progressive resistance training, when done properly, optimizes the adaptations to the stress of training, while at the same time reducing the likelihood of injury.

Interval Training—Interval training consists of periods of very intense exercise followed by periods of rest. Interval training can be used to improve all energy sources. In general, the *intensity* of the exercise during interval training should approximate or even exceed the intensity required in the activity of interest or the performance goal (e.g., running 400 meters in 45 seconds or swimming 200 meters in 2 minutes). During interval training, the desired intensity is typically produced over less time than the performance goal. For example, a workout using interval training for a 200-meter swim might be a set of 10 x 50-meter swims performed at maximal effort, each followed by one minute of rest. This type of workout allows the person to train with near maximal effort, but allows the effects of fatigue to dissipate during the rest period. In interval training, the intensity, repetition (number of repeats within a workout period or set), rest (after each repeat), and frequency (number of sets) can be varied depending on the activity of interest and fitness level of the individual. Interval training can enhance all the energy sources, although research has not yet revealed the specific physiological mechanisms responsible for the performance benefits. However, interval training appears to enhance the oxidative energy source through adaptations in skeletal muscle mitochondria (Brooks and Fahey, 1987).

Endurance Training—Another type of training technique particularly useful for enhancing oxidative energy sources is endurance training, also called long-distance training. This training involves continuous activity at moderate intensity over long time periods. The intensity of the activity typically is below the performance goal such that the individual's heart rate is between 60 percent and 80 percent of maximal heart rate (Wilmore and Costill, 1988). For example, an individual (i.e., non-elite runner) wishing to improve their oxidative energy source can do so by running continuously for 20 to 30 minutes, three times per week, at a pace that produces about 75 percent of the maximal heart rate.

SUMMARY

1. The laws of thermodynamics govern metabolism within the body.
2. The potential energy stored within carbohydrates, fats, and protein can be converted to kinetic energy during metabolism to perform the work necessary for muscle contraction, building new tissue, transmission of nerve impulses, and a variety of other important activities.
3. The major fuel for muscle contraction and other activities is adenosine triphosphate (ATP).
4. ATP is produced from three energy sources: immediate, non-oxidative, and oxidative.
5. A variety of training techniques can be used to improve these three energy sources.

CARDIOVASCULAR CHANGES WITH EXERCISE

The body's cardiovascular responses to exercise are **acute** (short term or immediate) and **chronic** (long term). The acute responses to a bout of exercise, such as running up a flight of stairs or running one mile, are manifested in changes in several cardiovascular parameters, whereas training protocols designed to improve the oxidative energy source will induce long-term or chronic changes in the cardiovascular system. Both the acute and chronic changes of the cardiovascular system due to exercise are explored in the next section.

ACUTE AND CHRONIC CHANGES WITH EXERCISE

The acute or short-term effects of exercise on the cardiovascular system depend on the type and the intensity of the activity. Activity that requires a large muscle mass with great intensity, such as running uphill, places a much greater stress on the cardiovascular system than lifting a bucket of water, for example. More long-term or chronic adaptations of the cardiovascular system occur as a result of endurance training (see last section). Let us now examine some of the important acute and chronic cardiovascular responses to exercise and physical activity.

Oxygen Consumption—In general, the utilization of oxygen for demanding activities is directly proportional to the intensity of the exercise (Skinner and McLellan, 1980). As the intensity of the activity increases, so does the demand for oxygen. Oxygen consumption is usually denoted as **VO2**. Resting at the bottom of the hill requires a VO2 of about 4 milliliters of O2 per kilogram of body weight per minute for the average young adult. As one runs up the hill, VO2 steadily increases, as shown in Figure 4.6.

Oxygen consumption continues to increase until it reaches some maximum called **VO2max**, where it begins to level off, even with further exercise.[3] The ability to consume and utilize O2 does not go beyond one's VO2max. Thus, to continue exercising with greater intensity beyond VO2max requires the body to obtain energy from non-oxidative energy sources. As the non-oxidative energy sources are used up, lactic acid builds up in the blood and exercising muscle, the painful effects of fatigue may set in, and the individual becomes less capable of sustaining the activity. The point at which lactic acid begins to build up in the blood is referred to as the onset of blood lactate accumulation (**OBLA**—also called lactate threshold) and often is expressed as a function of the percentage of the individual's VO2max. Individuals with greater endurance tend to have OBLAs at a higher percentage of their VO2max.

VO2 is determined by heart rate, stroke volume of the left ventricle in the heart, and the amount of oxygen extracted from the blood expressed as follows:

$$\textbf{VO2 = HR} \times \textbf{SV} \times \textbf{a} - \textbf{v O2 difference} \qquad (4.2)$$

where **VO2** = oxygen uptake in milliliters per kilogram of body mass per minute, **HR** = heart rate in beats per minute, **SV** = volume of blood pumped from the left ventricle in each beat (in milliliters), and **a – v O2**

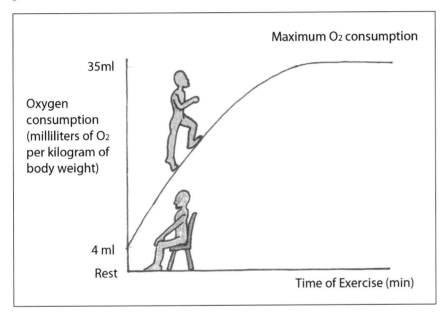

Figure 4.6

Oxygen consumption as a function of the duration of exercise.

difference = difference in oxygen content between the arteries and the veins, expressed in milliliters of O2 per 100 ml of blood.

Recall that cardiac output is equal to **HR x SV** (see Chapter 3); therefore, oxygen consumption is equal to cardiac output multiplied by the difference in oxygen content between the arteries and veins.

Perhaps the single most important criterion of an individual's cardiovascular fitness is the measure VO2max. As one's cardiovascular fitness increases, so does VO2max. Long-term endurance training results in increases in VO2max. Most published studies indicate improvements of VO2max of 20–30 percent are common (see Brooks and Fahey, 1987, for a discussion); however, increases of up to 93 percent have been reported (Pollack, 1973). The degree to which VO2max increases depends on the type of training program (intensity and duration), present fitness level, and age of the individual.

Heart Rate—Figure 4.8 illustrates how heart rate might change as one runs up the long hill and then stops to rest at the top of the hill. At rest, the average heart rate for a young adult is approximately 72 beats per minute. As shown, heart rate increases directly with exercise intensity until it reaches some maximum value at the point near exhaustion. The maximum heart rate attained by an individual depends on one's age. A rough estimate of maximum heart rate can be obtained by subtracting one's age from 220 (beats per min). For example, it would be expected that the maximum heart rate attained by a 20-year-old would be approximately 200 beats per

HIGHLIGHT

Oxygen Debt: Paying Back What You Owe!

If you were to run quickly up to the top of a small hill, you would notice that your heart rate and breathing rate would drastically increase from rest. Your oxygen consumption would be much higher during the recovery period than at rest. After some period of time, depending on the intensity of the exercise, both your heart rate and breathing rate would gradually return to normal resting values, as would your oxygen consumption. During a brief intense bout of exercise such as this, anaerobic energy systems are primarily used to supply the required energy. The use of these energy systems has been thought of as a debt to be paid for during the recovery period in the way of an increase in oxygen consumption over that required at rest. The term **oxygen debt**, coined by exercise physiologist A.V Hill (1927) and coworkers after the turn of the last century, has been used to reflect the increase in oxygen consumption over rest during the recovery period. Figure 4.7 diagrams the relationship between oxygen consumption and the time of exercise.

Also notice another term in Figure 4.7, **oxygen deficit**, which represents the difference between the oxygen consumption required during the exercise and the oxygen actually consumed.

An early theory by A. V. Hill had postulated that the oxygen debt was related to the metabolism of lactate that had built up in the blood as a result of the use of anaerobic energy sources during the exercise. However, subsequent research showed that the degree of oxygen debt was not related to lactate metabolism (e.g., Bang, 1936). Rather, it is likely that the elevated consumption of oxygen following exercise is due to several factors related to returning the body's physiological state back to normal resting conditions (see Brooks, Fahey, and White, 1996, for an excellent discussion).

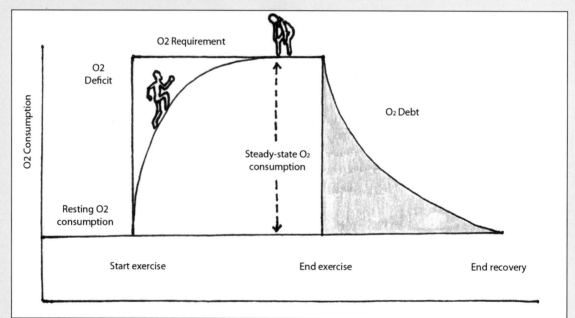

Figure 4.7

Oxygen deficit, oxygen requirement, and oxygen debt prior to, during, and after submaximal, steady-state exercise, (adapted from *Training for Sport and Physical Activity: The Physiological Basis of the Conditioning Process*, third edition by J. H. Wilmore and D. L. Costill, 1988, by Wm. C. Brown).

minute. Maximum heart rate decreases with age (from early adulthood to old age) limiting oxygen consumption. The decrease in maximum heart rate with age limits our ability to deliver oxygen to the tissues. It is an important reason why our cardiovascular endurance capabilities significantly decrease due to aging processes.

Endurance training lowers resting heart rate. Elite endurance athletes can have resting heart rates of 40 beats per minute or lower. Endurance training can also lower heart rates attained at submaximal levels of exercise. Reduced heart rates at rest and submaximal levels of exercise indicate that the heart is performing less work at the same work load, which means that it becomes more efficient as a result of endurance training. Maximal heart rates, however, are not changed as much by endurance training. Some research suggests that the lower heart rate effect is due to an increase in parasympathetic activity and a decrease in sympathetic activity in the heart.

Stroke Volume—The amount of blood pumped from the heart by the left ventricle in each heart beat is the **stroke volume (SV)**. At rest, the stroke volume can range from 60 to 100 milliliters of blood per beat and can increase to over 200 milliliters at maximal exercise, depending on the size and physical conditioning of the individual. As shown in Figure 4.9, the increase in stroke volume with exercise intensity reaches a maximum well before VO2max (usually around 40–50 percent of VO2max) and then begins to level off. Thus, additional increases in cardiac output beyond this point are accomplished by increases in heart rate.

Stroke volume significantly increases as a result of exercise training. Stroke volume increases at rest, at submaximal, and at maximal exercise. Essentially, this increase means that the heart becomes much stronger with endurance training. There appear to be several reasons behind the stroke volume increases:

1. more complete filling of the heart during diastole, resulting in greater end-diastole blood volume;
2. small increase in the mass of the left ventricular walls (hypertrophy);
3. increase in the volume of the ventricular chambers;
4. increase in the blood volume at rest.

Arterial-Venous Oxygen Difference in the Blood—At rest, the oxygen content of blood is around 20 ml of O2 per 100 ml of arterial blood, whereas the oxygen content is 14 ml of O2 per 100 ml of venous blood. The difference between them (a – v O2 difference) represents the extent to which oxygen is extracted from the blood

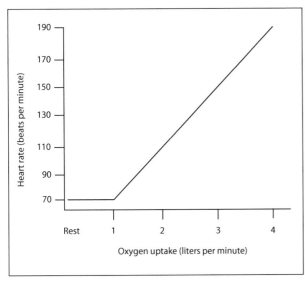

Figure 4.8

Heart rate at rest and as a function of oxygen uptake during exercise.

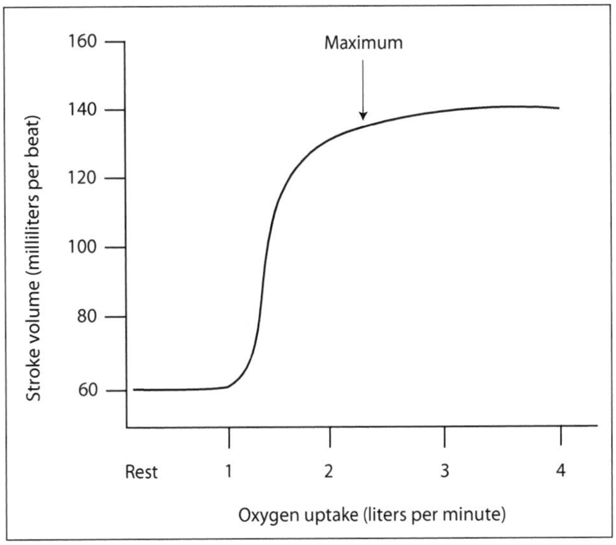

Figure 4.9

Stroke volume as a function of oxygen uptake at rest and during exercise.

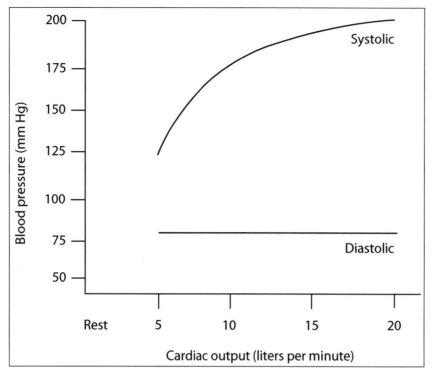

Figure 4.10
Systolic and diastolic blood pressure as a function of cardiac output during exercise.

as it passes through the body. With an increase in exercise intensity, the a – v O2 difference increases due to a *decrease* in venous oxygen content. This decrease occurs because, with exercise, the tissues extract more oxygen from the blood. The oxygen content in the arterial blood remains unchanged with exercise.

The a – v O2 difference is slightly increased with endurance training. This increase suggests that the affected tissues are capable of more efficiently extracting oxygen from the blood. These changes appear to be due to increases in mitochondrial adaptations, elevated myoglobin concentrations, and muscle capillary density.

Blood Pressure—Blood pressure is related to cardiac output and to total peripheral resistance in the blood vessels. With exercise, cardiac output increases and the total peripheral resistance generally decreases, resulting in a large increase in blood pressure. Systolic blood pressure rises from 120 mm Hg to 180 mm Hg or more as a function of exercise intensity. The increase in cardiac output causes a direct increase in systolic blood pressure, but very little change in diastolic blood pressure. In fact, significant increase in diastolic blood pressure with exercise usually indicates some abnormality or presence of coronary artery disease (Sheps, 1979). Figure 4.10 illustrates the changes in systolic and diastolic blood with increases in exercise intensity.

Endurance training tends to reduce blood pressure during rest. It is not clear what mechanism is responsible for this reduction. It may be that lowering of blood pressure is due to decreases in body fat with endurance training. Endurance training has been used as one type of treatment for hypertension but it is clear that it is not a cure for the disease. Other treatments of hypertension include changing to a low-fat diet, reducing sodium intake, and taking certain medications.

Blood Flow—During physical activity, blood is directed toward the exercising muscles and away from the viscera and non-active muscles. The change in direction of blood flow to different body parts is called **blood shunting**. Blood shunting is accomplished by the sympathetic nervous system, responsible for increasing the diameter of the blood vessels to increase blood flow, called **vasodilation**, and for decreasing the size of the blood vessels to reduce blood flow, called **vasoconstriction**. As we run up a hill, for example, the blood vessels supplying the legs vasodilate, effectively increasing blood flow to the active muscles. The blood vessels supplying many of the visceral organs, such as the stomach, undergo vasoconstriction, reducing blood flow. Blood flow to the brain remains relatively constant during exercise. Figure 4.11 illustrates blood shunting at rest, moderate exercise, and intense exercise.

There is evidence to suggest that coronary blood flow increases as a result of endurance training. In fact, recent evidence suggests that the size of the coronary artery increases as a result of long-term endurance training (see below). In addition, the number of capillaries appears to increase in the trained muscle. Both of these are major contributing factors to improvements in blood flow as a result of endurance training.

SUMMARY
- There are a number of significant changes in the cardiovascular system as a function of acute exercise and as a function of chronic endurance training.
- Oxygen consumption, heart rate, stroke volume, oxygen extraction from the blood, and blood pressure increase as a function of exercise intensity.
- Blood flow increases to the exercising muscles and decreases to other regions such as the visceral organs, a phenomenon called blood shunting.
- Blood flow to the brain remains relatively the same during exercise and rest.
- Oxygen consumption reaches a maximal level, called VO2max, and remains constant even with further increases in exercise intensity.
- Lactate in the blood accumulates, contributing to muscle fatigue.
- Endurance training causes a number of changes in the cardiovascular system.

Ventilatory Changes with Exercise

As one exercises, changes in the mechanical act of breathing, called ventilation, can be observed. In this section, the ventilatory changes that accompany exercise and the factors responsible for these changes are described.

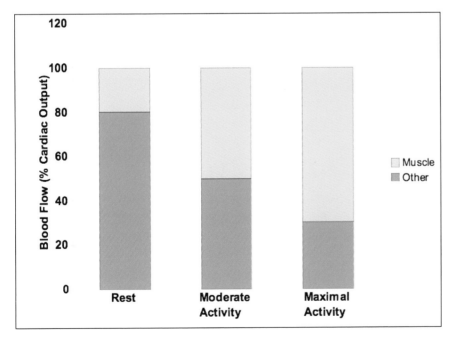

Figure 4.11
The distribution of blood flow in the exercising skeletal muscles and other organs at rest, moderate, and maximal exercise. Not shown is blood flow to the brain that remains relatively constant at rest and during exercise.

In addition, the effects of exercise training on ventilatory responses will be illustrated. Finally, the concept of the anaerobic threshold will be discussed as it relates to the onset of blood lactate accumulation during exercise.

Ventilatory Changes Accompanying Exercise—Shown in Figure 4.12 is a typical ventilatory response before, during, and after exercise. As one anticipates a bout of exercise, say, a vigorous run up to the top of a hill, an increase in ventilation will occur. This increase in ventilation *prior* to exercise is due to neural factors regulated by the interaction of a variety of brain structures. Once exercise begins, ventilation is regulated by both neural and chemical events in a complicated set of processes yet to be fully understood. Ventilation begins to level off at submaximal exercise (as shown) and once exercise stops, ventilation rate declines rapidly. The shape of the ventilation rate curve depends on the type, duration, and intensity of exercise.

Effects of Endurance Training on Ventilation—The influence of endurance training on ventilation rate is illustrated in Figure 4.13. In both trained and untrained individuals, ventilation rate in liters per minute (VE) increases linearly with exercise intensity. However, trained individuals have a lower ventilation rate than untrained individuals at lower workloads (< 75 percent of VO2max). At higher workloads (> 75 percent of VO2max), trained individuals may produce higher ventilation rates because of an increase in exercise capacity.

Anaerobic Threshold and the Onset of Blood Lactate Accumulation—As discussed previously, an increase in exercise intensity is matched with linear increases in VO2. However, onset of the accumulation of lactic acid in the blood (OBLA or lactate threshold) does not occur until about 60 percent of VO2max. After this point, lactic acid quickly builds up in the blood. For many years, it was thought that the onset of blood lactate accumulation would be accompanied by similar increases in ventilation rate. The rationale was that an increase in ventilation rate would indicate an inefficient delivery of O2 to the muscles. The point at which ventilation rate dramatically increases and departs from linear increases has been termed the **anaerobic threshold**. However, problems have been noted with this conceptualization. First, it has been shown that the anaerobic threshold and OBLA do not always occur at the same time (see Brooks, Fahey, and White (1996). Second, and most damaging, is research showing that McArdle syndrome patients, who are incapable of catabolizing glycogen and forming lactic acid, still show the nonlinear increases in ventilation (Hagberg, Coyle, Carroll, Miller, Martin, and Brooke, 1982). Thus, there does not seem to be a direct link between the anaerobic (ventilatory) threshold and the inefficient delivery of O2 to the muscles.

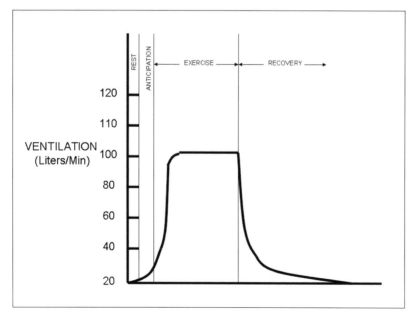

Figure 4.12

Ventilation rate at rest, during, and after exercise (adapted from *Fundamentals of Human Performance*, G. A. Brooks, and Fahey, T. D., 1987, Macmillan).

SUMMARY

- An increase in ventilation rate from resting values may be observed in anticipation of a bout of exercise due to neural factors.
- Further increases in ventilation rate occur with exercise due to both chemical and neural factors.
- Endurance-trained individuals have lower ventilation rates at low workloads and higher ventilation rates at high workloads compared to untrained individuals.

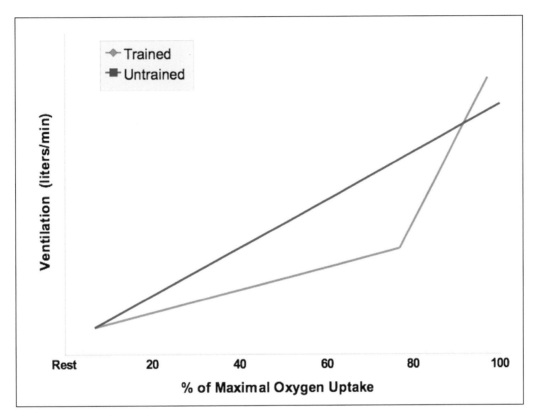

Figure 4.13

Ventilation rate in trained and untrained individuals as a function of the percentage of VO2max at rest and during exercise. (adapted from *Fundamentals of Human Performance*, G. A. Brooks, and Fahey, T. D., 1987, Macmillan).

The Importance of Physical Activity and Exercise

An interesting hypothesis was developed by Pearl (1928) called the Rate-of-Living Theory. The theory reflected the attitudes of some people in the medical profession at the time toward the value of exercise. The Rate-of-Living Theory stated that the greater the rate of energy expenditure and oxygen utilization, the *shorter* the life span. Similarly, Seyle and coworkers (Seyle, 1960; Seyle and Prioreschi, 1970) argued that vigorous exercise, along with infections, trauma, and nervous tension, can be considered a stress that imposes long-term *harmful* effects on the body. Is there any support for the notion that engaging in a long-term exercise program is harmful to the body?

After years of research in laboratories all over the world, no evidence has accrued that exercise produces serious deleterious effects on one's health or causes the body to "wear out" more rapidly, thereby reducing life span (Holloszy, 1983). In contrast, most evidence suggests that exercise causes the body to adapt in several *positive* ways that may improve one's state of health. There may be a variety of health benefits gained from engaging in a regular exercise program over an extended period of time. In this section, some of these benefits are summarized. It should be emphasized that much more research is needed in several areas to determine more precisely the scope and limits of the importance of exercise to one's health. Discussed in this section are the negative effects of physical inactivity and the benefits of exercise in reducing the risk of cardiovascular disease.

The Negative Effects of Inactivity—Several studies and critical reviews have been conducted on the effects of extended bed rest on a variety of physiological functions (Booth, Nicholson, and Watson, 1982; Greenleaf and Koslowski, 1982; Lamb, Stevens, and Johnson, 1965; Sandler and Vernikos, 1986). In the scientific investigations conducted, subjects were confined to rest for extended periods. Following these periods, physiological measurements were taken to determine whether the functional capacity of the individual changed in any way. The results of these studies almost uniformly indicated that functional capacity deteriorated as a result of the lack of physical activity. Particularly damaging effects of bed rest occurred on the cardiovascular system as shown in the following parameters:

- decrease in maximal stroke volume
- decrease in coronary blood flow
- decrease in VO2max
- decrease in systolic blood pressure and increase in diastolic blood pressure
- increase in resting heart rate.

All of these negative effects suggest a diminished work capacity of the heart. In addition, extended periods of bed rest have negative effects on the muscles and bones. As a result of inactivity, muscle mass decreases, reducing muscular strength and endurance. The bones, particularly the weight-bearing bones, undergo a process of demineralization, making them highly vulnerable to injury.

The negative effects of inactivity on the cardiovascular and musculo-skeletal system have serious implications, not only on the lifestyles of sedentary individuals but also on patients recuperating from certain types of surgeries or illnesses. Most medical experts now agree that, rather than extended bed rest after surgery or a bout with illness, individuals should be encouraged to participate in light to moderate levels of physical activity or exercise. When properly administered, the increase in the level of physical activity or exercise following surgery or an illness should improve the work capacity of the heart and cardiovascular system. In addition, exercise will strengthen both the muscles and bones.

In summary, in contrast with the Rate-of-Living Theory that exercise wears down the body, most available evidence indicates that the body undergoes positive adaptations as a result of the increased stress imposed upon it by adequate levels of exercise or physical activity.

Exercise and Cardiovascular Disease—Heart disease is the number one killer in Western countries such as the United States and Great Britain. Nearly one-half of the deaths in these countries are attributed to diseases of the cardiovascular system (see Figure 4.14). Heart disease manifests itself in a buildup of fatty deposits (or lipids) in the walls of the arteries supplying the heart, a condition called **atherosclerosis**. As these fatty deposits increase, blood flow through the artery decreases, thereby reducing the supply of nutrients and oxygen to heart muscle. Atherosclerosis can also impair blood flow to other organs such as the brain, liver, and kidneys.

There are many variables, called **coronary risk factors**, that increase one's chances of developing heart disease. These factors, developed by the American Heart Association (1990), are distinguished by those factors that cannot be changed by the individual, those that can, and other risk factors that possibly can be changed by the individual (see Table 4.4).

One of the coronary risk factors that is, to a large extent, controllable by the individual is physical inactivity. What evidence is there that participating in an exercise program over extended time periods might reduce the likelihood or incidence of certain diseases of the cardiovascular system? The answer to this question comes from two major lines of epidemiological research studies. **Epidemiology** is the study of the incidence, distribution, and control of a disease. The first line of research has compared the incidence of heart disease in physically active people with sedentary individuals. One hypothesis predicts that people who have been physically active for many years should have less incidence of heart disease than sedentary individuals. An example of this research is a study by Paffenbarger and Hale (1975) ,who compared the death rates from coronary heart disease in groups of professions that were labeled as either *physically active* (e.g., postal deliverers, active longshoremen) or *sedentary* (e.g., bus drivers and postal clerks). The results of the study showed that death rates from heart disease were significantly higher in the sedentary professions than in physically active professions. Many other studies have also found a similar relationship between the incidence of heart disease and physical activity level. However, it is important to point out that these types of studies are only *suggestive* of a causal link between the development of heart disease and physical activity (or exercise). It is possible that the physical activity demands in the two groups of professions (sedentary, physically active) do not determine the development of heart disease. It could be that people who choose the more physically demanding professions are more naturally resistant to the development of heart disease, for whatever reasons.

A second line of research investigating the hypothesis that physical activity and exercise reduces the incidence of heart disease are those studies that have experimentally manipulated the level of exercise in different groups of subjects over extended time periods, and then compared the groups for signs of atherosclerosis. A classic experiment of this type was performed by Kramsch, Aspen, Abramowitz, Kreimendahl, and Hood (1981) on monkeys. Three groups of monkeys were investigated: a control group was fed a normal low-fat diet, a non-exercising group was fed a high-fat diet known to induce heart disease, and an exercising group that also ate the

Table 4.4
Coronary Risk Factors

Factors that cannot be changed (genetic)
  Heredity
  Age
  Sex
  Race

Factors that can be changed
  Cigarette Smoking
  Hypertension
  Blood Cholesterol

Other factors that may be changed
  Body Weight
  Physical inactivity
  Stress
  Diabetes

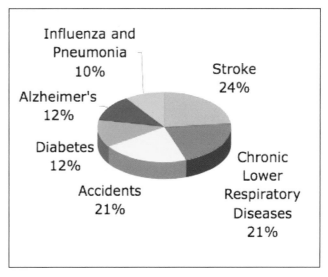

Figure 4.14
The leading causes of death in the United States.

high-fat diet. After the training period, all monkeys were assessed for signs of cardiovascular disease. The group that remained sedentary and ate the high-fat diet developed atherosclerosis, whereas the group who exercised and ate the high-fat diet actually showed an increase in the size of their coronary arteries and substantially less atherosclerosis. Figure 4.15 illustrates a comparison of the left main coronary artery in the sedentary and the exercising monkeys. While these results do not *prove* that exercise prevents coronary artery disease[4], they strongly suggest that exercise can improve blood flow to the heart, thereby reducing the likelihood of the development of severe atherosclerosis.

Taken together, these two lines of research suggest that exercise plays an important role in the prevention of coronary artery disease. But because this disease is thought to be brought on by several factors, much more research is needed to verify this hypothesis (see Holloszy, 1983, for detailed arguments).

SUMMARY

- Physical inactivity leads to a number of cardiovascular complications.
- Lack of physical activity has negative effects on the bone and muscle tissue.
- Research suggests that the incidence of atherosclerosis can be reduced by regular participation in physical activity and exercise.

CHAPTER SUMMARY

- The body performs many activities by converting one form of energy to another in accordance with the laws of thermodynamics.
- Kinetic energy is produced when the chemical bonds of various molecules are either broken (catabolism) or assembled (anabolism).
- The potential energy in carbohydrates, fat, and protein is measured in units called calories.
- The major fuel for the production of mechanical energy during muscle contraction is adenosine triphosphate (ATP).
- Energy is released when ATP is broken down allowing for muscle contraction.
- ATP is produced by the immediate, non-oxidative and oxidative energy sources.
- These three energy sources may be improved with certain types of exercise training.
- Cardiovascular and ventilatory responses to exercise may be acute (short term) and chronic (long term).
- Cardiovascular changes accompanying exercise may be seen in oxygen consumption, heart rate, stroke volume, oxygen extraction from the blood, blood pressure, and blood flow.
- The rate of ventilation changes with exercise intensity and endurance-trained individuals show different ventilation rates than untrained individuals.
- Ventilation rate does not correlate well with increases in blood lactate.
- The lack of physical activity leads to a number of cardiovascular, muscle, and bone complications.
- Physical activity and regular bouts of exercise have been shown to reduce the incidence of cardiovascular disease.

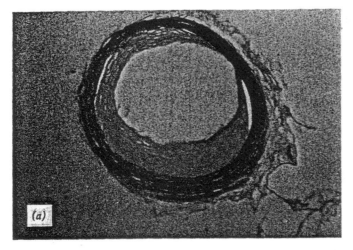

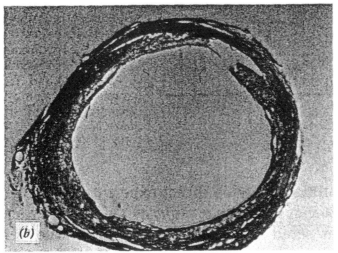

Figure 4.15

Cross section photographs of comparable sections of the left main coronary artery in sedentary *(a)* **and exercising** *(b)* **monkeys on a high-fat diet (from Kramsch, D. M., Aspen, A. J., Abramowitz, B. M., Kreimendahl, T., and Hood, W. B. 1981. Reduction of coronary atherosclerosis by moderate conditioning exercise in monkeys on an atherogenic diet.** *New England Journal of Medicine, 305,* **1483–89).**

IMPORTANT TERMS

A – v O2 difference – difference in oxygen content between the arteries and veins

Acute and chronic changes to exercise – short term and long term adaptations, respectively, in the body due to exercise

Adenosine triphosphate – an important molecule that provides energy to all physiological processes

Anabolism – a type of metabolism that creates stored energy

Anaerobic metabolism – metabolism in the absence of oxygen

Atherosclerosis – a major type of cardio-respiratory disease

Basal metabolism – the amount of energy needed to maintain all physiological processes at rest

Blood flow – is affected by the pressure gradient and resistant in a blood vessel

Blood pressure – changes during the contraction and relaxation of the heart

Blood shunting – blood that is directed to an exercised muscle

Calories – a unit of energy

Cardiovascular disease – disease that affects the heart and vascular system

Catabolism – a type of metabolism that provides energy needed for various physiological functions

Coronary risk factors – factors that help predict the incidence of heart disease

Electron transport chain – a process of oxidative metabolism that helps produce large amounts of ATP

Endurance training – sustained training at sub-maximal levels over long time periods

Energy - is governed by the first and second laws of thermodynanics

Epidemiology - the study of the incidence, distribution, and control of a disease

Glycogenolysis – a process that converts glycogen to glucose

Glycolysis – a process that breaks down glucose to ATP

Heart rate – the frequency of contractions in the heart

Hydrolysis – a process of breaking down ATP to produce energy

Hypertrophy - an increase in muscle tissue as a result of exercise training

Immediate energy sources – the energy that can sustain activity for a short time period

Interval training – repeated bouts of intense exercise following by rest

Kinetic energy – energy produces during movement

Krebs cycle – a process that produces large amount of ATP

Laws of Thermodynamics – physical laws governing the nature of energy

Maximum oxygen uptake - the maximum capacity of an individual's body to transport and use oxygen and is thought to be the best indicator of the cardiovascular fitness of an individual

Metabolism – the process of utilizing energy in the body

Muscular fatigue – a process that inhibits muscle activity

Non-oxidative energy sources – processes that help produce energy without the presence of oxygen

Nutrients – carbohydrates, protein and fat

Onset of blood lactate accumulation – the time during exercise when lactic acid appears

Overload principle - states that properly exercising the body above and beyond what it is accustomed to (stress), leads to physiological adaptations that result in a greater functional capacity

Oxidative metabolism – a process that produces large amounts of ATP through the use of oxygen

Oxygen consumption – the amount of oxygen used during exercise

Oxygen debt - is a measurably increased rate of oxygen intake following strenuous activity

Potential energy – stored energy

Power stroke – movement of the myosin head cross bridges during muscle contraction

Progressive resistance training - is the gradual increase of the stress level during training over some extended time period

Rate-of-living theory – a disproven theory that the greater the rate of energy expenditure and oxygen utilization, the shorter the life span

Rephosphorylation – a process in the immediate energy store producing small amounts of ATP

Specificity of training – an exercise principle that states that only those systems or body parts affected by the stress of training will make adaptations

Stroke volume – the amount of blood pumped out of the left ventricle per beat

Vasoconstriction – a process that reduces the diameter of a blood vessel

Vasodilation - a process that increases the diameter of a blood vessel

INTEGRATING KINESIOLOGY: PUTTING IT ALL TOGETHER

1. How many calories do you consume and expend on an average daily basis? For three consecutive days, keep track of all the calories you eat. In addition, estimate how many calories you expend over these same three days. To facilitate this assignment, obtain the following reference and use the tables provided on the nutritive value of commonly used foods, alcoholic beverages and nonalcoholic beverages, and specialty and fast-fast foods (Appendix B, pp. 715–62) and energy expenditure in household, recreational, and sports activities (Appendix D, pp. 769–81) to do your calculations:

McArdle, W. D., Katch, F. I., and Katch, V. L. (1996). Exercise Physiology: Energy, Nutrition, and Human Performance. *Philadelphia: Lea and Febiger.*

- Using these tables and the information provided in Chapter 9, Human Energy Expenditure During Rest and Physical Activity, it is possible to determine the amount (and percentage) of carbohydrate, fat, and protein caloric intake on a daily basis.
- Remember to include your basal metabolism in estimating energy expenditure. A discussion of how to obtain this estimate is in McArdle et al. (1996) on page 151.
- Also, remember to convert both caloric intake and energy expenditure to kcal.
- This is a challenging, but interesting, assignment. Good luck!

2. If you have the desire to improve your cardiorespiratory fitness level, try developing your own exercise training schedule using the exercise training techniques discussed in this chapter. In developing such a program, keep in mind the following considerations:
- What is the specific purpose or goal of the program? For example, do you want to compete in a running event, or do you just want to get in better physical shape?
- How do you plan to reach this goal?
- What specific training principles will you use and why?
- What physical activity(ies) will you incorporate into your program (e.g., swimming, running, cycling, etc.)?
- What will be the duration and intensity level of each workout?
- Consider keeping a log of your workouts.
3. Which of the three energy sources provides the major sources of energy for the following activities?
- Running a marathon
- Lifting a heavy box in one attempt
- A 50-meter sprint in swimming

KINESIOLOGY ON THE WEB

- http://journals.lww.com/acsmmsse/pages/collectiondetails.aspx?TopicalCollectionId=1—This Web site contains several important position statements from the American College of Sports Medicine on the importance of exercise and physical activity.
- http://americanheart.org—This is a Web site of the American Heart Association on risk factors in heart disease.
- www.eatright.org/public/—This is the Web site of the American Dietetic Association that provides information on good nutrition.

REFERENCES

American Heart Association (1990). Heart and stroke facts, Dallas, pp. 18–20.

Bang, O. (1936). The lactate content of the blood during and after muscular exercise in man. *Scand. Arch. Physiol. 74* (Suppl. 10): 49–82.

Booth, F. W., Nicholson, W. F., and Watson, P. A. (1982). Influence of muscle use on protein synthesis and degradation. *Exercise and Sport Sciences Reviews, Vol. 10*, R. L. Terjung (ed.). Philadelphia: Franklin Institute, pp. 27–48.

Brooks, G. A. (2001). Lactate doesn't necessarily cause fatigue: Why are we surprised? *Journal of Physiology, 536* (1).

Brooks, G. A., Fahey, T. D., and White, T. P. (1996). *Exercise Physiology: Human Bioenergetics and Its Application*. Mountain View, CA: Mayfield.

Brooks, G. A., and Fahey, T. D. (1987). *Fundamentals of Human Performance*. New York: Macmillan.

Greenleaf, J. E. and Kozlowski, S. Physiological consequences of reduced physical activity during bed rest. *Exercise and Sport Sciences Reviews, Vol. 10*, R. L. Terjung (ed.). Philadelphia: Franklin Institute, pp. 84–119.

Hagberg, J. M., Coyle, E. F., Carroll, J. E. , Miller, J. M., Martin, W. H., and Brooke, M. H. (1982). Exercise hyperventilation in patients with McArdle's disease. *Journal of Applied Physiology: Respiration, Environment and Exercise Physiology, 52*, 991–94.

Hill, A. V. (1927). *Muscular Movement in Man: The Factors Governing Speed and Recovery from Fatigue.* New York: McGraw-Hill.

Holloszy, J. O. (1983). Exercise, health, and aging: A need for more information. *Medicine and Science in Sports and Exercise, 15*, 538–42.

Huxley, H. E. (1958). The contraction of muscle. *Scientific American*, 1958.

Kramsch, D. M., Aspen, A. J., Abramowitz, B. M., Kreimendahl, T., and Hood, W. B. (1981). Reduction of coronary atherosclerosis by moderate conditioning exercise in monkeys on an atherogenic diet. *New England Journal of Medicine, 305*, 1483–89.

Lamb, L. E., Stevens, P. M., and Johnson, R. L.. (1965). Hypokinesia secondary to chair rest from 4 to 10 days. *Aerospace Medicine, 36*, 755–63.

Luttgens, K., Deutsch, H., and Hamilton, N. (1992). *Kinesiology: Scientific Basis of Human Motion.* Dubuque, IA : W. C. Brown, 8th ed.

McArdle, W. D., Katch, F. I., and Katch, V. L. (1996). *Exercise Physiology: Energy, Nutrition, and Human Performance.* Philadelphia: Lea and Febiger.

Paffenbarger, R. S., and Hale, W. E. (1975). Work activity and coronary heart mortality. *New England Journal of Medicine, 292*, 545–50.

Pearl, R. (1928). *The Rate of Living.* New York: Knopf.

Pollack, M. L. (1973). Quantification of endurance training programs. *Exercise and Sport Science Reviews, 1*, 155–88.

Rayment, I., Rypniewski, W. R., Schmidt-Base, K., Smith, R., Tomchick, D. R., Benning, M. M., Winkelmann, D. A., Wesenber, G., and Holden, G. M. (1993). Three-dimensional structure of myosin subfragment-1: A molecular model. *Science, 261*, 50–58.

Rayment, I., Holden, H. M., Whittaker, M., Yohn, C. B., Lorenz, M., Holmens, K. C., and Milligan, R. A. (1993). Structure of the actin-myosin complex and its implications for muscle contraction. *Science, 261*, 58–65.

Sandler, H., and Vernikos, J. (1986). *Inactivity: Physiological Effects.* Orlando: Academic Press.

Seyle, H. (1970). On just being sick. *Nutrition Today, 5*, 2–7.

Seyle, H., and Priorechi, P. (1960). Stress theory of aging. In *Aging: Some Social and Biological Aspects.* N. W. Shock (ed.), Washington, DC: American Association for the Advancement of Science, p. 261.

Sheps, D. (1979). Exercise-induced increase in diastolic pressure: Indicator of severe coronary artery disease. *American Journal of Cardiology, 43*, 708–12.

Skinner, J. S., and McLellan, T. H. (1980). The transition from aerobic to anaerobic metabolism. *Research Quarterly for Exercise and Sport, 51*, 234–48.

U.S. Department of Health and Human Services (1996). *Physical Activity and Health: A Report of the Surgeon General.* Atlanta, GA: U.S. Department of Health and Human Services, Centers for Disease Control and Prevention, National Center for Chronic Disease Prevention and Health Promotion.

Vale, R. D. (1993). Measuring single protein motors at work. *Science, 260*, 169–70.

Wilmore, J. H., and Costill, D. L. (1988). *Training for Sport and Activity: The Physiological Basis of the Conditioning Process.* Dubuque, IA :
Wm. C. Brown.

FOOTNOTES

1. The process of the actin-myosin interaction is of great debate. As reported by Vale (1993), the understanding of this process is "one of the great puzzles in biophysics for the last century, and, contrary to what might be inferred from reading a biology textbook, the mechanism of myosin motility is far from being solved." (p. 169). Indeed, Brooks, Fahey, and White (1996) have stated that "... the solution of this problem is probably worthy of a Nobel Prize ...". Thus, it should be made clear that the description given in this text of actin-myosin interaction, developed originally by H. E. Huxley (1958), is one of several viewpoints (see Vale, 1993, for a brief overview). Other evidence has been provided for the sliding filament theory (also called the cross bridge hypothesis of muscle contraction) by detailed electron microscopy and image analysis (see Rayment, Rypniewski, Schmidt-Base, Smith, Tomchick, Benning, Winkelmann, Wesenberg, and Holden, 1993; Rayment, Holden, Whittaker, Yohn, Lorenz, Holmes, and Milligan, 1993).

2. Unused glucose combines with other substances to form a much larger molecule called glycogen (the stored form of glucose) in a process called glycogenesis. The body can store nearly 500 g (1.1 lb.) of glycogen, about 80 percent in skeletal muscle and the rest in the liver.

3. Technically, there is a distinction between VO2max and a parameter **VO2peak**. Often, the maximal oxygen consumption obtained during an exercise test is not necessarily the highest *possible* level. For example, exercise physiologists often find that untrained subjects have greater difficulty reaching their "true" VO2max because of their inability or unwillingness to perform at maximal levels. In addition, VO2max is difficult to obtain when smaller muscles groups are primarily used in the exercise test. For example, VO2max tests using only the arms typically yield maximal oxygen consumption values far less than obtained during exercise using the leg muscles (e.g., exercise treadmill). Thus, the maximal oxygen consumption value obtained during an exercise may represent VO2peak and not the individual's true level of cardiovascular fitness.

4. It is generally recognized in science that no experiment or study can prove or disprove a hypothesis. Rather, experiments and studies are thought to show support or nonsupport for a hypothesis or a theory. Accepting or rejecting a hypothesis is typically a long, tedious process requiring the accumulation of results from several investigations.

Chapter Five
Biomechanical Foundations

Types of Human Motion

Linear

Angular

General

Summary

Describing Human Motion

Linear Kinematics

Distance and Displacement

Speed and Velocity

Acceleration

Vectors and Scalars

Projectile Motion

Angular Kinematics

Angular Distance and Angular Displacement

Angular Speed and Angular Velocity

Angular Acceleration

CHAPTER FIVE

Biomechanical Foundations

"Animals are bodies and their vital operations are either movements or actions which require movements. But bodies and movements are the subject of mathematics."

– Giovanni Alfonso Borelli (1679)

STUDENT OBJECTIVES

1. To distinguish the three major types of human motion.
2. To understand the differences between linear and angular kinematics in describing human motion.
3. To delineate the various explanations of human motion including linear kinetics and angular kinetics.
4. To understand how Newton's Laws apply to human motion.

In this chapter, we explore the subfield of **biomechanics**, defined as the study of internal and external forces acting on the human body and the effects produced by these forces (Hay, 1985). As indicated in the previous chapter, the production of movement depends on the body's ability to generate muscular tension. Muscular tension is produced when stored potential energy is converted to mechanical energy. Many of the movements we make, from wiggling our finger to walking up a flight of stairs, are produced by the internal forces generated by muscular contraction. External forces also influence the type of motions we can make. For example, the force of the earth's gravity affects all movement. Gravity is a force that tends to pull all objects toward the earth's center. The frictional forces of air molecules also are influential in many movements we make. A golf ball hit in Denver one mile above sea level will travel farther than one hit in the same manner in Miami at sea level. This ball flight difference is expected from biomechanics because air resistance is less in Denver than in Miami. The effects of gravity are also slightly less in Denver because of its altitude.

One of the goals in biomechanics is to account for all the relevant forces, both internal and external of the human body, responsible for a given movement or action. According to biomechanical principles, the total of all the forces both within the body[1] and imposed on the body determines the type of observed motion.

Figure 5.1

An example of straight-line or rectilinear translation (adapted from Hay, 1985, Figure 2).

Biomechanics is an extremely useful tool for helping solve applied and practical problems, such as in the field of ergonomics, which investigates the design of facilities, tools, equipment, and tasks that match human capabilities. For example, Andersson, Ortengren, and Schultz (1980) used biomechanical techniques to measure the effects of loads on the lumbar spine during sitting at a table. Biomechanical studies by Nashner and his coworkers (e.g., 1977; 1980; Nashner and McCollum, 1985) and Lichtenstein and coworkers (Lichtenstein, Burger, Shiavi, et al., 1990; Lichtenstein, Shields, Shiavi, et al., 1988; Lichtenstein, Shields, Shiavi et al., 1989) have been used to help clinicians evaluate the neuromuscular systems of patients. Biomechanical knowledge is useful for physical and occupational therapists for evaluating and treating a variety of musculoskeletal and neurological problems (e.g., Donatelli and Wooden, 1989). The proper construction of prosthetic devices and artificial limbs (e.g., Shaperman and Setoguchi, 1989), splints, braces, and other orthotic devices is largely based on biomechanical principles (e.g., Duncan, 1989). Certain types of corrective surgery, such as treatments for scoliosis, a condition that results in exaggerated curvature of the spine, rely on the use of biomechanics (e.g., Ghista, Viviani, Subbaraj, et al., 1988). Biomechanical principles have been extensively used in the analysis of sports skills (e.g., Yeadon and Challis, 1994). Biomechanics is also used as an important research tool at several levels of analysis (i.e., behavioral, systems, cellular) by a number of different subfields within kinesiology.

How can we categorize and describe movements using biomechanical concepts? What are the various ways movements can be described? What is the definition of a force? How are internal and external forces quantified in biomechanics and how do they account for observed body motion? These are some of the questions explored in this chapter. Let us begin by first describing the types or general forms of motions that can be produced by the human body.

TYPES OF MOTION

The human body is capable of producing a wide array of movements. The types of motions that can be produced may be categorized into three basic forms: translational, rotational (or angular), and general motion (a combination of translational and rotational). **Translational motion** occurs when all parts of the body move in the same direction, same distance, and in the same time. An example of translational motion is shown in Figure 5.1. as she skates across the ice, all parts of her body move in the same direction, same distance, and same time, thereby satisfying the criterion of translational motion. Because the motion is in a straight line it is called **rectilinear translation**. An example of a different type of translational motion is shown in Figure 5.2. In this case, the skydiver's motion is following a curvilinear path, yet the body still satisfies the criterion of translational motion. Thus, the skydiver's body motion is called **curvilinear translation**.

Figure 5.3 →

Examples of angular or rotational motion (adapted from Hay and Reid, 1988, Figures 78, 79, and 80).

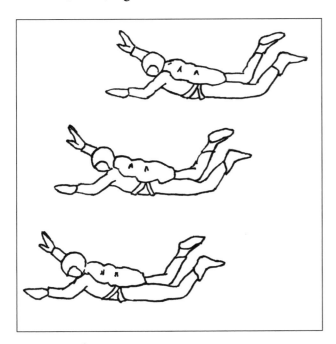

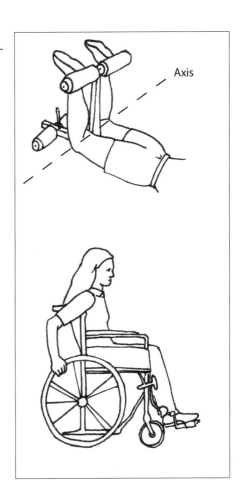

Figure 5.2 ↑

An example of curvilinear translation (adapted from Hay, 1985, Figure 3).

A second major form of motion is **rotational (or angular) motion**. Rotational motion occurs when a body moves in a circular path about some axis so that all parts of the body traverse the same angle in the same direction over the same time. Figure 5.3 illustrates two examples of rotational motion around the knee joint (above) and the motion of the wheel (bottom).

Finally, the third major form of motion is **general motion**, a combination of both translational and rotational (see Figure 5.4). A good example of general motion is the so-called "Thomas Flair" gymnastic skill. General motion is often difficult to analyze because of its complexity. Most biomechanists deal with this complexity by focusing on smaller components of the task. That is, the linear or angular aspects of the motion are examined individually to gain insight into this complexity.

Summary

- There are three major forms of motion: translational, rotational, and general.
- Translatory motion involves the movement of all parts of the body in the same direction, same distance, and in the same time.
- Rotational motion is the circular movement of a body around some axis.
- General motion involves a combination of translatory and rotational motions.

Figure 5.4

The "Thomas Flair"—an example of complex general motion (adapted from Hay and Reid, 1988, Figure 81, pg. 113).

DESCRIBING HUMAN MOTION

Linear Kinematics

The area of biomechanics that is used to *describe* human motion is called **kinematics**. There are two categories of kinematics: linear and angular. Let us first examine **linear kinematics**, considered to be description of motion along a linear path.

In kinematics, a number of questions can be asked to help describe the motion of interest, such as, how far did the body (or body part)1 move, in what direction, how fast, and how fast did it change speeds? First, let us examine two types of descriptions of human motion: distance and displacement.

Distance and Displacement. In Figure 5.5 is an illustration of the motion of the hand. If the elbow joint is flexed such that the hand moves from position A to position B, a vertical distance of 10 cm is covered. If the hand is moved from position A to position B and then back to A, the total vertical distance moved would be 20 cm. The **distance** moved represents the total length of the movement path, regardless of direction. The **displacement** of a body (denoted by the Greek letter θ) refers to a change of position of the body with reference to its starting

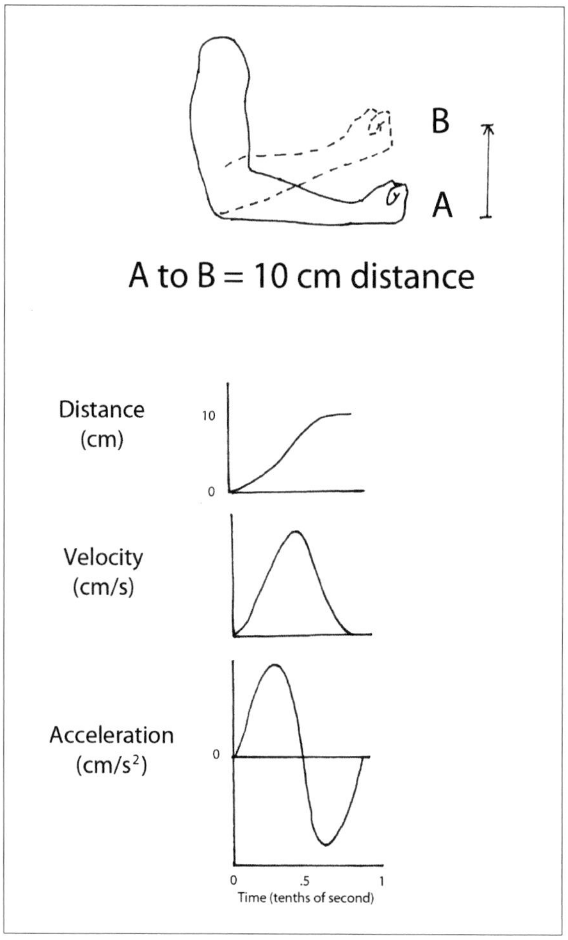

A to B = 10 cm distance

Distance (cm)

Velocity (cm/s)

Acceleration (cm/s²)

Time (tenths of second)

Figure 5.5

An upward, vertical motion of elbow flexion of both a linear distance and displacement of 10 cm (from position A to B). Plotted from top to bottom is the instantaneous displacement, velocity, and acceleration of the motion as a function of time.

position, taking into account direction of movement. For example, in the above example, a movement from position A to B results in a displacement of 10 cm in the vertical direction. In this case, both the distance and the displacement results in the same changes. But a movement from position A to B and then back to A results in a total distance of 20 cm, but a displacement of 0 cm! Using another example, a run around a 400 m track and back to the starting position results in a distance of 400 m, but a displacement of 0 m.

Speed and Velocity. The **average speed** of a body refers to the *distance* traveled relative to some time period. The speed limit of 55 miles per hour on many of the roadways in the United States means that a car going at this speed will travel 55 miles in one hour. The speed limit of 100 km / hr in Canada means that a car going this speed will travel 100 km in one hour. The general formula for the average speed is as follows:

$$s = l\,/\,t \qquad\qquad (5.1)$$

where s is speed, *l* is length or distance, and *t* is time. In the example above, if the person moves the hand from position A to position B in, say, 1 second, then the speed of the movement would be 10 cm / s. What would be the speed of a similar movement if the time *t* is .5 sec? (Answer: 20 cm/s). The **average velocity** (denoted by the Greek letter ω) of a body is the displacement (q) of a body relative to the total time of the displacement. The formula for the average velocity is as follows:

$$\omega = \theta\,/\,t \qquad\qquad (5.2)$$

In the example above, a displacement from position A to B in 1 second would result in an average velocity of 10 cm / s in the vertical direction. If the movement went from A to B and then back to A in one second, then the average velocity would be 0 cm / s. To obtain the **instantaneous velocity** of a body at any given moment in time during the movement, we can use formula 5.2 at very small increments of the movement. For example, we can calculate the instantaneous velocity of the movement from A to B by determining the displacement of the hand, say, every tenth of a second. Figure 5.5 illustrates the position at each .1 second and the results of our calculations. One can see that the instantaneous velocity of the hand increases from 0 cm / s to some maximum, then decreases to 0 cm/s when the hand stops at position B.

Acceleration. The velocity of a body can change during the course of a movement, such as the running of a 100-meter sprint. At the beginning of the race, the runner's velocity is zero while waiting in the starting blocks for the starting signal. Once the runner leaves the starting blocks, running velocity can increase up to some maximal level and may hold steady through the duration of the race. Acceleration (denoted by the Greek letter α) is defined as the *change* in velocity of a body over a given time period:

$$\alpha = \theta\,\mathrm{f} - \theta\,\mathrm{s}\,/\,\mathrm{t} \qquad\qquad (5.3)$$

where θ s is the starting velocity, θf is the final velocity and *t* is the time. Because velocity is represented in m / s (for example) in the numerator and is divided by time, represented in seconds in the denominator, acceleration is expressed in m per second squared (i.e., m / s2), or cm / s2 and so on.

As with instantaneous velocity, it is quite common in biomechanics to determine acceleration changes during small increments of a movement, called **instantaneous acceleration**. Figure 5.5c shows the instantaneous acceleration of the vertical movement from position A to position B. Notice that the highest point of acceleration occurs nearly one quarter of the way into the movement. Acceleration is zero halfway through the movement, when maximal velocity is reached. From this point, the movement slows down and the acceleration values become negative, a term commonly referred to as negative acceleration or de**celeration**. Acceleration is again

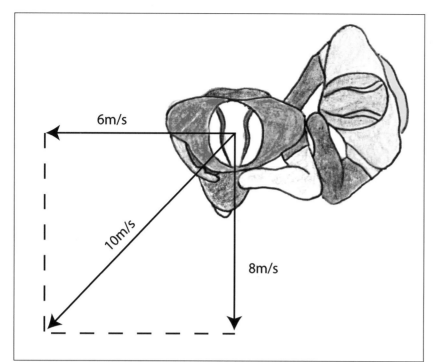

Figure 5.6
A tackler imparts a velocity of 6 m / s on a runner who is moving forward with a velocity of 8 m / s. At that given instant, a resultant velocity of 10 m / s in the direction shown can be determined using the parallelogram of vectors analysis and Pythagorus's theorem (adapted from Hay and Reid, 1988, Figure 84, pg. 121).

zero when the movement stops at position B. Thus, in a discrete movement (a brief movement with a clear beginning and end point) such as this, the hand accelerates until maximal velocity and then decelerates as the hand slows down.

As mentioned earlier, **gravity** is a force that attracts all objects toward the earth's center. If released above the earth's surface, a body will fall at a constant acceleration of approximately 9.81 m / s2, a quantity usually abbreviated as *g*.

Vectors and scalars. Some of the above quantities, like distance and speed, are completely defined when their magnitudes are expressed. These types of quantities are called **scalars**. Other quantities called **vectors**, such as displacement, velocity, and acceleration, require both their magnitudes and direction to be expressed. Vectors can be illustrated by arrows such as those shown in Figure 5.6. A football player is running down the field at a velocity of 8 m / s. At a given instant, a defensive player tackles the runner imparting a velocity of 6 m / s in a direction perpendicular to the runner's motion down the field. Using a geometric analysis known as the parallelogram of vectors (see HIGHLIGHT below), it is possible to determine the instantaneous result of these two interacting vectors, called the **resultant vector**.

Projectile Motion. Projectile motion is the movement of a body through the air. There are many physical activities that require the whole body to move through the air, such as jumping and sky diving. There are other activities that require throwing an object, such as a ball or javelin. Other similar activities involving projectiles include the striking of an object that results in the object being propelled through the air, such as hitting a baseball or golf ball. To completely predict projectile motion, it is necessary to take into account the projectile speed, the projection angle at the moment of release, and the relative projection height. Relative projection height is the difference in the height from which the body is initially projected compared to the height at which it lands or is stopped (Hall, 1991). Air resistance is also another factor affecting projectile motion, but we will ignore air resistance in the following example.

Let us consider a hypothetical case of a projectile thrown under conditions of micro gravity, such as in deep outer space. In deep space, the effects of earth's gravity are absent. If a body or object is propelled in micro gravity situations, the projectile continues indefinitely with the same velocity vector as that at the moment of release. In the hypothetical example shown in Figure 5.7, a javelin starting with a horizontal velocity of 10.7 m / s and a vertical velocity of 9 m / s under no gravity conditions would continue to travel 1.07 m horizontally and .9 m vertically for *every* one tenth of a second of its flight. However, in the presence of the earth's gravity, the projectile is subjected to a downward and constant vertical acceleration of 9.81 m / s2. But gravity has no influence on the projectile's horizontal velocity vector. As under the micro gravity conditions, the projectile travels the same horizontal distance for every second of flight. Because of gravity, the projectile's vertical position falls progressively below the vertical position under the micro gravity conditions. As a result, the trajectory (or path) of the projectile follows a parabolic shaped path (see Figure 5.7), unless its projectile angle is perfectly vertical (i.e., straight up).

Angular Kinematics. A wide variety of human movements involve rotation of the body (or body part) around a fixed point. Figure 5.8 provides two additional examples of this type of movement. For example, the elbow joint shown earlier in Figure 5.5 allows for rotational movement of the lower arm. There are several measures of rotational movement, called **angular kinematics**, that we will now examine.

Angular distance and displacement. The rotational movement of a body (or body part) is typically measured in units called degrees (°). The rotational movement around the elbow joint from position A to position B in Figure 5.5 results in an angular distance of 25°. A movement from A to B and back to A results in an angular displacement of 0°.

Angular speed and velocity. As with linear kinematics, the **average angular speed** of a rotational movement depends on the distance moved by the body or body part over time:

Average angular speed =
Final angle (°) – Initial angle (°) / Time (s)   (5.5)

Rotational movement around the elbow joint from position A (0°) to position B (25°) in 1 second results in an average angular speed of 25° / s. The average angular velocity (denoted by the Greek letter ω) is similarly calculated by using the rotational displacement of the body or body part over the time of the displacement:

Average angular velocity (ω) = Angular displacement (°) / Time (s)   (5.6)

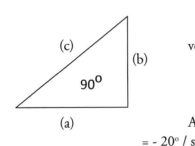

If the gymnast (in Figure 5.9) moves her legs 120° from position 1 to 2 and then 150° back to position 3 in 1.5 seconds, her average angular velocity is

– 20° / s:

= 120° – 150°

Average angular velocity = -30° displacement /s

= - 20° / s

The negative velocity expressed by the quantity – 20° / s means that, on average, velocity was greater during the movement from position 2 to 3.

Average Angular Acceleration. By subtracting the final angular velocity from the starting angular velocity and dividing by the time of the movement, it is possible to calculate the **average angular acceleration** of the body or body part (expressed by the Greek letter α) . Let us say we would like to determine the average angular acceleration of angular motion around the elbow joint between two points in time, at .25 seconds and .5 seconds, during, say, a 1.5-second movement. If the angular velocity at .25 second is 50° / s and the angular velocity at .5

second is 75° / s, then the average angular acceleration between these two points in time is 100° / s2. (Solution: 75° / s – 50° / s divided by .5 s - .25 s).

Angular Motion Vectors. Creating vectors for angular motion is not as straightforward as it is for linear motion, because vectors are denoted by straight lines. One system that has been used to help alleviate this problem is the

---

HIGHLIGHT

Calculating Resulting Vectors

To calculate a resulting vector, it is necessary to use a geometric analysis called the parallelogram of vectors and a theorem developed by the ancient Greek mathematician, Pythagorus (580 B.C. to around 500 B.C.). In the example provided in Figure 5.6, it is necessary to draw a parallelogram with the vectors of 6 m/s and 8 m/s as sides. Only if the two vectors form a 90-degree angle, resulting in a right triangle, may we use the theorem of Pythagorus to find the unknown vector. The unknown vector is the hypotenuse of the right triangle ABC. Pythagorus's theorem is stated as follows:

$$c^2 = a^2 + b^2 \qquad (5.4)$$

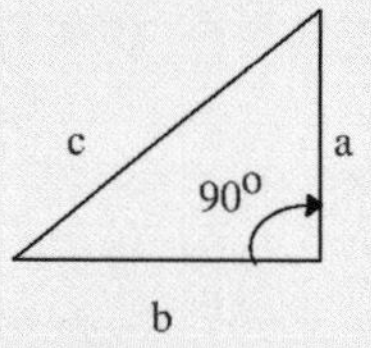

where c is the hypotenuse of the right triangle of ABC, and a and b are the other two sides of the right triangle ABC. If a and b are known quantities, which they are in our example, then it is possible to solve the above equation 5.4 to find c, the hypotenuse and the resultant vector:

$$c^2 = (8 \text{ m} / \text{s})^2 + (6 \text{ m} / \text{s})^2$$
$$= (64) + (36)$$
$$= 100$$

applying the square root to each side of the equation

$$\sqrt{c} = \sqrt{100}$$
$$c = 10 \text{ m} / \text{s}$$

Therefore, the instant at which the defensive player imparts a velocity of 6 m / s on the ball carrier who is running down the field at 8 m / s, a resultant vector of 10 m / s is created. At this instant, the ball carrier moves with velocity of 10 m / s in the direction shown in Figure 5.6.

Figure 5.7

In the absence of gravity, the javelin would travel equal horizontal and vertical distances in equal periods of time. A javelin released at 14 m / s at an angle of 40o to the horizontal would travel 1.07 m horizontally and .9 m vertically, for each .1 s of its flight. However, gravity causes a projectile to deviate from the no-gravity path. The actual distance fallen from the no-gravity path for selected times of flight can be calculated using the following formula: l = 1/2 g t2 where l is the vertical distance the javelin falls below the no-gravity path, g is gravity, and t is time. The lower panel shows the entire trajectory of the javelin, which is shaped like a parabola under the influence of gravity (adapted from Hay and Reid, 1988, Figures 86 (pg. 125), 87 (pg. 126) and 164 (pg. 249).

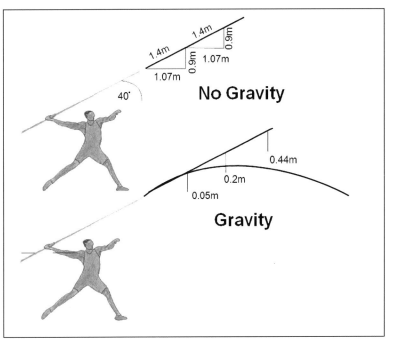

Figure 5.8

Two additional examples of rotational motion (Figures 6 and 7). Copyright © 2011 Depositphotos Inc./Andrei Ureche.

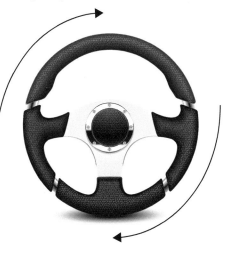

right-hand thumb rule. This rule states that if you curl your fingers in the direction of the rotation, the thumb points in the direction of the angular motion vector (see Figure 5.10).

A *resultant* angular motion vector can be created if the body is engaged in two simultaneous rotational movements using the parallelogram of vectors and Pythagorean theorem. An example of this can be seen in Figure 5.11. A diver who imparts both a forward somersault with a twist will show a vector specific for *each* angular movement and a *resultant* angular motion vector.

Summary

- Kinematics is the area of biomechanics that describes body motion.
- The two major areas of kinematics are linear and angular.
- Within linear kinematics, there are several specific measures of motion: linear distance and displacement, speed and velocity, and acceleration.
- Vectors of motion represent the body's magnitude of speed and its direction.
- Angular motion also can be described using comparable measures of rotational movement, including angular motion vectors.

Copyright © 2009 Depositphotos Inc./Maksym Yemelyanov.

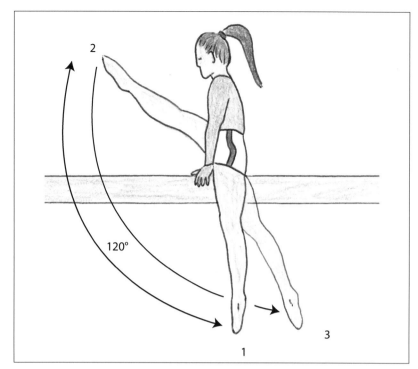

Figure 5.9
An example of angular distance and displacement in the execution of an exercise on the balance beam (adapted from Hay and Reid, 1988, Figure 92, pg. 138).

EXPLAINING HUMAN MOTION

Another major branch of biomechanics is **kinetics**. Kinetics deals with the causes of motion. Why does a body move a certain distance and with a particular speed and acceleration? What causes the velocity vectors of a body in motion? As with kinematics, kinetics can be used to study both linear and angular motion. The area of kinetics formally began with the development of Isaac Newton's three laws of motion. Newton's laws dealt with a number of properties of a body such as inertia, mass, force, and momentum. Before investigating Newton's laws, let us first define these four properties.

Linear Kinetics

Inertia. The inertia of a body is related to the body's resistance to a change in its present condition. For example, a heavy box on the floor tends to resist any attempts to move it. A ball thrown into the air tends to remain in motion. In general, a body at rest tends to remain at rest and when in motion, the body tends to remain in motion.

Mass. The quantity of a body's matter is called its **mass** and is directly related to its inertia. The greater its mass, the greater the body's resistance to a change in its present condition. It is easier to move someone's finger than it is to move someone's arm, because the mass of a finger is less than the mass of the arm. An adult's body mass (and inertia) is greater than a child's, and therefore, it is much easier to alter the motion of a running child compared to an adult.

Force. The pushing or pulling effect that one body may exert on another is called a **force**. As noted earlier, a body hurled through the air would tend to continue in a straight line if it weren't for the earth's gravity, which is a force that pulls the body toward the earth's center. The heavy box on the floor will tend to remain stationary unless it is acted upon by an external force, such as the force your muscles produce in an attempt to pick the box up.

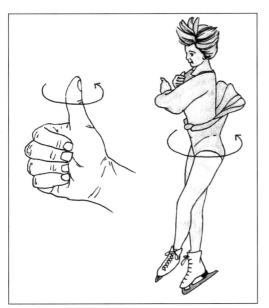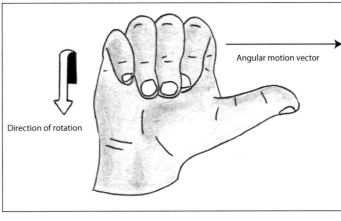

Figure 5.10
Angular motion vectors can be represented by arrows with the aid of the right-hand thumb rule (adapted from Hay and Reid, 1988, Figure 93).

Forces are vector quantities that have both a direction and a magnitude. If two or more forces are imparted to an object, a **resultant force vector** can be calculated using the parallelogram of vectors calculation and the Pythagorean theorem.

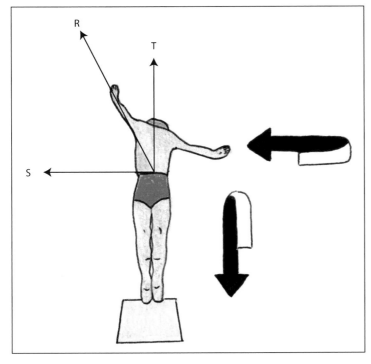

Figure 5.11
The right-hand thumb rule applied to a forward somersaulting and twisting dive (adapted from Hay and Reid, 1988, Figure 94c, pg. 140).

Momentum. The **momentum** of an object is equal to its mass times its velocity:

$$M = m \times v \qquad (5.7)$$

where M is momentum, m is the mass of the object and v is its velocity. Let us compare the momentum of an adult with a child during the running of a 100-meter sprint. If the child's mass is 30 kg and runs the 100-meter in 15 seconds, then the child's average momentum is:

$$M = 30 \text{ kg} \times 100 \text{ m} / 15 \text{ s}$$
$$= 200 \text{ kg m} / \text{s}$$

If the adult's mass is 90 kg and runs the 100-meter in 15 seconds, then the adult's average momentum is:

$$M = 90 \text{ kg} \times 100 \text{ m} / 15 \text{ s}$$
$$= 600 \text{ kg m} / \text{s}$$

or three times larger than the child's average momentum. Even though the velocity of the child and adult is the same, the adult's momentum during the 100-meter run is much greater because of the adult's greater mass. One implication of this fact is that it would be much more difficult to alter the adult's velocity vector toward the finish line with an outside force than the child's. For your understanding, to equal the adult's momentum, what 100-meter-dash time would be required of the child? Do you think it would be possible? (Answer: 5 seconds—impossible!)

Newton's Three Laws of Motion. Isaac Newton formulated three laws that describe the factors affecting a body at rest and in motion. Newton's three laws of motion are the law of inertia, the law of acceleration, and the law of action and reaction.

Newton's first law, **the law of inertia**, states:

A body will continue in its state of rest or uniform motion unless acted upon by an external force.

As mentioned earlier, a propelled object would continue indefinitely in a straight line path at a constant velocity if it weren't for the force of gravity that affects the object's vertical velocity vector. Every male ballet dancer will tell you that the hardest part of lifting the ballerina is generating enough force just to get her off the ground! Whether a body is at rest or in motion, the law of inertia makes it difficult to alter a body's present state. Recall that the inertia of a body is directly related to its mass.

Newton's second law, **the law of acceleration**, states:

The acceleration of a body is proportional to the force imparted to it and inversely proportional to the body's mass,

and can be formulated as follows:

$$a = F / m \qquad (5.8)$$

where a is acceleration, F is force, and m is the mass of the object. Thus, if the force imparted to an object increases and the mass of the object remains constant, the acceleration will also increase. If a ball of a given mass is thrown with greater force, it will accelerate more. These results are expected from the law when the mass is unaltered. However, if two balls of different masses are thrown with the same amount of force, the object with the smaller mass will accelerate more.

Another, perhaps more familiar, formulation of the law is the following:

$$F = m \times a \qquad (5.9)$$

This reformulation of Newton's second law allows for the prediction of force generated when the mass and the acceleration of the body are known. The basic measure of force is a quantity appropriately called a Newton (N), defined as the force required to accelerate one kilogram of mass at the rate of one meter per second squared (1 kg m / s2). If two bodies are accelerated equally, the body with the greater mass will have required a greater force (expressed in N). If two boxes of different masses are picked up from the floor with the same acceleration, more force will be required to move the box with the greater mass. Similarly, if a box is lifted upward with the same acceleration whether it is unfilled or filled with dirt, effectively doubling its mass, then twice as much force is required to lift the filled box. A baseball player swinging with the same acceleration using two bats of different masses on separate occasions will need to impart greater force when swinging the heavier bat, all other things being equal. Clearly, there are many applications of Newton's second law related to human movement.

Newton's third law, the **law of action and reaction** states:

When one body exerts a force on another, there is an equal, opposite, and simultaneous reactive force imparted on the first body from the second.

The equal, opposite, and simultaneous reactive force is sometimes called the **counterforce**. Examples of Newton's third law will now be provided. In the first example, consider a runner (Figure 5.12). To leave the ground on each step requires the runner to contract the extensor muscles of the legs so that force is generated downward into the ground. An equal and opposite force is directed upward that drives the runner up in the air. Consider another situation in which a baseball is tossed using an overhand throw. If the individual is on solid ground, the force generated forward by the arm also produces an equal and opposite force backward that tends to rotate the rest of the body backward. But, because the feet are planted securely, a backward force is directed to the ground. However, if the individual is on ice, the counterforce of the throw causes the body to rotate backwards. Similarly, when a diver tucks his or her head in the air to produce a forward flip, a counterforce is produced that causes the hips to flex.

Newton's Law of Gravitation. The story goes that Newton discovered gravity when an apple fell from a tree and landed on his head! The story may not be believable but the law of gravitation certainly is. We all know the adage, "What goes up, must come down," but Newton converted this generality on the effects of gravity to a precise mathematical formulation. The law of gravitation is an example of a different kind of force, the force of *attraction* of two bodies separated in space. Specifically, the law of gravitation states:

All bodies attract one another with a force proportional to the product of their masses and inversely proportional to the square of the distance between them,
and can be formally expressed as follows:

$$Fa = G\ m1 \times m2\ /\ d^2 \qquad (5.10)$$

where G is a gravitational constant, Fa is the attractive force exerted on each body, m1 is the mass of the first body, m2 is the mass of the second body and d is the distance between the geometric centers of the bodies. The masses of bodies on the earth is small compared to the earth and therefore the attractive force between such bodies is small and can be discounted for all practical purposes. For example, the attractive force between a person's left and right legs is so small that it can be ignored as a factor contributing to the mechanics of walking. However, the attractive force between any body (say m1) on the earth and the earth (m2) itself is large because the earth's mass is so large. The attractive force of the earth on all bodies on, or near, the earth's surface is called the earth's **gravity**. Gravity continuously pulls all objects in a direct line from the center of gravity of each object (see below) toward the center of the earth. The effect of the earth's gravity on the mechanics of human motion is quite significant.

The **weight** of an object is defined as its mass times the effect of gravity, or

$$W = m \times g \tag{5.11}$$

where W is weight of the object, m is its mass and g is the effect of gravity (9.81 m / s2). If the earth was shaped like a perfect ball, then an object would be exactly the same distance from the center of earth regardless of its position on the earth's surface. We know that the earth is flattened at its poles such that the distance from the center of the earth to the poles is less than to the earth's equator. As a result, the same object with a given mass will weigh more on the poles than at the equator because the distance between the object and earth's center is less when the object is at the north or south pole than when it is at the equator (see formula 5.10). It is important to remember that the weight of a body involves the pull of the earth's gravity on the body's mass.

Another important concept in biomechanics is a body's **center of gravity**, defined as the point around which a body's weight is equally balanced in all directions (Hall, 1991). The center of gravity of the human body can be located inside or outside of the body, depending on the body's configuration or alignment with respect to the ground. For example, standing straight up, an individual's center of gravity is located within the body. If the individual bends forward at the waist, the center of gravity also moves forward. If the individual bends backward, the center of gravity moves backward. An individual is stable as long as the center of mass stays between the individual's so-called **base of support**. which is often the area bounded by the two feet. If the center of gravity moves outside the base of support, the individual loses balance.

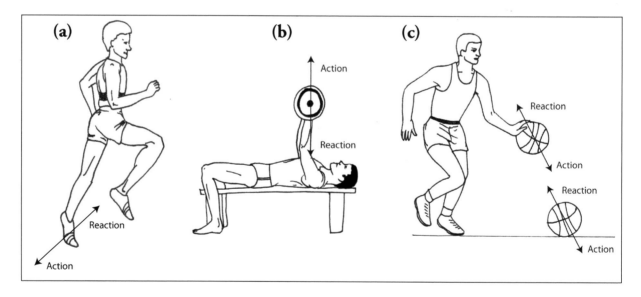

Figure 5.12

Examples of Newton's third law: (a) in running, (b) in performing a bench press, and (c) in dribbling a basketball (adapted from Hay, 1985, Figure 34).

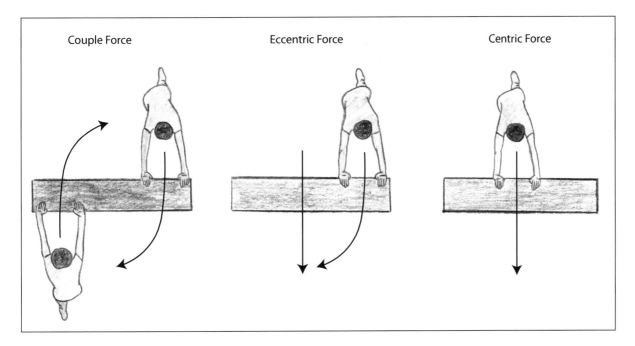

| Couple Force | Eccentric Force | Centric Force |

Figure 5.13

Examples of a centric force (right), eccentric forces (middle), and (left) a couple force (adapted from Hay and Reid, 1988, Figure 112 and from Hay, 1985, Figure 53).

Angular Kinetics

Most of the major movements performed by the human body involve rotation around the joints. In this section, we will examine the biomechanical causes of angular or rotational motion of the human body. A number of important quantities unique to angular motion will be covered first, followed by a discussion of how Newton's laws apply to angular kinetics.

The Direction of Forces Imparted on a Body. A force can be imparted on the center of a body or away from its center. If a sufficient force is directed at its center, called a **centric force**, then the body will undergo pure translational motion. If the force is not directed through the center of the body, then the body will undergo both rotational and translational motion (or general motion). This type of force is called an **eccentric force**. If two oppositely directed forces are imparted on a body that cause rotation of the body, then these forces are called a **couple**. Couples cause pure rotational motion of a body. Examples of a centric force, eccentric force and a force couple are shown in Figure 5.13.

Torque. In angular kinetics, it is necessary to introduce a new term called **torque**, which is the force necessary to cause rotation of a body (torque is also called a moment, or moment of force). Specifically,

$$\text{Torque} = \text{F} \times \text{MA} \tag{5.12}$$

where F is the applied force (N), and MA is the moment arm, that by definition, is the perpendicular distance from the center of the body's axis of rotation to the line of application of the force. Consider Figure 5.14, which shows a person supporting a heavy book in two conditions. In the first condition (a), the book rests near the axis of rotation of the elbow joint on the forearm. If the book exerts a downward force of 20 N and is located .1 m from the elbow joint, the torque produced is:

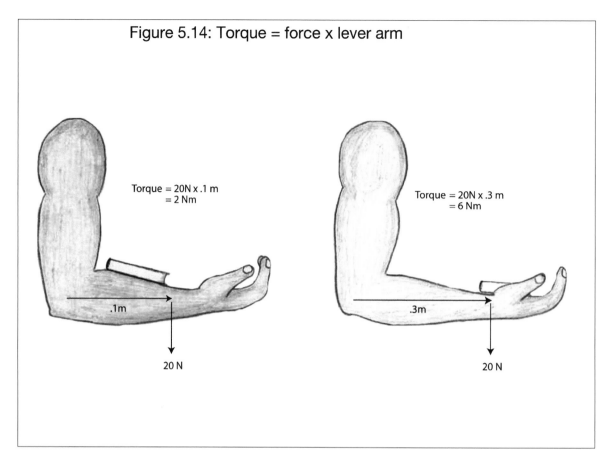

Figure 5.14

The torque imparted to the left arm is greater when the book is placed in the palm of the hand (right) compared to being placed near the elbow joint (left).

$$Torque = 20 \text{ N} \times .1 \text{ m}$$
$$= 2 \text{ Nm.}$$

This situation would require the person to contract the appropriate muscles so that the hand would exert an upward torque of 2 Nm in order to counterbalance the book. But if the same book were placed in the palm of the hand, say .3 m from the elbow joint, the resulting downward torque would be:

$$= 20 \text{ N} \times .3 \text{ m}$$
$$= 6 \text{ N m.}$$

In this case, the person would have to produce much more muscular effort to counterbalance the book.

Lever Arms. The example above illustrates the case when a load, such as a book, is located near or far from the axis of rotation (e.g., the elbow joint). As such, the elbow joint can be designated as one type of lever. A **lever** is a rigid body that has an axis of rotation. In fact, all levers have an axis of rotation (A), a resistance (R) or load which must be balanced or overcome and a force (F) that is applied to the lever. There are actually three major classes of levers that differ because of the relative positions of A, R, and F. Figure 5.15 illustrates the three

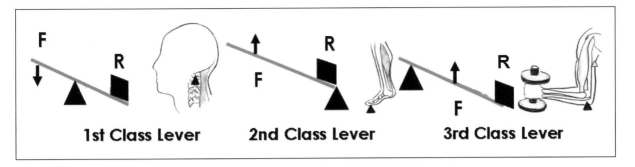

Figure 5.15

Examples of first-, second-, and third-class levers (adapted from Teeple and Misner, 1984, Figure 3-10).

classes of levers and an example of each within the human body. A first-class lever has A in the middle of the lever, with R and F at the ends of the lever. A second-class lever has R in the middle of the lever arm, with F and A at the ends of the lever. Finally, a third-class lever has the F in the middle of the lever with A and R at the ends of the lever.

The distance of the lever arm between A and F is called the **force arm** (FA) or moment arm (MA) and the distance of the lever arm between A and R is called the **resistance arm** (RA). The following formula, called the **Principle of Levers**, is useful in various leverage calculations:

$$F \times FA = R \times RA \tag{5.13}$$

The formula shows that when F x FA is equal to R x RA, then the lever arm is perfectly balanced. A good example of this can be seen in Figure 5.16 where two children of the same weight and equal distances from the axis of rotation are perfectly balanced on the seesaw (a first-class lever).

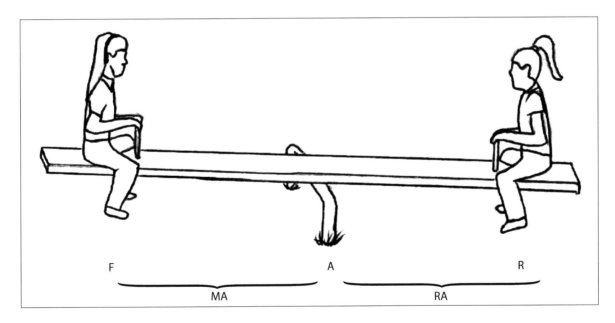

Figure 5.16

The moment arm (MA) and resistance arm (RA) of a first-class lever (adapted from Hinson, 1981, Figure 2.21).

Another useful formulation is that which defines the **mechanical advantage** of a lever system:

$$\text{Mechanical Advantage} = FA / RA \qquad (5.14)$$

When the FA is equal to RA, then the mechanical advantage is one. This means F must equal R to perfectly counterbalance the lever arm. The seesaw example above provides a good example of this situation.

If FA is less than RA, then the mechanical advantage is less than one, which means that F must be greater than R in order to counterbalance the lever arm. Figure 5.17 provides an example of this situation. In this case, a person is attempting to counterbalance a weight on their foot by producing an upward force through contraction of the quadriceps muscle. The axis of rotation (A) is the knee joint, the mass of 8 kg (80 N) is R, and F is the upward force (F) required by contracting the quadriceps muscle and synergist muscles. The force arm (FA) is 7 cm and is measured as the distance from the knee joint to the insertion of the quadriceps muscle. The resistance arm (RA) is 35 cm and is measured from the knee joint to the point on the lower leg directly below the center of the weight. This is an example of a third-class lever, where the mechanical advantage is calculated to be:

$$\text{Mechanical advantage} = FA\ (7\ \text{cm}) / RA\ (35\ \text{cm})$$
$$= 1/5.$$

Given this mechanical advantage, it requires 5 units of muscular force (F) generated by the quadriceps muscle (and other synergist muscles) for every unit of resistance force to balance the load (R).

There are a variety of examples of first-, second-, and third-class levers within the human body. However, the third-class lever is predominant. But because the moment arm in third-class levers is shorter than the resistance

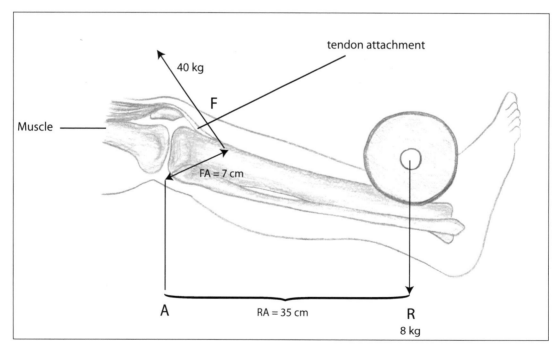

Figure 5.17

A third-class lever with a mechanical advantage of one fifth (adapted from Hinson, 1981, Figure 2.24).

arm, the mechanical advantage of a third-class lever is poor, and greater force is required to overcome the resistance. Given this fact, one may wonder why third-class levers are the most predominant levers in the human body. The answer is that third-class levers allow for a greater range and speed of movement, a quality that is required for so many skills performed by the human body. Even if the mechanical advantage of third class levers within the human body is poor, the muscles are more than capable of generating sufficient force to overcome the resistances we typically encounter.

Moment of Inertia. An important concept in angular kinetics is the moment of inertia of a rotating body. As discussed earlier, a body resists change in linear motion, a characteristic defined as inertia. Likewise, a body resists change in *angular* motion—a concept defined as the **moment of inertia**. Recall from Chapter 3 (Figure 3.3) that a given body, such as the human body, can rotate around three principal axes. The degree to which a body resists change in angular motion around a given axis of rotation depends on 1) all the particles of mass of the body; and 2) their individual distances from any particular axis of rotation. More formally,

$$\text{The moment of inertia (I)} = \Sigma \, (m \times r^2) \qquad\qquad (5.15)$$

where m is the mass of each particle of the body, and r is the distance each individual body part is away from a given axis of rotation. Essentially, this formulation says that the moment of inertia (resistance to change of rotation around a particular axis of rotation) is greater when the mass of the body is greater and when the various parts of the body are farther from the axis of rotation. Figure 5.18 shows that the human body's moment of inertia around the transverse axis is greater in a standing position than in a tucked position. Even though the mass of the body is the same in the two positions, the summed distances of the various body parts in the standing position is greater than in the crouched position. As a result, gymnasts and divers find it much easier to perform somersaults in a tucked position because the moment of inertia is less.

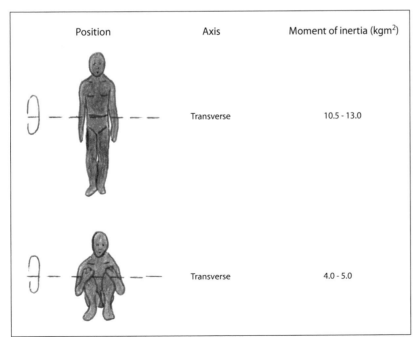

Position	Axis	Moment of inertia (kgm²)
	Transverse	10.5 - 13.0
	Transverse	4.0 - 5.0

Figure 5.18

The moment of inertia of the human body around the transverse axis of rotation is greater while standing than in a tucked position (adapted from Hay and Reid, 1988, Figure 134).

Angular Momentum. Another important quantity in angular kinetics is the **angular momentum** of a body. Recall that the momentum of a body in linear motion is its mass times its velocity (5.7). A body's angular momentum can be similarly calculated in the following way:

$$\text{Angular momentum} = I \times \omega \qquad (5.16)$$

where I is the body's moment of inertia around a particular axis of rotation, and w is the body's angular velocity. A body's angular momentum is greater if its moment of inertia is greater or if a body's angular velocity is greater.

Newton's Laws Applied to Angular Motion. As in linear motion, Newton's three laws of motion also apply to angular motion. In this section, these three laws will be presented and examples of how these laws apply to various movements will be provided.

Newton's first law in angular form can be stated as follows:

A rotating body will continue to turn about its axis of rotation with constant angular momentum unless acted upon by an external couple or eccentric force.

This particular law is also called the **law of conservation of angular momentum**. A good example of this law is a diver who alters his or her radius of rotation by extending the arms away from the body during the diving. If the diver extends the arms away from the body during the dive, the moment of inertia around the transverse axis will increase, but the rotational velocity will decrease. Thus, the angular momentum is conserved or unchanged because no *external* force was applied to the diver's rotating body.

Newton's second law in angular form states:

The angular acceleration of a body is proportional to the torque causing it and takes place in the direction in which the torque acts. This law can be quantitatively expressed in the following manner:

$$\alpha = T / I \qquad (5.17)$$

where a is the angular acceleration, T is the torque and I is the moment of inertia, all relative to a specified axis of rotation.

Newton's second law in angular form can explain situations where greater limb accelerations are achieved by simply reducing the moment of inertia (I), *without* increases in torque (T). For example, it is observed that the knee joint bends more as locomotion speed increases. Flexing the leg to bring the foot and shank mass closer to the hip joint reduces the moment of inertia. As a result, greater acceleration is produced even without additional torque that could be produced by increased hip flexor muscle contraction.

Recall that in linear motion, the force exerted by a body is equal to its mass times its acceleration. The angular form of the second law can be restated as:

$$T = I \times \alpha \qquad (5.18)$$

which says that torque (T) around a specified axis of rotation is equal to the product of the moment of inertia (I) and angular acceleration (a). One way to increase angular acceleration of the lower arm around the elbow joint is to generate more torque by increasing contraction of the flexor muscles of the elbow joint.

Newton's third law in angular form is stated in the following manner:

For every torque exerted by one body on another, there is an equal, opposite, and simultaneous torque exerted by the second body on the first.

A good example of this law can be seen in Figure 5.19, in which a long jumper swings his legs with an upward torque in a clockwise direction. Consistent with Newton's third law, the upper body produces a downward torque in a counterclockwise direction.

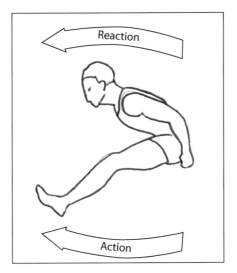

Figure 5.19

Newton's third law of action and reaction applied to the angular motions of the long-jump landing (adapted from Hay, 1985, Figure 90).

SUMMARY

- Linear kinetics is the area of mechanics that attempts to explain the causes of linear motion, namely the forces that are produced by a body and those that are imparted to the body.
- The inertia of a body is that characteristic which is related to a body's resistance to a change in its present state of motion. The greater the body's mass, the greater its inertia.
- Force is a pulling or pushing effect.
- The momentum of a body is equal to its mass times its velocity.
- Newton's first law states that a body will continue in its state of rest or uniform motion unless acted upon by an external force.
- Newton's second law states that the acceleration of a body is proportional to the force imparted to it and inversely proportional to its mass.
- Newton's third law states that when one body exerts a force on another, there is an equal, opposite, and simultaneous reactive force imparted on the first body by the second.
- Newton's law of gravitation states that all bodies attract one another with a force proportional to the product of their masses and inversely proportional to the square of the distance between them.
- The weight of an object is defined as its mass times the effect of gravity.

- The center of gravity of a body is defined as the point around which a body's weight is equally balanced in all directions.
- Angular kinetics is the area of mechanics that deals with the causes of the rotational motion of a body.
- Torque is the force necessary to cause rotation of a body and is equal to the force times the length of the lever arm.
- A lever is a rigid body that has an axis of rotation.
- The mechanical advantage of a lever system is equal to the ratio of the moment arm over its resistance arm.
- The characteristic of a body to resist change in *angular* motion is called its moment of inertia.
- A body's angular momentum is equal to its moment of inertia times its angular velocity.
- All three of Newton's laws of motion can be applied to angular movement.

CHAPTER SUMMARY

- The three forms of motion are translational, rotational and general.
- Body motion is described by the field of biomechanics, called kinematics.
- The areas of kinematics are linear and angular.
- The specific measures of motion in kinematics are distance, displacement, speed, velocity, and acceleration.
- Vectors represent the magnitude and a direction of a quantity under investigation.
- Angular motion is described by comparable measures of rotational movement, including angular motion vectors.
- Linear kinetics explains the causes of linear motion, namely the forces that are produced by, and imparted on, a body.
- Inertia is related to a body's resistance to a change in its present state.
- Force is a pulling or pushing effect.
- Momentum of a body is equal to its mass times its velocity.
- Newton's three laws of motion govern both linear and angular motion.
- The weight of a body is defined as its mass times the acceleration of gravity.
- Angular kinetics deal with the causes of rotational motion of a body.
- Torque causes the rotation of a body and is equal to the force times the length of the lever arm.
- A lever is a rigid body that has an axis of rotation.
- Mechanical advantage of a lever system is equal to its force arm over its resistance arm.
- The moment of inertia is a body's resistance to a change in angular motion.
- Angular momentum is equal to a body's moment of inertia times its angular velocity.

IMPORTANT TERMS

Acceleration – the change of velocity

Angular kinematics – an area of biomechanics that describes rotational movment

Angular kinetics – an area of biomechanics that describe rotational force

Angular momentum – the product of mass and velocity of a rotating body

Average angular acceleration – the average change in velocity of rotational movement

Average angular speed – the average change of distance over time in rotational motion

Average speed – the average change in distance over time

Average velocity - the displacement of a body relative to the total time of the displacement

Base of support - the area bounded by the two feet while standing

Center of gravity - the point around which a body's weight is equally balanced in all directions

Centric force – the force imparted directly to the center of a body

Counterforce – an equal, opposite, and simultaneous reactive force

Couple – two forces applied on the same body at the same time

Curvilinear translation - all parts of the body move in the same direction, same distance, and in the same time in a curvilinear path

Deceleration – the slowing down of a body in motion

Displacement - a change of position of the body with reference to its starting position, taking into account direction of movement

Distance - represents the total length of the movement path, regardless of direction

Eccentric force – a force imparted on a body away from the body's center

Force - the pushing or pulling effect that one body may exert on another

Force arm - the distance of the lever arm between the applied force and the axis of rotation

General motion – a type of motion that includes both translation and rotational

Gravity - a force that tends to pull all objects toward the earth's center

Inertia - the body's resistance to a change in its present condition

Instantaneous acceleration – the acceleration of a body between two points in time

Instantaneous velocity – the velocity of a body between two points in time

Kinematics – the area of biomechanics that describes movement

Kinetics – the area of biomechanics describing the forces produced in a movment

Law of acceleration – Newton's second law that states that the acceleration of a body is proportional to the force imparted to it and inversely proportional to the body's mass

Law of action and reaction – Newton's third law that states when one body exerts a force on another, there is an equal, opposite, and simultaneous reactive force imparted on the first body from the second

Law of conservation of angular momentum – Newton's first law in angular form that states that a rotating body will continue to turn about its axis of rotation with constant angular momentum unless acted upon by an external couple or eccentric force

Law of gravitation – according to Newton, all bodies attract one another with a force proportional to the product of their masses and inversely proportional to the square of the distance between them

Law of inertia – Newton's first law that states a body will continue in its state of rest or uniform motion unless acted upon by an external force

Lever - rigid body that has an axis of rotation

Lever arms – the distance between either the applied resistance or force of a lever and the axis of rotation

Linear kinematics – an area of biomechanics that describes the motion in a straight path

Mass – the quantity of a body's matter

Mechanical advantage - the ratio of the moment arm over its resistance arm in a lever system

Moment of inertia - the characteristic of a body to resist change in angular motion

Momentum – is the product of the mass of a body times its velocity

Newton – the formulator of the three laws of motion

Newton's laws applied to angular motion – Newton's three laws can be applied to both rectilinear and rotational motion

Newton's three laws of motion – form the foundation of biomechanics

Principle of Levers - F x FA = R x RA, where F is the force, FA is the force arm, R is the resistance and RA is the resistance arm

Projectile motion - the movement of a body through the air

Rectilinear translation – a straight path of a body's motion

Resistance arms – the distance between the resistance and axis of rotation in a lever system

Resultant force vector – the force vector that results from the interaction of two or more forces

Resultant vectors – are the results of the interaction of two or more vectors

Right-hand thumb rule – a method used to estimate a resultant angular motion vector

Rotational (angular) motion – the motion of a body around an axis of rotation

Scalars – the magnitude of a quantity

Torque – is the product of a body's force and the length of the force's lever arm

Translational motion - occurs when all parts of the body move in the same direction, same distance, and in the same time

Vectors – the magnitude and direction of a quantity

Weight – the product of a body's mass and the effect of gravity

INTEGRATING KINESIOLOGY: PUTTING IT ALL TOGETHER

1. A group of children decides to play a game on a teeter-totter that is 3 meters in total length. One of the children weighs 30 kg, and sits on one end of the teeter-totter. The game is to require each of the other children to take turns sitting somewhere on the other end of the teeter-totter to obtain perfect balance. If one of the other children weighs 45 kg, where would that child have to sit on the teeter-totter to obtain perfect balance with the first child? [Hint: Use the Principle of Levers.]

2. Could any child weighing less than 30 kg obtain perfect balance with the first child? Be able to explain your answer.

3. A swimmer orients herself perpendicular to the parallel banks of a river. If the swimmer's velocity is 2 m / s and the current is 0.5 m / s, what will be the swimmer's resultant velocity? [Hint: Use Pythagorus's theorem.]

4. The relative angle at the knee changes from 180 degrees to 95 degrees during the knee flexion phase of a squat exercise. If 10 complete squats are performed, what is the total angular distance and the total angular displacement undergone at the knee?

5. How much force must be applied by a kicker to give a stationary 2.5kg ball an acceleration of 40 m / s^2? [Hint: Use one of Newton's laws.]

6. Select three sports or daily living implements and explain the ways in which you can modify each implement's moment of inertia with respect to the axis of rotation.

KINESIOLOGY ON THE WEB

- http://en.wikipedia.org/wiki/Sports_biomechanics—This Web site contains extensive information on how biomechanics can apply to sport and other activities.

REFERENCES

Andersson, B. G., Ortengren, R., and Schultz, A. (1980). Analysis and measurement of loads on the lumbar spine during work at a table. *Journal of Biomechanics*, *13*, 513–20.

Barham, J. N. (1978). *Mechanical Kinesiology*. St. Louis: C.V. Mosby.

Borelli, G. A. (1679). *On the Movements of Animals*. Translated by P. Maquet. New York: Springer-Verlag, 1989.

Donatelli R., and Wooden, M. J. (1989). *Orthopaedic Physical Therapy*. New York: Churchill Livingstone.

Duncan, R. M. (1989). Basic principles of splinting the hand. *Physical Therapy*, *69*, 1104–16.

Ghista, D. N., Vivani, G. R., Subbaraj, K., et al. (1988). Biomechanical basis of optimal scoliosis surgical correction. *Journal of Biomechanics, 21,* 77–88.

Hay, J. G. (1985). *The Biomechanics of Sports Techniques.* Englewood Cliffs, NJ: Prentice Hall.

Hay, J. G., and Reid, J. G. (1988). *Anatomy, Mechanics, and Human Motion.* Englewood Cliffs, NJ: Prentice Hall.

Hinson, M. M. (1981). *Kinesiology.* Dubuque, IA: Wm. C. Brown, 2nd ed.

Kirby, R., and Roberts, J. A. (1985). Introductory Biomechanics. Ithaca, NY: Mouvement.

Lichtenstein, M. F., Burger, M. C., Shiavi, R. G., et al. (1990). Comparison of biomechanical platform measures of balance and videotaped measures of gait with a clinical mobility scale in elderly women. *Journal of Gerontology, 45,* M49–54.

Lichtenstein, M. F., Shields, S. L., Shiavi, R. G., et al. (1988). Clinical determinants of biomechanics platform measures of balance in aged women. *Journal of American Geriatric Society, 36,* 996–1002.

Lichtenstein, M. F., Shields, S. L., Shiavi, R. G., et al. (1989). Exercise and balance in aged women: A pilot-controlled clinical trial. *Archives of Physical and Medical Rehabilitation, 70,* 138–43.

Nashner, L. M. (1977). Fixed patterns of rapid responses among leg muscles during stance. *Experimental Brain Research, 30,* 13–24.

Nashner, L. M. (1980). Balance adjustments of humans perturbed while walking. *Journal of Neurophysiology, 44,* 650–64.

Nashner, L. M., and McCollum, G. (1985). Organization of postural human movements: A formal basis and experimental synthesis. *Behavioral and Brain Sciences, 8,* 135–73.

Norkin, C. C. and Levangie, P. K. (1992). *Joint Structure and Function: A Comprehensive Analysis.* Philadelphia: F. A. Davis.

Piscopo, J., and Baley, J. A. (1981). *Kinesiology: The Science of Movement.* New York: John Wiley and Sons.

Rasch, P. J., and Burke, R. K. (1978). *Kinesiology and Applied Anatomy: The Science of Human Movement.* Philadelphia: Lea and Febiger, 6th ed.

Shaperman, J., and Setoguchi, Y. (1989). The CAPP terminal device, size 2: A new alternative for adolescents and adults. *Prosthetics and Orthotics International, 13,* 25–28.

Teeple, J. B., and Misner, J. E. (1984). *Bioscientific Foundations of Exercise and Sport.* Champaign, IL: Stipes.

Yeadon, M. R., and Challis, J. H. (1994). The future of performance-related sports biomechanics research. *Journal of Sports Science, 12,* 3–32.

FOOTNOTE

1. Throughout this chapter, the term "body" is used to represent the entire human body or a part of the body. The term is also used to identify any object, like a ball.

Chapter Six
Motor Control and Motor Learning Foundations

CHAPTER SIX

Motor Control and Motor Learning Foundations

"The nervous system is not, in fact, like a lazy donkey which must be struck (or, to make the comparison more exact, must bite itself in the tail) every time before it can take a step. Instead, it is rather like a temperamental horse which needs the reins just as much as the whip."

— Erich von Holst (1939/1973)

STUDENT OBJECTIVES

1. To appreciate the various levels of the motor control system.
2. To understand the role of various brain structures in motor control.
3. To know the spinal cord's function in motor control.
4. To distinguish the various models of motor control.
5. To understand the factors affecting reaction time and movement time.
6. To differentiate between the types of feedback and their effects on the performance and learning of motor skills.
7. To describe the major theories of motor learning.

In the previous chapter, we examined the description and causes of human motion from a biomechanical perspective. In this chapter, we explore the control of human motion as well as some conceptual ideas about how this control is acquired. The subfield of motor control and motor learning draws upon the scientific knowledge, experimental techniques, and theoretical concepts in several other disciplines, including neurophysiology, biomechanics, and psychology. The subfield of motor control and motor learning attempts to answer a variety of questions, such as, how are reflexive and voluntary movements controlled? What is the brain's role in controlling human movement? Are certain areas of the brain responsible for different movement functions? If different brain areas are diseased or injured, what are the effects on the control of movement? What are the factors that affect the speed and accuracy of movement? What are the different models of motor control? And finally, how are motor skills learned? These are just some of the interesting and difficult questions to be examined in this chapter.

NEUROMUSCULAR CONTROL OF MOVEMENT

A prized possession of the human body is its ability to move in the environment. Most of us probably take for granted the vast array of skilled movements we perform every day of our lives. It is virtually impossible to list all the types of movements we are capable of producing. But whether it be brushing of our teeth, typing on a computer, or walking up a flight of stairs, researchers in the area of motor control believe that there are a number of different levels within the neuromuscular system responsible for the control and coordination of movement.

Levels of Motor Control

An understanding of motor control may begin with an appreciation that there are several contributing levels within the human motor control system. Figure 6.1 provides a simplified conceptual model of the different

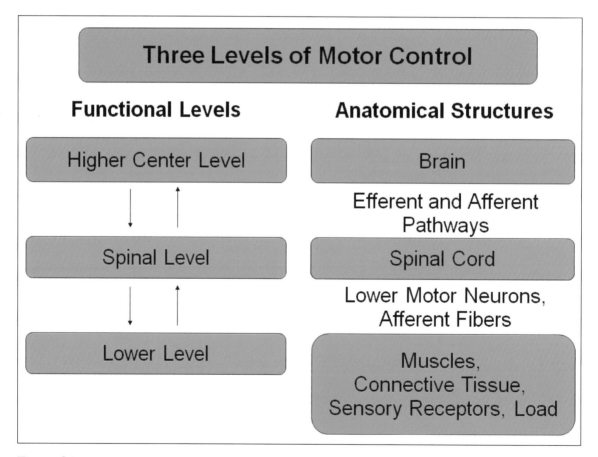

Figure 6.1
The three levels of motor control.

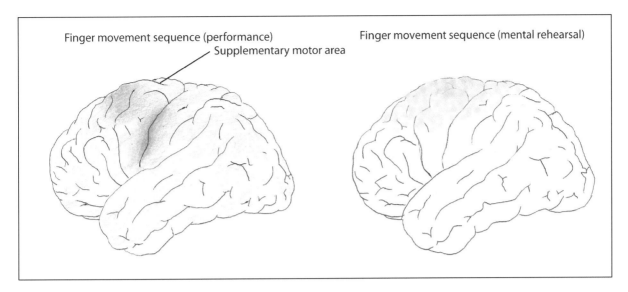

Finger movement sequence (performance)

Supplementary motor area

Finger movement sequence (mental rehearsal)

Figure 6.2

Cerebral blood flow during physical performance of a finger sequence (left panel) and during mental rehearsal of the same finger sequence (right panel). From Roland, Larsen, Lassen, and Skinhoj (1980).

motor control levels. Each level involves a variety of physiological structures contributing to the production of movement.

The *Higher Center Level* is composed of the brain, responsible for the planning and initiation of movement and the conscious sensation of movement. Ascending and descending neural pathways connect the Higher Center to the Spinal Level.

The *Spinal Level* is composed of the spinal cord, which contains sensory, motor, and interneurons. The Spinal Level serves as the intermediate relay station between the Higher Center Level and the Lower Level. A variety of reflexes also are housed within the neural networks of the spinal cord.

The *Lower Level* consists of the muscles, ligaments and tendons, and various receptors. The load that must be moved has considerable influence on the neuromuscular system and is included within the level. Sometimes the load is simply the weight of the arm or, on many occasions, includes the weight of an object that is grasped, transported or thrown. At this Lower Level, we will also examine the characteristics of muscle tissue and the contributions of the muscle spindle, an important sensory receptor located within the muscle.

It is tempting to conceptualize these levels in some hierarchical fashion by wrongly assuming that the Higher Center Level is more important than the others. Rather, it is more appropriate to deem each level as no more important than the others, and to understand that normal motor control depends on the proper functioning of each level.

The Higher Center Level

The brain contributes in a significant way to motor control. There are a variety of structures in the brain that affect the control of movement. Suppose that you decide to voluntarily flex your elbow joint. Evidence suggests that the idea for this act is partly formulated in an area of the brain called the **supplementary motor cortex** located just anterior to the motor cortex. This area becomes activated prior to movement, suggesting that it is highly involved in the planning of movements (Deecke, Scheid, and Kornhuber, 1969; Roland, Larsen, Lassen, and Skinhoj, 1980). Roland et al. (1980) required subjects to actually perform, and later to only imagine, a sequence of finger movements (see Figure 6.2). Using positron emission tomography (PET scan), they measured blood flow to both the supplementary motor cortex and the motor cortex under both conditions. They found that *both* the supplementary motor cortex and the motor cortex were highly active during performance of the finger movements. However, while *imagining* the finger sequence, only the supplementary motor cortex showed increased blood flow. This interesting finding suggests that the supplementary motor cortex is involved in the imaging, but not the initiation or execution of movement.

The initiation and intensity of activation of movement is largely affected by the **basal ganglia**. One movement disorder associated with damage to the basal ganglia is **Parkinson's disease**, which is caused by a deficit of the neurotransmitter called dopamine. Parkinson's patients exhibit involuntary shaking of the limbs, called *resting tremor*. They also typically have difficulty initiating movements, even when instructed, and their movements are produced much slower than normal individuals.

The **motor cortex** plays an important role in movement because it:

1. chooses the appropriate muscles to be used in the movement (Penfield and Rasmussen (1950);
2. helps determine the direction of motion (Georgeopoulos, Caminiti, Kalaska, and Massey, 1983);
3. helps develop the appropriate levels of force required for the movement (Evarts, 1981).

The motor cortex is *somatotopically* organized—that is, certain regions in the motor cortex provide muscular control to a particular body part or area. In general, muscles located in certain parts of the body that require very fine, coordinated movement (e.g., fingers, lips, and the tongue for speech) are represented in the motor cortex with more neurons than muscles responsible for gross movements (e.g., trunk and toe muscles).

The **somatosensory cortex** lies just posterior to the motor cortex and is responsible for the processing of sensory information from all parts of the body. Like the motor cortex, the somatosensory cortex is also somatotopically organized such that certain areas of the body have a greater neuron representation than others. When a movement is made, sensory feedback from sensory receptors for touch, vision, hearing, and proprioception is sent to the somatosensory cortex (as well as other brain structures at the Higher Center Level) informing them of changes of the body position in the environment resulting from the movement.

The **cerebellum**, located at the base of the brain, regulates or "fine tunes" voluntary movements. Damage to the cerebellum can leave the patient with diminished muscular strength, tone, and coordination. The cerebellum receives information from the motor cortex, other structures in the Higher Center and from the Spinal Level. On the basis of sensory information it receives, the cerebellum helps control posture and balance, and makes modifications in voluntary motor commands.

The **thalamus** is located in the deeper regions of the brain, and is considered to be a major relay station that connects a variety of brain structures to each other.

The plan to flex your elbow is sent from the supplementary motor cortex to the motor cortex where the appropriate muscles are chosen to perform the task. The basal ganglia provide the activation necessary for producing the movement. Signals from the motor cortex are sent to the cerebellum where adjustments to the motor commands can be made throughout the movement. The pathway from the motor cortex to the cerebellum and back to the motor cortex is called the **cerebrocerebellar loop**. It takes approximately 10 ms for the motor commands to be sent from the motor cortex to the cerebellum for adjustments, and back to the motor cortex. The motor commands to flex the elbow are then sent down efferent pathways to the spinal cord. These efferent pathways, as well as sensory pathways that transmit sensation about the movement to the brain, are described in the next section.

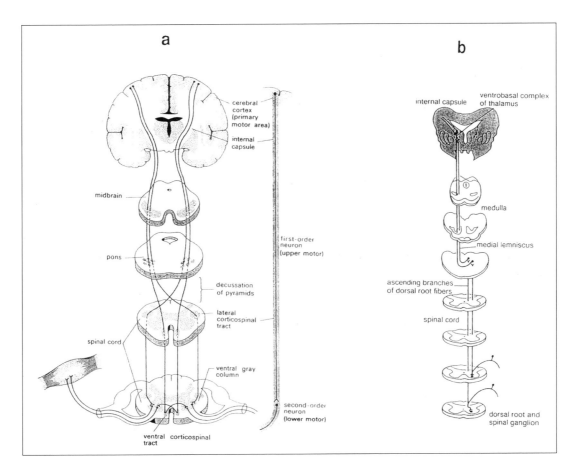

Figure 6.3

The corticospinal tract (a) and the dorsal column (b), adapted from Sage (1984), Figure 10.2 a and b, pg. 157.

Major Efferent and Afferent Pathways

Once a plan of movement is formulated and initiated, impulses from the motor cortex must be sent down to the Spinal Level and eventually to the appropriate muscles. In addition, sensory information from the muscles, joints, and skin must be sent to the Higher Center Level to indicate the position of the body (and body parts) before, during, and after movement.

A major efferent pathway for voluntary movement that connects the cerebral cortex to the Spinal Level is a pathway that is part of the pyramidal tract, called the **corticospinal tract** shown in Figure 6.3a. The corticospinal tract descends without synapse to the spinal cord originating primarily from the motor cortex. The axons of some of these motor neurons cross over just below the midbrain and synapse with alpha motor neurons at the spinal level on the opposite (or contralateral) side of the spinal cord. Axons of the other motor neurons of the cerebral cortex descend directly to the spinal cord, then cross over and synapse with alpha motor neurons. The upper motor neurons from the cerebral cortex to the spinal cord are called **first-order neurons**, and the lower motor neurons from the spinal cord to the muscles are called **second-order neurons** within the corticospinal tract pathway. Damage to this pathway typically causes a flaccid paralysis in the muscles. In addition, damage to one side of the motor cortex, for example, will affect the muscles on the contralateral side of the body, due to the motor axon fibers crossing over to the other side of the body.

Another motor pathway, called the **extrapyramidal tract**, begins in a variety of structures in the brain and descends to the spinal cord. It is quite complicated and includes all the motor axons not included in the pyramidal tract. Many of the descending motor neurons within the extrapyramidal tract are responsible for sending efferent signals to support balance and posture.

A major afferent pathway that sends sensory information from sensory receptors to the somatosensory cortex is called the **dorsal column**, shown in Figure 6.3b. Sensory information related to vision, hearing, touch, and proprioception is transmitted to the somatosensory cortex where it is consciously perceived. The dorsal column contains three order neurons. The axons of first-order neurons start at a sensory receptor, enter the spinal cord, and ascend without synapse to the brain stem, where they synapse with the second-order neurons. The axons of the second-order neurons cross the midline and ascend to the thalamus, synapsing with the third-order neurons. From there, the third-order neurons ascend to the somatosensory cortex.

The Spinal Level

At the Spinal Level, upper motor neurons from the Higher Center synapse with lower motor neurons that send axons to the muscles and glands and to interneurons. Sensory receptors from the muscles (see Lower Level) send axons to the Spinal Level, where they also synapse with lower motor neurons as well as interneurons. Thus, the Spinal Level is a major integration center for motor and sensory information.

Early neuroanatomical theories posited that the spinal cord was merely a type of relay station between the brain and the musculature. However, modern views hold that the spinal cord, in addition to being a major integration center of sensory and motor information, also houses a variety of reflexes (see Chapter 8 Motor Development Foundations), and plays a major role in the coordination of locomotion and other movements.

The spinal cord also plays a major role in the coordination of locomotion as confirmed by the interesting experiments of Shik (see HIGHLIGHT) and others (see Grillner, 1975). The act of locomotion, such as walking and running, requires precise timing of the contractions of the agonist and antagonist muscles in both legs. The left and right legs have to move in opposition, and the agonist and antagonist muscles of each leg have to contract in a reciprocal fashion for smooth, efficient locomotion. Coordination must occur both within and between the legs. Through a complex network of neurons, the spinal cord is capable of directing neural impulses in a coordinated fashion to control the agonist and antagonist muscles in each leg during locomotion. Thus, the higher centers of the brain are not totally responsible for the coordination of movement.

HIGHLIGHT

The Classic Shik Experiments

Shik and his colleagues (see Shik and Orlovskii, 1976, for a review) investigated the question of whether the spinal cord plays a significant role in locomotion by studying the leg movements of a spinalized cat. Shown in Figure 6.4 is a schematic drawing of the Shik preparation. The cat was supported by clamps because the surgical preparation prevented the cat from having *voluntary* control of posture, balance, or other self-directed movement. The surgical preparation performed by Shik involved the sectioning or cutting of the spinal cord just inferior (or below) the cat's midbrain. The preparation severed both the cat's sensory and motor pathways to and from the higher centers of brain, yet allowed for continuance of the vital functions of the midbrain (respiration, heart rate, etc.) necessary for survival. As a result, the cat could not consciously sense or produce voluntary movement.

Under these severe circumstances, what might be expected of the cat in the way of coordinated locomotor movement? If all coordinated movement is dictated by the higher centers of the brain, then we would expect little coordinated movement behavior from the cat under spinalized conditions. But very interesting results appeared when Shik performed the following experiments. In one experiment, Shik electrically stimulated the motor pathways just inferior to the surgical cut. As a result of this stimulation, the cat's legs started to move, and in a short time the cat started to produce coordinated locomotor movements with its four legs! When the stimulation was turned off, locomotion continued for a few seconds before stopping. In another experiment, Shik stimulated the cat's legs not by electrical stimulation but by turning on the treadmill under the cats suspended legs whose paws contacted the treadmill. After a few seconds of the paws dragging on the treadmill, the cat once again began to produce coordinated leg movements resembling locomotion! Thus, even though the cat could not produce voluntary locomotion, the cat's intact nervous system below the cut produced coordinated motion resembling locomotion if stimulated by outside sources. How can these results be explained? The explanation provided by Shik, and subsequently elaborated on by others (Grillner, 1975; Kupferman and Weiss, 1978), was that much of the coordination evident in locomotion must be housed in the neural circuitry of the spinal cord (called *central pattern generators*) and the neural connections within the musculature of the legs.

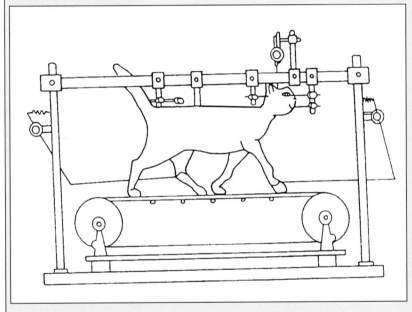

Figure 6.4
Mesencephalic (midbrain) cat supported on a treadmill as used in the study of spinal mechanisms in gait (adapted from Schmidt and Lee, 2005). Figure 6.5, pg. 170.

Central Pattern Generators. The neural circuitry in the spinal cord purported to be responsible for much of the coordination of locomotion has been termed the **central pattern generator** (or CPG). Let us briefly explore the hypothesized neural circuitry of a simple CPG shown in Figure 6.5. The circuitry shown here involves a total of four interneurons (1–4) within the spinal cord configured in such a way that allows an initial impulse of stimulation to travel from the first (1) interneuron stimulated to the last interneuron (4) in the circuit. Neurons 5 and 6 are alpha motor neurons originating within the spinal cord and stimulate an agonist and antagonist muscle, respectively, within a *single* limb. As the impulse travels around the circuit, the alpha motor neurons of the agonist and antagonist muscles are stimulated at different times, causing the agonist muscle to contract first, followed by the antagonist muscle. In this type of simple CPG, reciprocal innervation can be coordinated at the

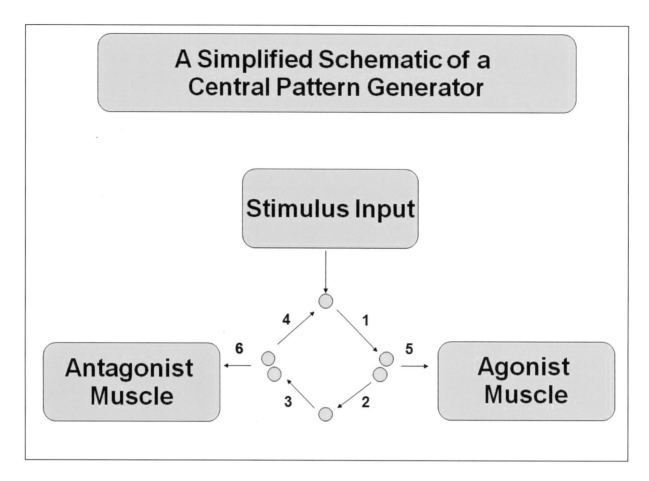

Figure 6.5

A schematic diagram of a simplified central pattern generator. The stimulus input from the higher centers or from sensory feedback innervates interneuron 1 and the stimulation is cycled through the neuronal network, first activating the agonist muscle, then the antagonist muscle, (adapted from Schmidt and Lee, 2005). Figure 6.4, pg. 169

spinal level. Other, more complicated CPG circuits could help explain how coordinated locomotor movement of the cat's legs in the Shik experiments might occur at the level of the spinal cord.

Lower Level

The Lower Level of motor control includes the muscles, connective tissue of tendons and ligaments, sensory receptors, and the load to be moved. In this section, we will examine certain characteristics of muscle tissue itself that affect motor control. Sensory receptors called **muscle spindles**, located within skeletal muscles, will also be investigated. Another important receptor is the **golgi tendon organ** that is sensitive to changes in muscle tension.

Length-Tension Relations in Muscle. As mentioned in Chapter 3, muscle tissue has certain characteristics called *extensibility* and *elasticity*, allowing the muscle to be stretched beyond its resting length and to return to its normal resting length following stretch, respectively. One important relation to know about muscle tissue is called the **length-tension relation**. Shown in Figure 6.6, this relation indicates that as a given skeletal muscle is passively stretched, the tension (or the restoring force) of the muscle increases in a nearly linear fashion

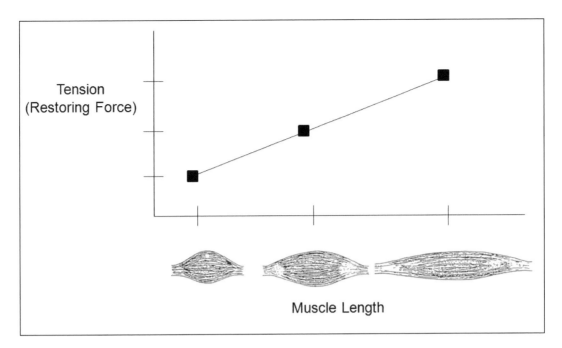

Figure 6.6

The relationship between muscle length and tension (or restoring force). The muscle can be likened to an ordinary spring.

(within a certain muscle length range). This type of characteristic is similar to that of an ordinary spring. One of the major theories of motor control (see below) argues that the length-tension relation in muscle plays a prominent role in motor control.

Muscle Spindles. We have all had the experience of visiting the doctor's office for a routine medical check-up. During the examination, the doctor may test our reflexes using the "knee-jerk response" by tapping

our knee with a small mallet. A normal response to this stimulus is a very quick, but small displacement of the foot. The knee jerk response is due to what is known as the "**knee-jerk**" **stretch reflex**. Let's examine how this reflex works by first considering the relevant neuroanatomical structures involved.

Within the skeletal muscles are located muscle receptors called **muscle spindles** that lie in parallel with the muscle fibers (called **extrafusal muscle fibers**). The polar ends of the spindle are composed of contractile muscle tissue called **intrafusal muscle fibers**. As with the extrafusal muscle fibers that make up a given muscle, intrafusal muscle fibers within the spindle contract when stimulated by motor neurons. Surrounding the middle portion or cell body of the spindle are **Ia afferent fibers**, which transmit sensory information about the condition of the spindle to the spinal cord. Muscle spindles provide the nervous system with information about changes in muscle length. Shown in Figure 6.7 is a simplified sketch of an individual muscle spindle receptor. In reality, the spindles' neuroanatomy is more complex than shown, but this sketch will suffice in helping to explain the knee-jerk reflex. It can be seen that muscle spindles lie parallel to the muscle fibers within a given muscle. Muscle spindles are sensitive to changes in muscle length, so when the muscle fibers around the muscle spindle are stretched or elongated, the spindles also are stretched. When the cell body or middle portion of the spindle stretches, for any reason, an *increase* in nervous impulses is sent back to the spinal cord via the Ia fibers. When the impulses reach the spinal cord, they jump a synapse between the Ia fibers and a motor neuron, called the **alpha motor neuron**. Alpha motor neurons originate in the spinal cord and their axons synapse with the muscles. Impulses from the Ia fiber are then transmitted to the alpha motor neurons to the same (or homologous) muscle that was originally stretched, causing the muscles fibers to contract. **Gamma motor neurons** originate in the spinal cord and synapse the intrafusal muscle fibers of the muscle spindle. Gamma motor neurons excite the muscle spindle primarily in voluntary movement.

The principle of **reciprocal inhibition**, discovered by Sherrington (1906), operates in the knee-jerk stretch reflex, because when the contraction of agonist muscle occurs, the antagonist relaxes. Figure 6.7 also shows how reciprocal inhibition is accomplished through the use of interneurons in the gray matter of the spinal cord that connect the axon terminals of the Ia afferents from the spindles of the quadriceps to the alpha motor neurons of hamstrings.

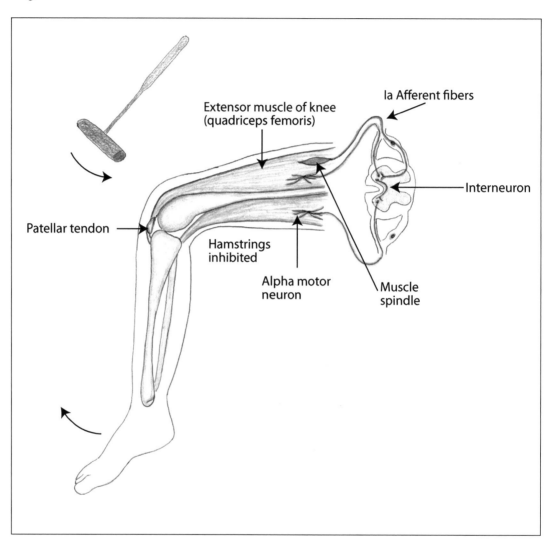

Figure 6.7

The muscle spindle lying in parallel with the muscle fibers of the quadriceps femoris. The tendon tap causes a stretch of the patellar tendon and the muscles fibers of the quadriceps, resulting in an increasing discharge of the Ia afferent fibers. The Ia fibers synapse directly with alpha motorneurons to the quadriceps, causing a contraction that results in lifting of the lower leg. In addition, the Ia fibers synapse with interneurons, causing a *relaxation* of the hamstrings in a process called reciprocal inhibition.

Reflexive and Voluntary Movement

Generally, movements can be reflexive, voluntary or some combination of the two. **Reflexive movement** is an involuntary muscular response due to some specific stimulus. The stimuli that invoke reflexes often originate in the environment and are detected by specific sensory receptors within the body. The sensory receptors send impulses to the spinal cord, and sometimes to the brain. Axons from the sensory receptors can synapse with motor neurons within the spinal cord. Motor neurons then send impulses directly to certain muscles causing them to contract and produce movement. **Voluntary movement** is initiated by the individual such that the stimulus to move is generated by certain areas of the brain. The impulses from the brain are sent to the spinal cord, which are then sent to the muscles.

Reflexive movement. Reflexive and voluntary movement can involve all three levels of the human motor control system. However, the *order* in which the levels are activated can be different. For example, the order of activation for some types of reflexive movement is:

Lower Level → Spinal Level -→ Lower Level -→ Movement
(sensory receptor)   (spinal cord)   (muscles)

As stated above, reflexive movement is an involuntary motor response to a particular stimulus. Reflexive movements can be quite simple, involving only a few muscles, but they can also be complex, involving several muscle groups. A simple reflex is the blinking reflex that results in the blinking of one's eyelids in response to a puff of air into the eye. An example of a complex reflex is the withdrawal reflex, elicited when one accidentally touches a hot stove. One's response to this painful stimulus is a rapid withdrawal of the hand from the hot stove, involving several muscle groups of the arm and shoulder and even trunk muscles. Reflexive movement involves the following components:

1. a sensory receptor that is sensitive to certain types of stimuli (Lower Level);
2. the receptor sends impulses over its axon to the spinal cord (Spinal Level) and the receptor's axon synapses with motor neurons and interneurons;
3. the motor neurons (Spinal Level) send impulses that cause specific muscles to contract (Lower Level);
4. an effector, such as a muscle or a gland.

In should be emphasized, in some reflexes, impulses sent to the spinal cord from the receptors continue to the brain before descending to spinal cord again, and then to the muscles. These type of reflexes are called **long-loop reflexes**.

Voluntary Movement. The order of activation for a simple voluntary movement is typically the following:

Higher Center Level → Spinal Level→ Lower Level →Movement
(brain)   (spinal cord)   (muscles)

When we produce voluntary movements, such as typing on a computer, running a 10K race, or reaching for a coffee mug, neural impulses from the higher centers of the brain are sent down to the Spinal Level.[1] These impulses then exit the spinal cord via the ventral roots and travel down the lower motor neurons (e.g., alpha motor neurons) to the skeletal muscles, causing the muscles to contract, and to the gamma motor neurons, activating the muscle spindles within the muscles in a process called the **alpha-gamma coactivation principle**

(Vallbo, 1970). The neural signals sent to the muscle spindle during voluntary movement allow the spindle to remain sensitive to changes in muscle length.

Summary

- This section has outlined some of the major neurological and muscular structures at the Higher Center, Spinal, and Lower Levels responsible for movement.
- In the Higher Center, a variety of brain structures are responsible for the planning and execution of movement.
- Major efferent and afferent neural pathways connect the brain with the spinal cord.
- At the Spinal Level, complex neural circuitry contributes to coordination of locomotion. Many reflexes also are controlled at the Spinal Level.
- Muscles at the Lower Level have springlike qualities that allow them to change tension, depending on their length.
- Muscles spindles located within the muscles are responsible for the stretch reflex and also contribute to feedback during voluntary movement.

Models of Motor Control

There are several conceptual models that deal with the control of movement. Some of the models focus on how sensory information is processed by the central nervous system to help control movement. Other models attempt to unravel questions related to the nature of descending or motor information that affects muscle contraction. Finally, other models focus on the nature of movement coordination. In this section, we will briefly outline some of the major theories and conceptualizations of how voluntary movement is controlled and coordinated. As we shall see, there is no one unifying theory of motor control.

Open- and Closed-Loop Concepts

One of the first conceptualizations to emerge in motor control was the realization that the control of movement depends on both efferent information from the Higher Center Level and afferent information from the sensory receptors located in the muscles, joints, and other body structures. If only efferent information is available, the motor system can provide what is termed open-loop control. Figure 6.8 illustrates an open-loop model of motor control. With **open-loop control**, the Higher Center is responsible for the planning of movement, selection of different movement parameters (e.g., force, direction, limbs, etc.), and the sending of efferent signals to the spinal cord, then are directed to the appropriate muscles causing movement.

Evidence for open-loop control came from studies by Taub (Taub and Berman, 1968; Taub, Perrella, and Barro, 1973), who performed an operation called deafferentation on monkeys. **Deafferentation** is a surgical procedure in which the major afferent pathways to the Higher Center Level are severed, but the efferent pathways are left intact. Theoretically, the animal should be able to send efferent commands to the muscles but not be able to perceive any sensory information resulting from the movement. Deafferentation forces the animal to perform open-loop, that is, without feedback. Taub found that deafferented monkeys could perform many skilled movements such as locomotion and climbing cages in the absence of sensory feedback, providing support for open-loop control. However, the movements performed were not as coordinated or "eloquent" as those produced by intact monkeys.

If both efferent and afferent information is available, the motor system can provide **closed-loop control**. Figure 6.8 shows a closed-loop model of motor control. Notice that in addition to efferent information from the

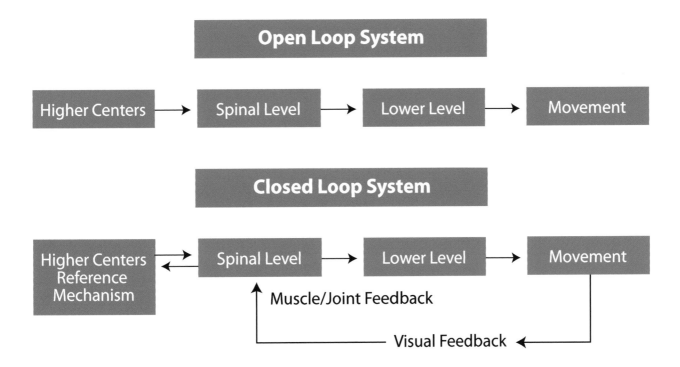

Figure 6.8

An open-loop (upper panel) and closed-loop (lower panel) system. In open-loop control, the higher centers send motor commands to the spinal and then the lower levels, causing movement without sensory feedback. In closed-loop control, sensory feedback can be used to help modify the movement. The sensory feedback can be compared to a reference mechanism to detect errors in the movement. Corrections can then be sent once again to the spinal and lower levels.

Higher Center, a closed-loop model has sensory feedback from the movement that is sent back to the muscles via the spinal cord to the Higher Center. In this way, feedback about the movement can be provided to the motor system to make corrections. The feedback loop to the muscles can take place between 30–60 ms, providing rapid but minor corrections to the existing movement. The muscle spindles are involved in this feedback loop. The feedback loop to the Higher Center takes 60 ms or longer, depending on the type of correction.

Evidence for closed-loop control of movement comes from a variety of sources, showing that sensory feedback is essential for successful movement execution. For example, years ago Woodworth (1899) demonstrated the importance of visual sensory information to movement control by simply requiring subjects to perform aiming movements with the eyes closed or open. He showed that when movements were performed slowly, errors increased under the eyes-closed condition. This result provided support for closed-loop control and the need for sensory information (in this case, vision) in controlling movement. However, when movements were made quickly, there was no difference in error rates between the eyes-open and eyes-closed conditions. This result suggests that fast movements might be primarily under open-loop control. Thus, the speed (or movement time) of movement seems to determine whether movements are controlled in an open-loop or closed-loop fashion. We will return to this issue when the notion of motor programming is discussed below.

Another important aspect of closed-loop control is the concept of a comparator or **reference mechanism** (see Figure 6.8). The reference is a type of perceptual-motor template of the desired movement (see Adams's theory below) that is stored in the Higher Center. For the motor system to detect errors in the movement execution, the feedback from the movement is compared to the reference mechanism. Any discrepancy between the feedback and the stored template of the desired movement generates an error. This error can then be corrected by the motor system, producing a more appropriate efferent signal to the musculature.

Information Processing Model

There are two conceptual models that deal with our ability to use sensory stimuli to help produce and control movement, the *information processing model* and the *ecological perspective model*. One important way these two models differ is in the proposed manner that information is obtained from the stimuli in the environment.

The **information processing** model can be called an *indirect* processing model, because in order for the performer to perceive and use certain stimuli in the environment, stimuli must first be processed by the central nervous system. In this model, first advanced by a Dutch physician, Donders, well over a century ago (1868/1969) and later refined by Sternberg (1969), sensory information from the environment (e.g., visual, auditory) as well as from our bodies (e.g., proprioception, touch), is processed in different stages within the central nervous system. Figure 6.9 illustrates two versions of the information processing model. Shown in Figure 6.9a is a simplified version of this model, with stimuli entering the central nervous system as input and

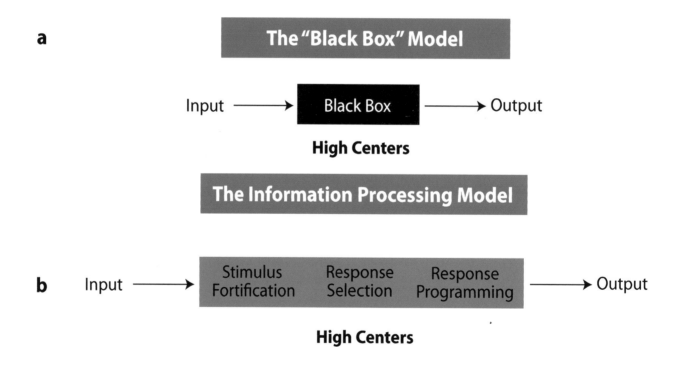

Figure 6.9

In a, the black box model. In b, the information processing model that recognizes the three stages of information processing: stimulus identification, response selection, and response programming.

commands to the muscles that result in movement as output. This type of model does not identify any specialized functions within the central nervous system, nor does it acknowledge the use of feedback. This simplified version has been labeled the "black box" model, because specialized functions within the central nervous system are not acknowledged.

Another more updated version of this model suggests that within the central nervous system, stimuli must be processed in three separate stages called stimulus identification, response selection, and response programming (Schmidt and Lee, 1999), shown in Figure 6.9b. In the **stimulus identification stage**, stimuli are first identified and distinguished from one another. For example, the brake light of the car in front of you has to be identified and distinguished from other stimuli before you can develop an adequate motor response. Once the stimulus has been identified, your central nervous system must now formulate an appropriate response, a process governed by the **response selection stage**. In this stage, the central nervous system must decide what response to make or to decide not to make a response at all. So, after the brake light is identified, a certain amount of time is required for this type of decision. For example, if you identify the brake light but the car is far away, you will probably decide not to apply your brakes, but if the car is close in front of you, a decision to apply the brakes will probably be made. The formulation of the actual efferent commands to be sent to the muscles is determined by the **response programming stage**. In this stage, *how* the movement is to be produced is determined. For example, if the car is close in front of you, it is likely that the response programming stage will decide to rapidly apply the brakes with a significant amount of force. This decision will also involve the selection

Figure 6.10: Reaction time, movement time and total respronse time

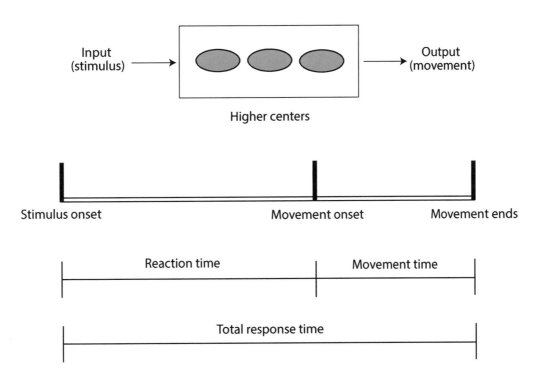

Figure 6.10

Total response time to a stimulus is composed of reaction time and movement time.
Reaction time is determined by how long it takes for the stimulus to be processed
in the three information processing stages

of the appropriate musculature (i.e., right foot versus left foot). Once this decision is made, the corresponding efferent commands are sent to the muscles producing the movement.

The amount of time it takes for information to pass through each of the above processing stages is proposed will vary depending on the situation. The total time from identification of the stimulus to the *initiation* of the response is called **reaction time**. The time from the initiation of the movement to its completion is called **movement time**. The total **response time** is the reaction time plus the movement time. Figure 6.10 shows a breakdown of the total response time in situations that call for a motor reaction to a stimulus.

Ecological Perspective Approach

The second major model that attempts to explain how stimuli are processed is called the **ecological perspective approach**. This model, originally advanced by J. J. Gibson (1966), can be called a direct **perception** model, because it proposes that stimuli be directly used by the perceptual-motor system without elaborate processing through separate stages by the central nervous system. Gibson believed that instead of the nervous system "extracting" information from the stimulus, information is already contained within the stimulus, requiring no further processing. He believed that the sense organs evolved such that they are sensitive to the information inherently contained in the stimulus.

An example of direct perception of information contained in a stimulus is the concept of the optical flow originally proposed by Gibson and later elaborated on by David Lee and others (e.g., Lee and Aronson, 1974; Lee, Young, Reddish, Lough, and Clayton, 1983; also see Tresilian, 1995, for a critical review). **Optical flow** is the *change* of ambient (reflected) light as a result of movement of the environment, the observer, or both. For example, if a ball is tossed toward an observer, the optical flow or reflected light from the ball changes on the observer's retina. Specifically, the size of retinal image increases as the ball approaches the observer (Figure 6.11). Thus, the size of the retinal image is *directly* related to the distance from the ball to the observer's eye. However, the rate of change of the retinal image of the ball is dependent on the speed of the ball as it approaches the eye of the observer. Lee proposed that the optical flow of an approaching object contains an optical variable, called **tau** (τ), which is the ratio of the size of retinal image over the rate of change of the retinal image. He argued that under conditions of a direct approach of the object toward the observer and under constant object speed conditions, τ is directly related to the time remaining before contact of the ball with the eye (i.e., time-to-contact). The simple calculations below illustrate Lee's idea:

τ = retinal image size / rate of change of image size = distance (d) / velocity (d/t)

$\tau = d / d/t = 1 / 1/t$

$$\tau = t \tag{6.1}$$

where t = time-to-contact.

Lee's formulation shows that t, an optical variable, can specify time-to-contact. Thus, consistent with Gibson's ecological perspective, Lee's concept of t is an example of direct perception of stimulus information that can be used to help control interceptive actions such as catching a ball, judging the arrival of an approaching vehicle, or even the timing of a long jump. There are some problems with the t concept, because it has difficulties predicting time-to-contact, particularly in indirect approaches of an object and an observer (Wallace, Stevenson, Weeks, and Kelso, 1992) and it may not be the only information used for time-to-contact judgments (Tresilian, 1995).

Motor Program Theory

One of the more popular theories of motor control is called **motor program theory**, whose biggest proponent over the years has probably been Richard A. Schmidt (e.g., Schmidt, 1980; Schmidt, 1982; Schmidt and Lee, 1999; Schmidt and McCabe, 1976). The emphasis in motor program theory is on the efferent signals sent to the muscles, and thus, it is considered to be primarily an open-loop theory. One of the strict versions of motor program theory was defined by Keele (1968) as a set of muscle commands allowing the sequence to be carried out uninfluenced by peripheral feedback (p. 387). While this version has been modified over the years, the notion that a motor program contains some set of muscle commands has essentially remained. One of the debates among motor control scientists has been over the nature of these commands regarding the type of information sent to the musculature.

Schmidt (1988) argued that these motor commands contain two types of characteristics: **variant and invariant features**. The variant features of a motor program are those characteristics of the movement that change

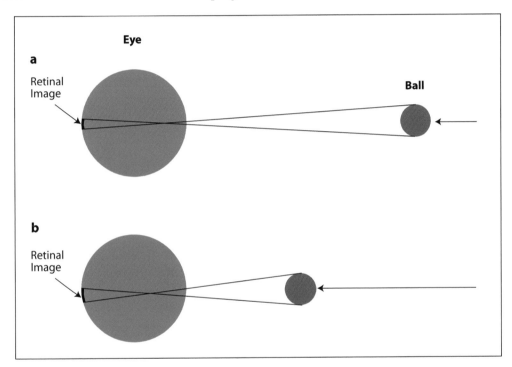

Figure 6.11

A ball tossed toward an observer's eye creates optical flow that is the change of reflected light. The distance of the ball from the eye determines the size of the retinal image. In (a), the size of the retinal image is smaller than in (b).

when variations of the movement are attempted. Invariant features are those characteristics of the movement that remain unaltered even when variations of the movement are produced. Schmidt developed these concepts to help explain how we can produce the seemingly endless variations of movements associated with different skills.

One question confronting Schmidt was, is every possible movement we make dictated by a separate motor program? Schmidt believed that the answer was clearly no, because this would result in a storage problem in the brain. Rather, Schmidt proposed that motor programs stored in the brain are much more flexible. For example, are there separate motor programs responsible for an overhand softball versus baseball throw? Clearly, the general pattern of the two motions is very similar, if not identical, but the softball is slightly heavier than the baseball, requiring more force generated by the muscles. Changes in the absolute levels of the force in two similar motions is an example of the variant feature of the motor program. But because the general patterns of the motions are the same, this would be an example of the invariant feature of the motor program. Thus, in Schmidt's conceptualization, the same motor program can generate different force commands, thereby greatly increasing the flexibility of motor control.

The Equilibrium Point Hypothesis

Motor programming theory argues that the motor system programs the correct level of force needed by the muscles to move a limb the correct speed and distance. In contrast, a hypothesis developed by Anatol Feldman, originally from the Soviet Union, argues that the motor system controls the *threshold* of activation of the motor neurons that innervate the muscles (Asatryan and Feldman, 1965; Feldman, 1986). During voluntary movement, the motor system reduces the threshold of the motor neuron activation to recruit more motor neurons, thereby increasing the contractile force of the muscle. Thus, the equilibrium point hypothesis asserts that force is not directly specified by the motor system; rather it is *indirectly* controlled by an alteration of the threshold of motor neuron activation.

Another important difference between the equilibrium point hypothesis and motor programming theory is that a limb moves, not by changes in force, but in the *relative* changes of tension generated by agonist and antagonist muscles. Recall that the tension generated by a muscle depends partly on its length. Figure 6.12 shows that if the tensions (or even more correctly, the torques) of the agonist and antagonist muscles around a joint are equal, the limb is stationary at a given point, called an **equilibrium point**. To *move* a limb to another equilibrium point, the motor system must change the relative tensions of the agonist and antagonist muscles. It is important to emphasize that sensory feedback from the joints and muscles is used to help stabilize the limb at an equilibrium.

Dynamic Pattern Theory

Another motor control theory that has emerged is the **Dynamic Pattern Theory** (Haken, Kelso, and Bunz, 1985; Jeka and Kelso, 1989; Schoner and Kelso, 1988; Wallace, 1996). This theory focuses on issues related to the *coordination* of two or more components within the motor system (i.e., joints, muscles, neurons). The theory is based on principles of *nonlinear dynamics*, an approach in the study of complex phenomena developed primarily in mathematics and physics, sometimes referred to as **chaos theory** (see Gleick, 1987). The details of dynamic pattern theory are much beyond the scope of the present chapter, but a few of its principles and some examples will hopefully give the reader an appreciation of its contribution to the field of motor control.

Rather than focusing on the forces exerted upon and produced by a single body as in Newton's laws, nonlinear dynamics attempts to understand how two or more bodies behave within a complex system, defined as any system composed of two or more interacting bodies. Nonlinear dynamics arose from the work of the French mathematician Jules Henri Poincaré in the late 19th century, who showed the impossibility of predicting the motions of two or more interacting bodies (such as the planets) over long time periods.

A major aspect of nonlinear dynamics incorporated into the dynamic pattern theory is the focus on *patterns* of coordination among the system's components and the processes associated with changing from one pattern to another. For example, if you are walking on a treadmill and the treadmill speed is systematically increased, at some particular speed you will change from a walking to a running pattern (the German physiologist Erich von

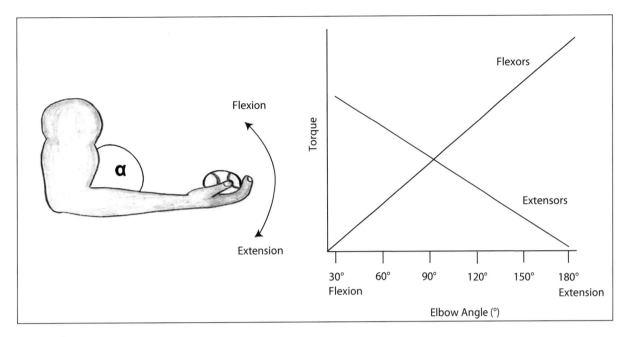

Figure 6.12

The equilibrium point model. (Left, muscles seen as springs; right, the length-tension diagrams for the flexors and extensors, with the intersection being the equilibrium point where the tensions, or torques, in the two muscle groups are equal and opposite), adapted from Schmidt and Lee, 2005. Figure 7.20, pg. 233.

Holst was one of the first scientists to study the processes by which patterns of coordination change in animals and man, [von Holst, 1939/1973]). This change can be, more or less, automatic in that you do not necessarily have to think about voluntarily switching to a new pattern of coordination. The quantitative description of these patterns is expressed by an **order parameter**. A change from one pattern to another means that the complex system has changed from one value of the order parameter to another. Often, the order parameter can be expressed as a quantitative relationship of how the components are moving relative to one another.

In the dynamic pattern theory, pattern change in a complex system like the human body can be due to the influence of an outside variable called a **control parameter**, on the interactive behavior of the components (in this case, the legs). As an analogy, increasing the temperature on a stove will cause the molecules of water in a pan to change from a random pattern distribution throughout the medium at room temperature to a bubbly or turbulent pattern at the boiling point. The control parameter, in this case temperature, causes a change in the pattern of molecules. According to nonlinear dynamics, the change is due to a *loss of stability* of the random pattern and an increase in the stability of the turbulent pattern. **Stability** can be defined as the tenacity of a pattern to resist change. Thus, a control parameter is any variable that causes a change in stability of the patterns within a complex system. In the treadmill example, increasing the speed of the treadmill (the control parameter) causes the walking pattern to become unstable and the running pattern to become more stable, causing a switch, or what nonlinear dynamical scientists call a phase transition from a less to a more stable pattern of locomotion.

In dynamic pattern theory, both efferent and afferent information, in addition to many other factors (biochemical, anatomical, psychological, etc.) play a role in stability of patterns of coordination. Thus, movement behavior is not completely prescribed by the higher centers, as in motor programming theory, but rather it is an emergent property of a complex system like the human body, which contains many interacting components.

As we have seen, there are numerous models of motor control. In spite of these conceptual advancements, much remains to be learned about how movement is controlled and coordinated. In the next section, we examine a couple of important laws of motor control related to the speed and accuracy of movement.

Summary

- Open-loop control refers to the control of movement in the absence of sensory feedback, whereas closed-loop control involves the use of sensory feedback and a comparator or reference mechanism to detect errors.
- The information processing and the ecological perspective models differ on how information is obtained from sensory stimuli in order to help control movement
- In motor program theory, movements are primarily controlled by a set of muscle commands generated by the Higher Center, which allows the movement to be carried out uninfluenced by peripheral feedback.
- Variant features of a motor program are those that change across variations of the movement (e.g., absolute levels of force) while invariant features are those characteristics that remain constant (e.g., relative timing of the movement components).
- The equilibrium point hypothesis posits that the motor system controls the threshold of activation of motor units. In addition, a limb moves from one position to another by changes in the relative tensions of the agonist and antagonist muscles
- The dynamic pattern theory focuses on the coordination of two or more components of a movement. Control parameters are variables that can cause the motor system to change from one pattern of coordination to another. Patterns of coordination have different levels of stability.

Speed and Accuracy of Movement

We human beings clearly have limitations in performing motor tasks. In particular, we frequently cannot respond to sensory stimulation as quickly as we would like. For example, we have all had the experience of reacting to an unexpected stimulus such as a car suddenly braking in front of us. Typically, there is an inherent delay from the time the brake lights of the car in front of you are perceived to the completion of the braking movement. This delay is a result of the time required by the central nervous system to identify the stimulus, prepare an appropriate motor response, and to send efferent commands to the muscles to produce movement. Often these delays can significantly affect the performance of a skill. For example, in a sprint race like a 100-meter dash, the sprinter can win or lose the race in the blocks, depending on the sprinter's reaction to the starter's gun. The speed with which we can react to a stimulus is affected by several variables. In addition, there are several factors that affect the speed and accuracy of movement itself. In this section, we will examine some factors that affect how quickly we can respond to a stimulus, called **reaction time**, as well as those factors or variables that influence the speed, or more specifically, the time required in producing movement, called the **movement time**.

Reaction Time

Reaction time is defined as the time from the onset of a stimulus to the initiation of a response to that stimulus. A major factor affecting reaction time is the *number of stimulus-response alternatives*. As the number of possible alternative responses to the stimulus increases, reaction time also increases. A study done over a century ago by Merkel (1885) illustrates this effect. Merkel required subjects to move the appropriate finger in response to a stimulus that indicated which finger to move. The number of stimulus-response alternatives varied between one and ten. For example, in one set of trials the subject was told that only one stimulus-response alternative

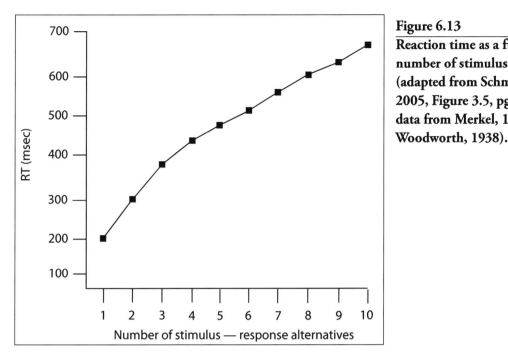

Figure 6.13
Reaction time as a function of the number of stimulus alternatives, (adapted from Schmidt and Lee, 2005, Figure 3.5, pg. 60, original data from Merkel, 1885, as cited by Woodworth, 1938).

would be used. In this condition (Condition 1 on the horizontal axis), the subject would know in advance which finger to move when the stimulus came on. But in Condition 10, *any* of the ten fingers might be required to respond to the corresponding stimulus. Merkel wished to know whether the reaction time to a particular stimulus would be affected by the number of possible stimulus-responses available to the subject.

Figure 6.13 shows the results of Merkel's experiment. On the horizontal axis is the number of stimulus-response alternatives for a set of trials, and on the vertical axis is the reaction time. It can be seen that reaction time increased as the number of stimulus-response alternatives increased. It should be noted that the reaction to only one stimulus-response alternative is called **simple reaction time**. and in Merkel's experiment this was shown to be around 180 ms. **Choice reaction time** is involved when more than one stimulus-response alternatives are presented.

Hick-Hyman Law

Why should choice reaction time be longer than simple reaction time, and why should choice reaction time increase as a function of the number of stimulus-response alternative? One answer to this question was provided by both Hick (1952) and Hyman (1953). Both of these investigators argued that as the number of stimulus-response alternatives increases, the amount of uncertainty as to which stimulus (and associated response) might appear also increases. The central nervous system must take longer to resolve the uncertainty as the number of stimulus-response alternatives increases. Hick and Hyman noticed, both in Merkel's and their experiments, that reaction time increased by a nearly constant amount (about 150 ms) every time the number of stimulus-response alternatives *doubled*, as seen in the original experimental data by Merkel (1885) shown in Figure 6.14. Both Hick and Hyman proposed that choice reaction time could be predicted from the following formula:

$$\text{Choice RT} = a + b\,[\text{Log}_2\,(N)] \tag{6.2}$$

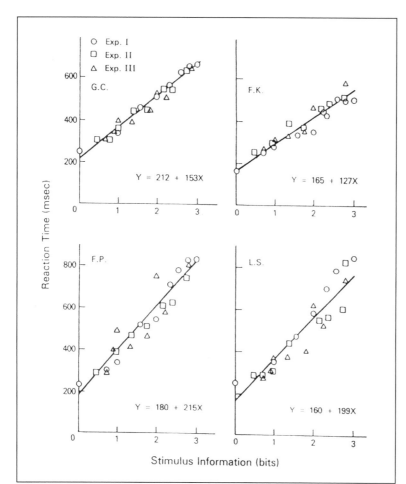

Figure 6.14
Choice reaction time as a function of bits of stimulus information. From Schmidt and Lee (2005), Figure 3.6, page 61. (Original data are from four subjects in the 1953 Hyman study). The graphs clearly show a near linear relationship between choice reaction time and stimulus information. These results indicate that reaction time increases as the number of possible stimulus-response alternative increases.

where a is the intercept of the line, b is the slope of the line, and N is the number of stimulus-response alternatives. This formulation has become known as the **Hick-Hyman Law**. Thus, choice reaction time is proposed to be strongly dependent upon the Hick-Hyman Law.

Movement time

Movement time is defined as the time from the initiation of a movement until its completion. There are several factors that affect the movement time. We may have all heard the adage "Haste makes waste." This saying implies that we become more inefficient at doing things if we attempt to do them too quickly. For example, if you write your name faster and faster (i.e., a shorter movement time) it becomes more and more unreadable. Similarly, you are much more accurate in threading a needle if you move slowly (i.e., a longer movement time). Thus, there appears to be a relationship between the speed and accuracy of the movement.

Fitts's Law

The first scientist to establish a quantitative relationship between the speed and accuracy of movement was Paul Fitts (Fitts, 1954; Fitts and Peterson, 1964). Fitts (1912–1965) performed experiments showing that the

Calculating the Amount of Information in the Hick-Hyman Law

Hick (1952) and Hyman (1953) argued that doubling the number of stimulus-response alternatives in a reaction time situation was analogous to doubling the amount of information the central nervous system has to deal with, where **information** is defined as the amount of uncertainty to be resolved. One bit (short for *b*inary dig*it*) of information is defined as the amount of uncertainty related to deciding between one of two alternatives, two bits of information is equal to four possible alternatives, three bits of information is equal to eight alternatives, and so on. For example, if you hide a marble in one of your hands and ask a friend to guess which hand the marble is in, the amount of uncertainty is equal to one bit of information, because there are two possible choices to make and the uncertainty is resolved after one choice is made *regardless* of which choice is made. However, if you can hide the marble in one of your hands *or* under one of your feet, it takes two bits of information for your friend to resolve the uncertainly. For example, your friend could first ask if the marble is in your hands or under your feet. Only one more choice is required to resolve the uncertainty. Therefore, two choices or two bits of information are required to solve the problem. The calculation of the amount of information contained in any number of stimulus-response alternatives can be determined as follows:

$$\text{The amount of information} = \text{Log}_2 (N) \qquad (6.3)$$

where N is the number of stimulus-response alternatives. As you may know, the logarithm of the base 2 of the number 1 is the power to which the base 2 must be raised to determine the number. Thus the $\text{Log}_2 (1) = 0$, $\text{Log}_2 (2) = 1$, $\text{Log}_2 (4) = 2$, $\text{Log}_2 (8) = 3$, and so on. Hick and Hyman figured that the constant increase of reaction time every time the number of stimulus-response alternatives is doubled must be due to the associated increase in the number of bits of information to be resolved by the central nervous system.

time to a complete movement (i.e., the movement time) was dependent on two factors: the movement distance and the size, or more specifically, the width of the target. Figure 6.15 illustrates the task used by Fitts that manipulated these two variables. Fitts found that the movement time in this task could be predicted by the following formula:

$$\text{Movement time} = a + b [\text{Log}_2 (2A/W)] \qquad (6.4)$$

where a is the y intercept, b is the slope of line, A is the movement amplitude (distance) and W is the width of the target. One can see that the formula predicts that when A is increased, movement time also increases, reflecting the physical principle that it takes more time to travel longer distances. Therefore, movement time is *directly* related to the distance traveled. In addition, if W is increased, movement time is predicted to *decrease*, reflecting the principle that larger targets allow individuals to be less accurate and go faster. Therefore, movement time is *inversely* related to target width. The expression $\text{Log}_2 (2A/W)$ was considered to be the **index of difficulty**. Fitts found that when the index of difficulty increased, so did the movement time. In fact, the relationship between the index of difficulty and movement time is predicted to be linear. The data from Fitts's original experiment is shown in Figure 6.16, and verifies the prediction. His findings demonstrated that when the index of difficulty increased due to increases in distance or decreases in target width, so did the movement time.

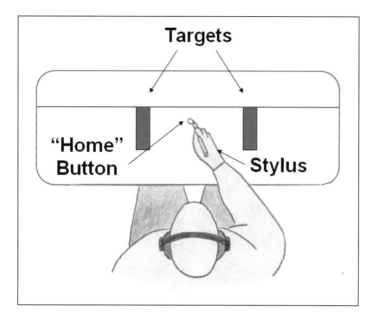

Targets

"Home" Button

Stylus

Figure 6.15

Overhead view of a subject performing a Fitts tapping task. This task requires the subject to alternately tap between the two targets as quickly as possible. The task is altered by changing the distance between and the width of the targets, (adapted from Shea, Shebilske, and Worchel, 1993), Figure 9.2, originally from Fitts, P. M. (1964). Perceptual-motor skills learning. In A. W. Melton (ed.). *Categories of Human Learning*, p. 258. Copyright 1964 by Academic Press.

The quantitative relation (6.4) has been shown to hold in a variety of situations and with different populations, and has become known as **Fitts's Law** (see Schmidt and Lee, 2005, for a discussion). The implications of Fitts's Law for the control of movement are quite clear. Individuals can increase the speed (i.e., reduce movement time), when the index of difficulty reduces. In other words, movement time reduces when the distance traveled is short or when the target width is large, or both. Violations of Fitts's Law result in more errors (or less accuracy). In essence, the performer must trade speed for accuracy.

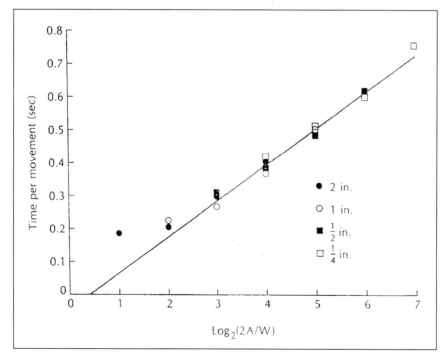

Figure 6.16

Average movement time for the Fitts tapping tasks of different difficulties, (adapted from Shea, Shebilske, and Worchel, 1993), Figure 9.4, pg. 187, original data from Fitts (1954).

SUMMARY

- Reaction time is defined as the time from the presentation of a stimulus to the initiation of the response.
- Movement time is defined as the time from the initiation of the movement to its completion.
- Response time is the reaction time plus the movement time.
- According to the information processing model of motor control, there are three distinct stages of information processing in reaction time situations: the stimulus identification stage, the response selection stage, and the response programming stage.
- The stimulus identification stage is affected by the intensity and clarity of the stimulus, the response selection is affected by the number of stimulus-response alternations, and response programming is influenced by the complexity of the movement.
- The Hick-Hyman law is stated as follows: Choice RT = a + b[Log_2 (N)].
- Information is defined as the amount of uncertainty to be resolved and can be calculated as follows: the amount of information = Log_2 (N).
- Fitts's Law governs the relationship between the speed and accuracy of movement and is expressed by the following formula:
- Movement time = a + b [Log_2 (2A/W)].

MOTOR LEARNING

Definition of Motor Skills

Motor skills are tasks requiring voluntary head, body, and/or limb movement for the purposes of achieving some goal (Magill, 2011). Earlier in the chapter, we described the difference between reflexes and voluntary movement. Motor skills consist of primarily voluntary movements that are learned. We have all attempted to learn new motor skills, such as riding a bicycle, typing, or downhill skiing to name just a few, and each clearly have well-defined goals or purposes. In fact, all of the motor skills we possess have been learned sometime during our lives.

Types of Motor Skills

How many motor skills are there? Probably countless! Because of the vast number of motor skills, researchers have tried to figure out ways to classify them and to identify their major elements. Probably the most well known classification scheme was one developed by Gentile (1987:2000). In this classification scheme or taxonomy, Gentile identified two key elements of all motor skills. The first element is the *action function*, which defines whether the person's body is stable (not moving) or is in transport (moving). The action function also defines whether an object is manipulated or not. The second element is the *environmental context* that defines certain characteristics of the environment in which the skill is performed. In some skills, the environmental conditions are relatively stationary, like when shooting a foul shot in basketball. The basket is 15 ft. away and 10 ft. high, and these distances never change when shooting a foul shot. In other skills, the environment is constantly moving or changing, such as in hitting a baseball. The so-called regulatory conditions of the environment can therefore be stationary or in motion. In addition, the environmental conditions may or may not change from one practice attempt to another. In free-throw shooting, the environmental context does not significantly change from one attempt to another, but it does if the baseball pitcher throws different types of pitches (fast ball, curve ball, slider, etc.) from one pitch to the next. After considering the type of action function and environmental context, Gentile proposed that there were 16 separate categories or types of motor skills. Gentile's taxonomy has been useful in teaching and evaluating motor skill development in a variety of contexts such as sports, physical education, and rehabilitation (see Magill, 2011).

Factors Affecting Motor Learning

How humans learn motor skills and some of the variables that affect the learning process is the topic of this section. It should be recognized that there are a number of factors that contribute to the learning of motor skills. Much of motor learning is influenced by various types of information provided to the learner either before, during, or after each practice attempt (Newell, 1981). For example, it has been shown that observing others perform a skill prior to physical performance of that skill can influence subsequent performance, a factor called **modeling** (e.g., McCullagh, Weiss, and Ross, 1989). During performance, **intrinsic feedback**, information gathered through our senses about the position of our bodies relative to the environment, is also important. Finally, the feedback we achieve after a performance attempt that provides information either about how the skill was executed, called **knowledge of performance**, or about how successful the skill was in achieving the goal, called **knowledge of results**, is also critical for motor learning. In addition, research has shown the importance of the structure of practice conditions. For example, how much rest should occur between practice attempts? Should only one variation of a skill be practiced at a time, or should the learner's practice conditions include more varied practice (see Schmidt and Lee, 2005, for a comprehensive review of this literature). To introduce the reader into the area of motor learning, our discussion will focus on feedback and its importance in motor learning. Then, the major theories of motor learning will be briefly described. These theories attempt to explain how and why motor learning occurs.

Feedback and Motor Learning

We are all familiar with the saying, "Practice makes perfect." This saying implies that the mere repetition of a skill will insure improvement. However, research in the area of motor learning has shown that practice alone will not guarantee significant improvement in the learning of a motor skill. In addition to practice, the learner must also be provided feedback about his or her performance. There are different types of feedback, and in this section we will describe them and provide some experimental evidence demonstrating their importance in the learning process.

Types of Feedback. As we are learning a motor skill such as a golf swing, we can receive a variety of feedback about our performance. As we perform the movement, we receive sensory information from our body such as touch and pressure feedback from gripping the club, proprioceptive information about the position of our arms and legs from joint receptors, muscle length changes from our muscle spindles, and visual feedback throughout the swing. This type of feedback is called **intrinsic feedback**, because it is generated from our bodies and it informs our central nervous system about how the movement was produced.

We may also receive information from other sources, such as from a teacher, coach, or physical therapist. This type of feedback is called **extrinsic feedback**, because it is generated from an outside source. Generally, extrinsic feedback can take one of two forms. The first form provides information about the outcome of movement, called **knowledge of results** (or KR). Following a swing, a golf coach might say "Nice shot! The ball hooked a little but traveled about 200 yards down the left side of the fairway." Another form of extrinsic feedback, called **knowledge of performance** (or KP), provides information about the execution of the movement (Gentile, 1972). During rehabilitation therapy of a patient suffering from paralysis of the legs, a physical therapist might

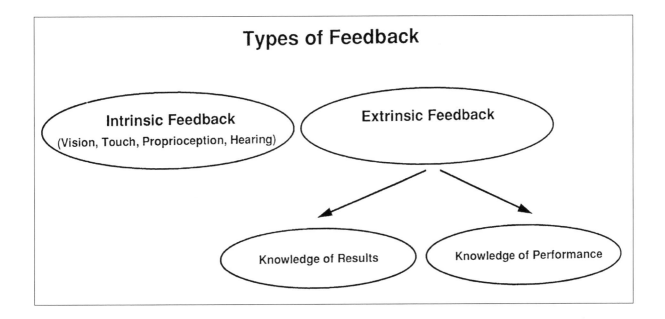

Figure 6.17

Types of feedback available during motor learning.

say, "That was a good effort! You were able to flex your big toe about 15 degrees." Figure 6.17 illustrates a breakdown of the types of feedback important in the learning of a motor skill.

The Motivational and Informational Function of Feedback. Receiving feedback from an outside source can be motivating or create feelings of anxiousness to the learner. Motivation can be defined as a positive emotional state leading to the tendency of the learner to continue the activity, while anxiety is a negative emotional state of discontinuing the activity. Aspiring ballerinas taking pointe lessons can receive feedback (or corrections) from their teacher that is encouraging or motivating such as statements like, "Great job, your turnout is looking much better!" Such statements following a performance attempt tend to facilitate performance. However, statements like, "Your turnout today looks pathetic! Why can't you do what I am telling you to do?" can be quite discouraging and may sometimes lead to negative or anxious reactions from the learner and can impair performance. However, the research on the *long*-term effects of positive and negative feedback is lacking.

In addition to its influence on the emotional state of the individual, feedback also provides *information* to the learner about how he or she is performing relative to some goal (Adams, 1971). The information provided by KR, for example, informs the learner about how close the outcome of the performance attempt was to the goal outcome or standard. KR can contain directional information that specifies on which side of the standard the movement was on. For example, a runner is training to run the 400-meter at a pace that represents 75 percent of her best time. After completing one attempt, the track coach might say, "Susan, you ran that too slowly." This type of KR would be somewhat helpful, but more informative KR can provide the direction and the *magnitude* of the error, such as, "Susan, that lap was five seconds too slow." Research has suggested that providing more precise error information, up to some point, leads to better performance and learning (e.g., Smoll, 1972; Trowbridge and Cason, 1932), with the results depending on the skill level of the performer (see Magill, 2011, for a discussion).

Importance of KR and KP in Motor Learning. How do KR and KP affect the learning of a motor skill? To answer this question, researchers have generally required adult subjects to learn a novel or unfamiliar motor skill. Extrinsic feedback is provided to one group of subjects in the experimental group after performance attempts of the motor skill, while no feedback or impoverished feedback is provided to another group called the control

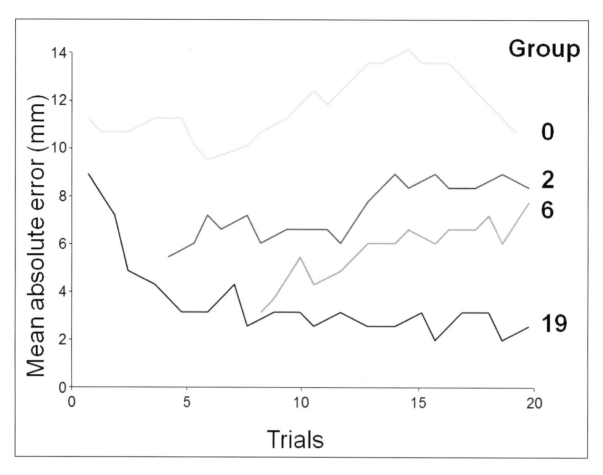

Figure 6.18

Absolute error (smaller errors = better performance) for linear positioning arm responses under different knowledge of results (KR) conditions. The group numbers indicate the number of trials for which KR was provided. Notice that without KR (Group 0), performance does not improve. Performance improves as the number of KR trials increases, (adapted from Shea, Shebilske, and Worchel, 1993), Figure 10.1, pg. 213, (original data from Bilodeau, Bilodeau, and Schumsky, 1959).

group. A comparison of the experimental and control group's performances can be made during the initial learning trials of the skill, called the *acquisition phase*, and during the final *retention phase* when both groups have to perform on their own under similar, no-feedback conditions. The retention phase is thought to reflect what the performer has learned, or retained from previous practice during the acquisition phase (Schmidt and Lee, 2005).

Many studies have shown that performance during the acquisition phase of a novel motor skill does not improve unless the performer receives KR (e.g., Trowbridge and Cason, 1932; Bilodeau, Bilodeau, and Schumsky, 1959; Newell, 1974; see Schmidt and Lee, 2005, and Magill, 2011, for a discussion). Figure 6.18 illustrates the results of the Bilodeau et al. experiment, in which subjects in a variety of groups attempted to learn a simple yet novel motor skill of producing an arm movement at a given distance while blindfolded. The blindfold prevented subjects from seeing the outcome of their movements. In the experimental groups, KR, in the form of specific error information, was given to the subjects by the experimenter after each performance attempt, while subjects in the control group (no KR) performed without extrinsic feedback. In another interesting experimental manipulation, KR was withdrawn from the experimental groups at different times during learning. It can be seen that the no-KR control group's performance did not improve over the entire practice period, refuting the adage that practice (alone) makes perfect! Rather, the results suggested a dependency of learning on KR. Improvement in performance occurred when KR was provided during the acquisition phase and performance tended to deteriorate during the retention phase when it was withdrawn. It is interesting to note, however, that the group receiving the most KR trials were able to sustain their performance even when KR was withdrawn. These and other similar results allow the conclusion that KR is essential for both improvements in performance and on the learning (or retention) of motor skill.

A study by Wallace and Hagler (1979) illustrated the importance of KP on the performance and learning of a motor skill. Subjects with little experience playing basketball attempted to learn a novel skill of shooting the basketball with their non-dominant hand. Both groups had full vision of each attempt and therefore received knowledge of the outcome of each shot (KR) on their own. In the KP group, subjects received specific

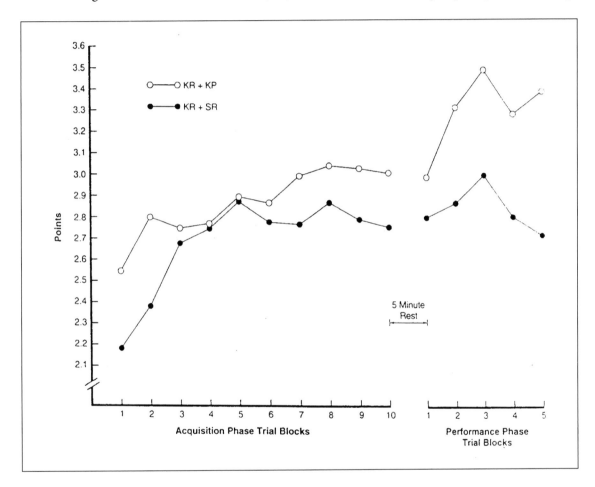

Figure 6.19

Results of the Wallace and Hagler (1979) study showing the benefits of knowledge of performance (KP) to motor learning. The open-circle group received both KR and KP while the closed-circle group received KR and social reinforcement (SR) or encouraging comments from the experimenter. Notice that at the end of the acquisition phase, the KR + KP group performed better than the KR + SR group. During the performance (or retention) phase, when all verbal feedback from the experimenter ceased, the KR + KP group retained the motor skill better. From Wallace, S.A. & Hagler, R.W. (1979). Knowledge of performance and learning of a closed motor skill. Research Quarterly, 50, 265–271.

information about the execution or form after each shot from the experimenter such as, "You didn't bend your knees," or "You didn't follow through." In the other group, subjects received verbal encouragement from the experiment such as "Good effort," or "Nice try." After 50 practice trials in the acquisition phase, both groups then performed 25 retention trials without any extrinsic feedback from the experiment. The results can be seen in Figure 6.19. Since both groups received KR, we would expect both groups to improve, which is exactly what happened. However, notice by the end of the acquisition phase, the KP group performed at a higher level than the other group. In addition, their performance during the retention phase, when no extrinsic feedback was provided to either group, was also superior, indicating that the KP group actually learned (or retained) the motor skill to a greater degree. While fewer KP studies have been done compared to KR studies, it is generally thought that both KR and KP are potent contributors to the learning of motor skills (Magill, 2011; Schmidt and Lee, 2005).

Theories of Motor Learning. How does the learning of a motor skill occur? There are three major theories of motor learning that attempt to answer this question, *closed-loop theory* (Adams, 1971), *schema theory* (Schmidt, 1975), and *dynamic pattern theory* (Kelso, 1995). In this section, we will briefly outline the important concepts of each theory and mention experiments that provide some support for each theory.

The first theory of motor learning to be discussed is the **closed-loop theory** developed by Jack Adams (Adams, 1971). When it was published, Adams's theory sparked a renewed interest among researchers in the processes associated with the learning of new motor skills. The theory was rather simple and it was testable, two important qualities of good theories. According to the theory, there are two main sources of information required by the performer to learn new motor skills: sensory (intrinsic) feedback and knowledge of results (KR). These two sources of information are used to develop a reference mechanism or memory representation of the desired movement to be learned. Adams argued that after each attempt (or trial) at learning the motor skill, the sensory feedback from the movement (vision, proprioception, etc.) is stored in the learner's memory.

The sensory representation of the movement is called the **perceptual trace**. Because the learner's initial attempts at producing the desired movement is likely to be incorrect, the perceptual trace resulting from these movements is weak. However, with repeated exposure to KR during the acquisition phase, the movements become more accurate and the sensory feedback from these movements helps to develop a stronger perceptual trace. Adams posited that after each attempt at producing the desired movement, the sensory feedback from the current movement is compared to the perceptual trace. Any discrepancy between the two results in the generation of an error, an error to be reduced by future practice with KR.

According to Adams, the perceptual trace provides the learner with an *image* of the desired movement to be produced, and this image is theorized to become stronger with extended practice accompanied by KR. One of the consequences of a stronger perceptual trace is that the learner should be better able to evaluate their own performance. If the perceptual trace is strong enough, the learner should be able to continue to evaluate performance (i.e., detect performance errors relative to the desired movement) even in the *absence of KR*.

In addition to a stronger perceptual trace, extended practice with KR strengthens the learner's ability to actually initiate and produce movement, necessary for correcting errors. Adams called the mechanism for initiating and producing the efferent commands for movement the **memory trace**. The memory trace is a motor program (see above) that is also strengthened with physical practice accompanied by KR. Equipped with a strong perceptual and memory trace, the learner is capable of continued improvement in performance and long-term retention or learning of the motor skill. This improvement is due to an increased ability to detect errors, the responsibility of the perceptual trace, and the increased ability to make physical corrections over practice trials, the responsibility of the memory trace.

Evidence supporting Adams's theory came from previous studies such as the Bilodeau et al. experiment discussed above. Adams's theory predicted that extended practice with KR should not only result in better performance improvement, but also an increased ability to maintain performance after KR is withdrawn. These predictions were consistent with the results of the earlier Bilodeau et al. study, and additional support came quickly with a study by Schmidt and White (1972). Other supportive studies followed.

The second theory of motor learning called **schema theory** was developed by Richard Schmidt (Schmidt, 1975). Schmidt's theory was similar to Adams's theory in acknowledging a mechanism like the perceptual trace for performance evaluation and a mechanism like the memory trace for initiating error corrections. However, Schmidt's theory argued that the manner in which these mechanisms or processes are represented in memory was quite different than that proposed by Adams. Recall in Adams's theory that the perceptual and memory trace were stored memory representations of the *correct* movement. One of Schmidt's arguments against this conceptualization was that it would prevent the learner from ever being allowed to produce *novel* variations of learned movement. For example, suppose someone wishes to learn to shoot a basketball from a variety of distances and positions around the basketball key area. A strict interpretation of Adams's theory would require the learner to develop a perceptual (and memory) trace for *every* possible distance and position on the court! Schmidt thought this type of learning would be burdensome for the central nervous system, given the infinite number of variations possible. Alternatively, Schmidt believed that learning must be more flexible. Specifically, he proposed that the learner develops rules (called schemata) for both evaluating one performance and actually producing movements in novel situations. The rule for evaluating one's performance was called the **recognition schema**, and the rule for producing movement was named the **recall schema**. The recognition schema is developed by forming a *relationship* between the sensory feedback of the movement and the KR provided to the learner. The recall schema is developed by forming a *relationship* between the efferent commands used to produce movement and the KR provided to the learner.

Let's briefly examine how a schema rule might be developed during the acquisition phase of learning a new movement. For simplicity, we will focus on the development of the recall schema responsible for producing the efferent commands to the musculature. A young child decides to learn how to shoot a basketball from various distances away from the basket. On each attempt, the child must generate the correct amount of force generated by efferent commands to propel the basketball the appropriate distance. To develop the recall schema, relationships must be formed between the efferent commands and the KR of the movements produced during the acquisition phase. According to schema theory, the best way to develop a rule or schema is through **variable practice**. With variable practice, the learner practices a variety of movements using the same motor program. Figure 6.20 illustrates how a rule or schema is better formed with variable practice as opposed to **constant practice**. An example of variable practice is allowing the child to practice shooting the basketball from a variety of distances from the hoop. This type of practice allows the learner to develop a relationship (shown by the straight line) between the force required and the KR. If the learner has developed this relationship after variable practice, he or she should be able to successfully produce a novel movement from a *novel* distance, namely, a distance never practiced at before. In contrast, constant practice does not allow for the development of a rule or schema, because it does not permit the learner to develop a good relationship between the force required and KR. The theory predicts that novel variations of the practiced movement can be produced better following variable practice compared to constant practice. Shapiro and Schmidt (1982) reviewed the literature and found considerable support for this prediction. These results provide evidence against Adams's conceptualization of the perceptual trace in favor of a schema- or rule-based representation of learned movement in memory.

The last theory to be discussed is **dynamic pattern theory** as it applies to motor learning. As discussed earlier in the chapter, dynamic pattern theory, strongly advocated by J. A. Scott Kelso, emphasizes the identification of patterns of coordination and in the quantification of the stability of the coordination patterns. A major assumption of the theory is that the learner brings previous coordination patterns developed by past experiences into the learning of a new motor task. These previous coordination patterns possess certain degrees of stability,

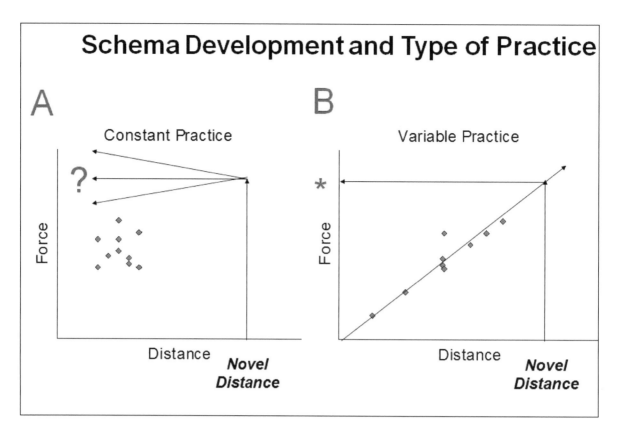

Figure 6.20

Schema development and the type of practice. Each circle represents a practice trial. In (A), if the learner only practices shooting a basketball from similar distances (constant practice), it is very difficult to develop a rule (schema) or relationship between the distance of the shot and the force required. When the learner transfers to a novel distance, say, further than that originally practiced, it is difficult for the learner to determine the required force. In (B), if, on the other hand, the learner uses variable practice, a schema rule can be established between the distance of the shot and the force required. With the rule established, it is possible to extrapolate the required force even at a novel distance.

or tenacity to resist change, as we discussed earlier. The patterns of coordination brought into a new motor task are called the learner's **intrinsic dynamics** or **coordination tendencies** (Kelso, 1995). In layman's terms, we could call these intrinsic dynamics the learner's unique "style" in performing the motor task, such as the walking pattern differences of people. As we all have observed, people seem to have a unique walking pattern or style in moving their body and limbs. According to dynamic pattern theory, the intrinsic dynamics of the learner must first be identified in order to assess the starting point of the learning process.

In addition to the intrinsic dynamics, environmental information is another important factor in motor learning. **Environmental information** defines the *required* coordination pattern involved in the new motor task to be learned. For example, environmental information could be the demonstration of the required pattern of coordination by an instructor or coach. According to the theory, if the intrinsic dynamics of the learner do not match the required pattern of coordination expressed by the environmental information, performance is inaccurate and inconsistent. Accuracy and consistency increase when the intrinsic dynamics start to match the environmental information. The theory states that to learn a new pattern of coordination required for a given motor skill, the learner may have to overcome his or her intrinsic dynamics if the intrinsic dynamics do not match the environmental information.

We have probably all had the difficult experience of ridding ourselves of an inappropriate pattern of coordination during the learning of a new motor skill. Most new golfers have the wrong coordination pattern for the golf swing. The timing of their arms, with respect to the shift in body weight and leg motion, is typically inappropriate. However, the inappropriate pattern (representing the learner's intrinsic dynamics) brought into the task is possibly quite stable (i.e., resistant to change). In the early motor learning stages of the golf swing, the learner's intrinsic dynamics are probably stronger than the environmental information provided by the instructor. Therefore, the learner's swing is likely to be inaccurate and inconsistent. However, with much practice, the inappropriate pattern(s) of coordination can change to the desired or required pattern. Zanone and Kelso (1992) found that not only can a learner acquire a new pattern of coordination, but that the intrinsic dynamics of the learner can change as well.

SUMMARY

- The two major types of feedback during motor learning are intrinsic and extrinsic feedback.
- Knowledge of results (or KR) is a type of extrinsic feedback that provides the learner with information about the outcome of the movement.
- Knowledge of performance (or KP) is a type of extrinsic feedback that provides the learner with information about the execution of the movement.
- Extrinsic feedback has both a motivational and informational function.
- Evidence suggests that both KR and KP improve the performance and the learning (or retention) of motor skills.
- The three major theories of motor learning are Adams's (1971) closed-loop theory, Schmidt's (1975) schema theory, and dynamic pattern theory (Kelso, 1995).

CHAPTER SUMMARY

- The three levels of motor control are the Higher Center Level within the brain, the Spinal Level in the spinal cord, and the Lower Level consisting of the muscles, sensory receptors, connective tissue, and the load to be moved.
- There are a variety of brain structures that have specific functions related to motor control.
- The two major types of movement are reflexive and voluntary movement.
- Voluntary movements can be under closed- or open-loop control—that is, they may or may not be affected by sensory feedback, respectively.
- There are several models of motor control.
- The Hick-Hyman Law states that reaction time depends on the number of stimulus-response alternatives.
- Fitts's Law states that movement time depends on the distance of the movement and the target size.
- The two major types of feedback during motor learning are intrinsic and extrinsic feedback.
- The two major types of extrinsic feedback are knowledge of results (KR) and knowledge of performance (KP). Both types are known to affect the motor learning process.
- The three major theories of motor learning are closed-loop theory, schema theory, and dynamic pattern theory.

IMPORTANT TERMS

Alpha and gamma motor neurons – lower motor neurons that stimulate extrafusal and intrafusal muscles fibers, respectively

Alpha-gamma coactivation principle – the concept that these motor neurons are activated simulataneously during voluntary movement

Basal ganglia – an area in the brain responsible for the initiation of movement

Central pattern generator – a hypothesis that rhythmical control of locomotion is located in the spinal cord by a network of neurons

Cerebellum – a brain structure that assists in the coordination of movement

Cerebrocerebellar loop – a neural pathway that connects the cerebral cortex and the cerebellum

Chaos theory – a theoretical approach to understanding complex systems

Choice reaction time – the time required to react to a stimulus when two or more responses are possible

Closed-loop theory – a theory proposed by Adams (1971) on how movements are learned that emphasized the importance of feedback

Constant practice – extended practice on only one variation of a skill

Control parameter – an outside variable that induces pattern change in complex systems

Coordination tendencies – patterns of coordination that making up an individual's intrinsic dynamics

Corticospinal tract – a major motor pathway involved in voluntary movement

Deafferentation – a lack of sensory feedback

Direct perception – a concept developed by J.J. Gibson that argues that we can pick up important features of the environment without major analysis by the brain

Dorsal columns – important sensory pathways that send information to the somato-sensory cortex

Dynamic pattern theory – a type of chaos theory on the coordination of movement

Ecological perspective approach – proposed by J.J. Gibson that stimuli be directly used by the perceptual-motor system without elaborate processing through separate stages by the central nervous system

Environmental information – in dynamical pattern theory, this information defines the required coordination pattern involved in the new motor task to be learned.

Equilibrium point – a stationary limb position where the torques of the agonist and antagonist muscles are equal and opposite

Equilibrium point hypothesis - a hypothesis developed by Anatol Feldman, arguing that the motor system controls the threshold of activation of the motor neurons that innervate the muscles

Extrafusal fibers – muscles fibers in skeletal muscles

Extrinsic feedback – information about a movement from outside the body

Feedback – information both within and outside the body that is used to help make adjustments of a movement

First-, second-, and third-order neurons – nerve cells sequentially linked with each other within a pathway

Fitts' Law – movement time is directly related to the distance of the movement and inversely related to the target size

Golgi tendon organs – sensory receptors sensitive to tension

Hick-Hyman Law – a law governing choice reaction time

Higher center level – areas of the brain involved in producing movement

Ia afferent fibers – sensory pathways from the muscle spindles

Index of difficulty – a ratio of the distance moved and the target size in Fitts' Law

Indirect perception – describes how information is analyzed in the information processing model

Information – the amount of uncertainty

Information processing model – a concept proposing that information is processed in a series of stages within the brain

Intrafusal fibers – muscle fibers within the muscle spindle

Intrinsic dynamics – similar to coordination tendencies in dynamical pattern theory

Intrinsic feedback – sensory information within the body used to help control movement

Knee-jerk stretch reflex – an involuntary response that occurs when the patellar tendon and quadriceps muscles are stretched

Knowledge of performance – information about how a movement is executed

Knowledge of results – information about the outcome of a movement

Length-tension relation – how the length of the muscle influences its tension

Long-loop reflexes – a component of the stretch reflex involving the higher centers

Lower level – a part of the motor system involving the muscles, sensory receptors and the load of a movement

Memory trace – a hypothetical construct in Adams (1971) theory that controls the selection and initiation of a movement

Modeling – the demonstration of a movement

Motor cortex – a part of the brain responsible for selecting the appropriate muscles for a movement

Motor skill – a learned movement that is goal oriented

Motor program theory – a theory proposing that details of the movement are stored in certain parts of the brain

Movement time – the duration of a movement

Muscle spindles – sensory receptors within the muscles

Open- and closed-loop control – types of movement control that do not or do use feedback, respectively

Optical flow – the change of reflected light on the retina

Order parameter – a variable in dynamic pattern theory that describes the various patterns of coordination

Parkinson's disease – a movement disorder due to damage of the basal ganglia

Perceptual trace - a hypothetical construct in Adams (1971) theory that is responsible for error correction

Pyramidal and extrapyramidal tracts – major motor pathways

Reaction time – a voluntary response to an unpredictable stimulus

Recall schema – a rule in Schmidt's schema theory responsible for selecting movement and making correction

Reciprocal inhibition – the relaxation of an antagonist muscle during the contraction of the an agonist muscle around the same joint

Recognition schema - a rule in Schmidt's schema theory responsible evaluation performance

Reference mechanism – a hypothetical construct in closed loop theory that represents the correct movement

Reflexive movement – involuntary responses to stimuli

Response programming stage – the stage in the information processing model responsible for selecting various movement parameters

Response selection stage - the stage in the information processing model responsible deciding the proper movement

Response time – reaction time plus movement time

Resting tremor – involuntary shaking of a limb

Schema theory – a theory developed by Schmidt (1975) that proposed the motor learning is dependent on the acquisition of 'rules' that govern movement error detection and correction

Simple reaction time – the time from the stimulus to the voluntary initiation of the movement

Somatosensory cortex – an area of the brain responsible for conscious sensory information

Spinal level – represents the spinal cord contributions to movement

Stability – the resistance to change of a coordination pattern

Stimulus identification stage – the first stage in the information processing model

Supplementary motor cortex – one of the areas of brain assisting in the planning of movment

Tau – a perceptual variable that represents the time to contact

Thalamus – an integrative relay station in the brain

Variable practice – extended practice with several variations of a skill

Variant and invariant features – aspects of a motor program that control changes in the movement and rhythmical timing, respectively

Voluntary movement – willed movement

INTEGRATING KINESIOLOGY: PUTTING IT ALL TOGETHER

1. A person decides to reach forward and pick up a coffee cup. At each level of the neuromuscular system (higher center level, spinal level, and lower level), describe some of the neuromuscular structures that would be involved from the planning through the execution of the movement.
2. For this same movement, what types of bones and joints may be involved? [Hint: Refer to Chapter Three for help.]
3. How many degrees of freedom of joint movement can you count for this movement?
4. Try to teach yourself to juggle three balls at once. Go to the Web site below for some help in learning how to juggle. See how long it takes for you to juggle three balls for 10 seconds. Once you have reached this goal, don't practice for a week. Then, perform a retention test and see how well you have retained the skill.

KINESIOLOGY ON THE WEB

What is your reaction time? These Web sites allow you to measure your visual reaction time:

- www.mathsisfun.com/games/reaction-time.html
- www.exploratorium.edu/baseball/reactiontime.html
- www.bbc.co.uk/science/humanbody/sleep/sheep/reaction_version5.swf

Do you want to learn the complex motor skill of juggling? Go to this Web site!

- www.thejimshow.com/juggle/

REFERENCES

Adams, J. A. (1971). A closed-loop theory of motor learning. *Journal of Motor Behavior, 3*, 111–50.

Asatryan, D. G., and Feldman, A. G. (1965). Functional tuning of the nervous system with control of movements or maintenance of a steady posture. I. Mechanographic analysis of the work of the limb on execution of a postural task. *Biophysics, 10*, 925–35.

Bilodeau, E. A., Bilodeau, I. M., and Schumsky, D. A. (1959). Some effects of introducing and withdrawing knowledge of results early and late in practice. *Journal of Experimental Psychology, 58*, 142–44.

Deecke, L., Scheid, P., and Kornhuber, H. H. (1969). Distribution of readiness potential, premotion positivity, and motor potential of the human cerebral cortex preceding voluntary finger movements. *Experimental Brain Research, 7*, 158–68.

Donders, F. C. (1969). On the speed of mental processes. In W. G. Koster (ed. and trans.), *Attention and Performance II*: Amsterdam: North Holland. (Original work published in 1868.)

Evarts, E. V. (1981). Role of motor cortex in voluntary movements in primates. In V. B. Brooks (ed.), *Handbook of Physiology* (Sec. 1, Vol. 11, pp. 1083–1120). Bethesda, MD: American Physiological Society.

Feldman, A. G. (1986). Once more on the equilibrium-point hypothesis (l model) for motor control. *Journal of Motor Behavior, 18*, 17–54.

Fitts, P. M. (1954). The information capacity of the human motor system in controlling the amplitude of movement. *Journal of Experimental Psychology, 67*, 103–12.

Fitts, P. M. and Peterson, J. R. (1964). Information capacity of discrete motor responses. *Journal of Experimental Psychology, 67*, 103–12.

Gentile, A. M. (1972). A working model of skill acquisition with application to teaching. *Quest, 17*, 3–23.

Gentile, A. M. (1987). Skill acquisition: Action, movement, and neuromotor processes. In J. H. Carr, R. B. Shepard, J. Gordon, A. M. Gentile, and J. M. Hinds (Eds.), *Movement Science: Foundations for Physical Therapy in Rehabilitation* (pp. 93–154), Rockville, MD: Aspen.

Gentile, A. M. (2000). Skill acquisition: Action, movement, and neuromotor processes. In J. H. Carr, and R. B. Shepard (Eds.), *Movement Science: Foundations for Physical Therapy*, (2nd ed., pp. 111–87). Rockville, MD: Aspen.

Georgeopoulos, A. P., Caminiti, R., Kalaska, J. F., and Massey, J. T. (1983). Spatial coding of movement direction by motor cortical populations. In J. T. Massion, J. Paillard, W. Schultz, and M. Wiesendanger (eds.), neural coding of motor performance. *Experimental Brain Research, Supplement 7*, 327–36.

Gibson, J. J. (1966). *The Senses Considered as Perceptual Systems*. Boston: Houghton-Mifflin.

Gleick, J. (1987). *Chaos: Making a New Science*. New York: Viking Penguin.

Grillner, S. (1975). Locomotion in vertebrates: Central mechanisms and reflex interaction. *Physiological Reviews, 55*, 247–304.

Haken, H., Kelso, J. A. S., and Bunz, H. (1985). A theoretical model of phase transitions in human hand movements. *Biological Cybernetics, 51*, 347–56.

Hick, W. E. (1952). On the rate of gain of information. *Quarterly Journal of Experimental Psychology, 4*, 11–26.

Hyman, R. (1953). Stimulus information as a determinant of reaction time. *Journal of Experimental Psychology, 45*, 188–96.

Jeka, J. J. and Kelso, J. A. S. (1989). The dynamic pattern approach to coordinated behavior: A tutorial review. In S. A. Wallace (ed.), *Perspectives on the Coordination of Movement* (pp. 3–45). Amsterdam: North Holland.

Kelso, J. A. S. (1995). *Dynamic Patterns: The Self-Organization of Brain and Behavior*. Cambridge, MA: MIT Press.

Keele, S. W. (1968). Movement control in skilled motor performance. *Psychological Bulletin, 70*, 387–403.

Klapp, S. T. (1977). Response programming, as assessed by reaction time, does not establish commands for particular muscles. *Journal of Motor Behavior, 9*, 301–12.

Kupfermann, I., and Weiss, K. R. (1978). The command neuron concept. *The Behavioral and Brain Sciences, 1*, 3–39.

Lee, D. N., and Aronson, E. (1974). Visual proprioceptive control of standing in human infants. *Perception and Psychophysics, 15*, 529–32.

Lee, D. N., Young, D. S., Reddish, P. E., Lough, S., and Clayton, T. M. H. (1983). Visual timing in hitting an accelerating ball. *Quarterly Journal of Experimental Psychology, 35A*, 333–46.

McCullagh, P., Weiss, M. R., and Ross, D. (1989). Modeling considerations in motor skill acquisition and performance: An integrated approach. In K. Pandolf (Ed.), *Exercise and Sport Science Reviews*, Vol. 17, 475–513, Baltimore: Williams and Wilkins.

Magill, R. A. (2011). Motor learning: Concepts and Applications. Madison, WI: Brown and Benchmark. 9th ed.

Merkel, J. (1885). Die zeitlichen Verhaltnisse der Willensthatigkeit. *Philosophische Studien, 2*, 73–127.

Newell, K. M. (1974). Knowledge of results and motor learning. *Journal of Motor Behavior, 6*, 235–44.

Newell, K. M. (1981). Skill learning. In D. Holding (Ed.), *Human Skills*. New York: Wiley and Sons.

Penfield, W., and Rasmussen, T. (1950). *The Cerebral Cortex of Man: A Clinical Study of Localization of Function*. New York: Macmillan.

Roland, P. E., Larsen, B., Lassen, N. A., and Skinhoj, E. (1980). Supplementary motor area and other cortical areas in organization of voluntary movements in man. *Journal of Neurophysiology, 43*, 118–36.

Sage, G. H. (1984). *Motor Learning and Control: A Neuropsychological Approach*. Dubuque, IA: Wm. C. Brown.

Schmidt, R. A. (1975). A schema theory of discrete motor skill learning. *Psychological Review, 82*, 225–60.

Schmidt, R. A. (1980). Past and future issues in motor programming. *Research Quarterly for Exercise and Sport, 51*, 122–40.

Schmidt, R. A. (1982). More on motor programs. In J. A. S. Kelso (ed.)., *Human Motor Behavior: An Introduction*. Hillsdale, N.J. : L. Erlbaum.

Schmidt, R. A., and Lee, T. D. (2005). Motor Control and Learning: A Behavioral Emphasis. Champaign, IL: Human Kinetics. 4th ed.

Schmidt, R. A., and McCabe, J. F. (1976). Motor program utilization over extended practice. *Journal of Human Movement Studies, 2,* 239–47.

Schmidt, R. A. and White, J. L. (1972). Evidence for an error detection mechanism in motor skills: A test of Adams's closed-loop theory. *Journal of Motor Behavior, 4,* 143–53.

Schoner, G., and Kelso, J. A. S. (1988). Dynamic pattern generation in behavioral and neural systems. *Science, 239,* 1513–20.

Shapiro, D. C., and Schmidt, R. A. (1982). The schema theory: recent evidence and developmental implications. In J. A. S. Kelso and J. E. Clark (eds.), *The Development of Movement Control and Coordination,* 113–50. New York: Wiley.

Shea, C. H., Shebilske, W. L., and Worchel, S. (1993). *Motor Learning and Control,* Englewood Cliffs, NJ: Prentice Hall.

Sherrington, C. S. (1906). *The Integrative Action of the Nervous System.* New Haven: Yale University Press.

Shik, M. L. and Orlovskii, G. N. (1976). Neurophysiology of locomotor automatism. *Physiological Reviews, 56,* 465–501.

Smoll, F. L. (1972). Effects of precision of information feedback upon acquisition of a motor skill. *Research Quarterly, 43,* 489–93.

Sternberg, S. (1969). The discovery of processing stages: Extensions of Donders's method. *Acta Psychologica, 30,* 276–315.

Sternberg, S., Monsell, S., Knoll, R. L., and Wright, C. E. (1978). The latency and duration of rapid movement sequences: comparisons of speech and typewriting. In G. E. Stelmach (ed.), *Information Processing in Motor Control and Learning* (pp. 117–52). New York: Academic.

Taub, E., and Berman, A. J. (1968). Movement and learning in the absence of sensory feedback. In S. J. Freedman (ed.), *The Neuropsychology of Spatially Oriented Behavior.* Homewood, IL: Dorsey.

Taub, E., Perrella, P., and Barro, G. (1973). Behavioral development after forelimb deafferentation on day of birth in monkeys with and without blinding. *Science, 181,* 959–60.

Tresilian, J. R. (1995). Perceptual and cognitive processes in time-to-contact estimation: Analysis of prediction-motion and relative judgment tasks. *Perception and Psychophysics, 57,* 231–45.

Trowbridge, M. H., and Cason, H. (1932). An experimental study of Thorndike's theory of learning. *Journal of General Psychology, 7,* 245–58.

Vallbo, A.B. (1970). Slowly adapting muscle receptors in man. *Acta Physiologica Scandinavica, 78,* 315–33.

Von Holst, E. (1939/1973). Relative coordination as a phenomenon and as a method of analysis of central nervous function. In R. Martin (ed. and trans.), *The Collected Papers of Erich von Holst: Vol. 1. The Behavioral Physiology of Animals and Man* (pp. 33–135). Coral Gables, FL: University of Miami Press. (Original work published 1939.)

Wallace, S. A. (1996). The dynamical pattern perspective of rhythmic

movement: an introduction. In H. N. Zelaznik (ed.), *Advances in Motor Learning and Control,* Champaign, IL: Human Kinetic, 155–94.

Wallace, S. A., and Hagler, R.W. (1979). Knowledge of performance and learning of a closed motor skill. *Research Quarterly, 50,* 265–71.

Wallace, S. A., Stevenson, E., Weeks, D. L., and Kelso, J. A. S. (1992). The perceptual guidance of grasping a moving object. *Human Movement Science, 11,* 691–715.

Woodworth, R. S. (1899). The accuracy of voluntary movement. *Psychological Review 3,* (2, Whole no. 13).

Woodworth, R. S.(1938). *Experimental Psychology.* New York: Holt.

Zanone, P., and Kelso, J. A. S. (1992). Evolution of behavioral attractors with learning: Non-equilibrium phase transitions. *Journal of Experimental Psychology: Human Perception and Performance, 18,* 403–21.

FOOTNOTE

1. While voluntary movement is thought to be initiated by the brain, and reflexive movement initiated by an environmental stimulus initially detected by a receptor, the beginnings of a movement are sometimes difficult to discern. For example, if you are riding a bicycle, both reflexive and voluntary movement pathways are sending impulses at the same time. Thus, the order of activation of the three levels of motor control in voluntary and reflexive movement may alter, depending on the type of movement or environmental context.

Chapter Seven
Psychological Foundations

CHAPTER SEVEN

Psychological Foundations

"The athlete who goes into a contest is a mind-body organism and not merely a physiological machine."

– Coleman Griffith (1925)

STUDENT OBJECTIVES

1. To know the major subdivisions of sport and exercise psychology.
2. To understand the factors influencing one's motivation to participate in physical activity and exercise.
3. To identify the relationship between arousal and performance.
4. To distinguish different types of cognitive strategies used to enhance performance.
5. To understand the problem of adherence to exercise and physical activity.
6. To identify the factors responsible for changes in the stress response.
7. To appreciate the relationship of stress to injury and the psychological factors in injury rehabilitation.

To this point, we have primarily focused on the anatomical, physiological, and biomechanical factors that contribute to human movement and performance. However, as Coleman Griffith (1925) stated, the performer is a mind-body organism and not merely a physiological machine. He was referring to the fact, of course, that *psychological* processes also play a significant role in human performance. In this chapter, we will identify and describe major concepts within the sport and exercise psychology subfield. Next, we explore the various factors contributing to one's motivation to participate in physical activity, exercise, and sport. Psychological factors that influence one's performance will be investigated, such as intrinsic and extrinsic motivation and arousal. Cognitive strategies that can be used to enhance motor performance will then be discussed. Finally, we examine the health psychology area and discuss the research related to the problem of

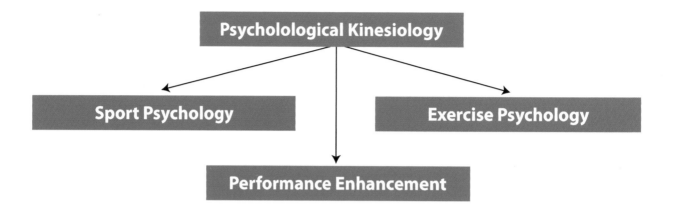

Figure 7.1

The three major subareas of sport and exercise psychology.

adherence to exercise and physical activity programs. In addition, the relationship of psychological stress to injury and the psychological factors influencing injury rehabilitation will be examined.

It should be pointed out that the subfield of sport and exercise psychology focuses primarily on the psychological level of analysis. However, attempts at describing psychological phenomenon often utilize measurements at other levels of analysis, such as physiological (e.g., Dishman, 1994), developmental (Brustad, 1998) and cross-cultural (Duda and Hayashi, 1998). There is also growing appreciation for a multi-level of analysis approach in studying psychological phenomena (Abernethy, Summers, and Ford, 1998; Cacioppo and Berntson, 1992; Kimiecik and Blissmer, 1998; Wiese-Bjornstal and Weiss, 1992).

MAJOR SUBDIVISIONS OF SPORT AND EXERCISE PSYCHOLOGY

The study of psychological phenomena related to human performance, as noted in Chapter 2, began with the studies of Triplett (1897), who noticed that individuals performed better in the presence of other competitors, compared to performing alone. Since that time, most of the work on the role of psychological processes in human performance has been studied within *sports*-related settings. Thus, it was natural to name this subfield of kinesiology "sport psychology," even though debates about this name continue (e.g., Dishman, 1983; Feltz, 1992). Today, because of continuing interest and research in exercise, the subfield is generally referred to as "sport and exercise psychology," even though other physical activity settings have been investigated, such as work and rehabilitation.

The Sport Psychology Area

The **sport psychology area** deals with many social and psychological factors that influence one's performance and behavior. One of the most important fields of study within the social-psychological area is motivation. Factors both internal and external to an individual contributing to the desire *or* the aversion to participate in certain activities or behaviors are examined within this important field (see Weiss and Chaumeton, 1992). The relationship between one's level of arousal (or activation) and performance is another interesting field within the social-psychological field. In the social-psychological area, the interrelated topics of arousal, motivation, anxiety, and performance have also been investigated.

The Performance Enhancement Area

The performance enhancement area focuses on cognitive techniques and strategies for improving one's performance. The influence of **mental imagery** and specific techniques such as mental practice on motor performance enhancement has also been investigated (e.g., see Murphy and Jowdy, 1992, for a review). An interesting issue in this field is whether the effects of mental imagery are similar across different motor skills.

Goal setting in advance of actual performance has been extensively studied in business and academic settings, and recently in motor skills settings. It is generally accepted that the setting of realistic goals enhances physical performance, but there are a number of considerations to be discussed in this chapter regarding the proper utilization of goal-setting procedures.

The Exercise Psychology Area

The health psychology area is involved with psychological factors contributing to one's health, such as exercise adherence and the relationship between stress and injury. The field of exercise adherence examines the factors contributing to one's desire to continue an exercise program, sport, or physical activity. Once one has participated in some physical activity, the chances of becoming physically injured naturally increase. Researchers in the field of stress and injury have uncovered evidence that psychological states of the individual, like one's anxiety level, can predispose one to injury. Once an individual is injured, however, there may also be social-psychological strategies incorporated into one's rehabilitation program to aid recovery.

As we can see, the foundation of sport and exercise psychology is broad, encompassing a wide range of social and psychological phenomena related to physical activity. And many of the areas of study within it overlap with one another, making the investigation of social-psychological phenomena a difficult, but exciting, enterprise. However, researchers have made significant advances in our understanding of the role of social-psychological factors in performance and behavior.

SUMMARY

- The foundation of sport and exercise psychology attempts to understand the sport psychological contributions to performance, exercise, and behavior.
- The major fields of study within sport and exercise psychology are: sport psychology, performance enhancement, and exercise psychology.
- The sport psychology field deals with the many social and psychological factors in performance such as motivation and arousal.
- The performance enhancement field examines various intervention strategies that improve one's performance such as goal setting and mental imagery.

- The exercise psychology field studies a number of psychological factors involved in a person's adherence to an exercise program, sport, or other physical activity. Exercise psychology also investigates the relationship between stress and injury, and attempts to uncover optimal psychological strategies within a physical rehabilitation program.

SPORT-PSYCHOLOGICAL FACTORS IN HUMAN PERFORMANCE

There are many social-psychological influences affecting our beliefs and actions. Shown in Figure 7.2 is a simple model of this relationship. In this model, important social and psychological antecedents to performance are shown as influencing performance or participation in a physical activity. Following the performance or physical activity are various outcomes (e.g., success, failure, or other achievements) that feed back to influence the social or psychological antecedents of the next performance or future participation in some physical activity. One such social-psychological antecedent is motivation, the next topic.

Motivation and Physical Activity

Before engaging in any physical activity, whether it be a 10K race, a swim workout, or a rehabilitative exercise, we must first be motivated. The term "motive" is derived from the Latin word *movere*, which means to move. Psychologists use the term motivation to account for changes in an individual's activity, as well as how long and at what intensity one participates in an activity (McKeachie, Doyle, and Moffett, 1976). Thus, **motivation** implies that the individual has the *desire* to direct his or her behavior in a given direction, with some duration and intensity.

Defining Important Terms Related to Motivation. Before describing various factors that may influence motivation, it is first important to distinguish some terms related to motivation, such as drive, motive, and cues. A **drive** is a physiological need that leads to an aroused state *but with no directionality*, whereas a **motive** is the arousal of an individual to strive for some goal (Tolman, 1951). A newborn baby is thought to be initially equipped with drives but no motives (McKeachie, Doyle, and Moffett, 1976). Certain **drive cues** or stimuli, such as hunger or thirst, allow the baby to know it is hungry or thirsty. But the baby does not really know what to do about it. Eating food and drinking water are pleasant experiences the baby eventually learns to expect. With experience, these expectations become motives that arouse the baby to strive for the goal of satisfying its needs for food or

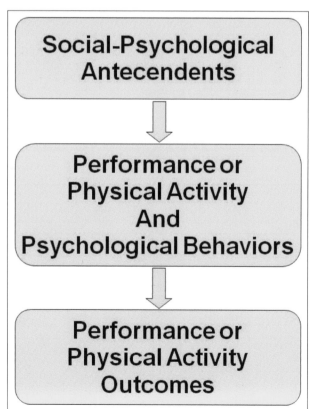

Figure 7.2

The relationship between social-psychological antecedents, physical activity, and physical activity outcomes.

water. Thus, when we observe an aroused individual in the process of striving for a goal, we may say that the individual is motivated.

Types of Motives. As we grow older, a number of motives may serve to change and guide our behavior. Figure 7.3 illustrates one theorist's view (Maslow, 1954) of the types of motives and their relative importance. Maslow identified survival and safety, affiliation, self-esteem, and competency as four basic motives and arranged them within a pyramid of relative importance. **Survival motives** are related to satisfying basic physiological needs, such as hunger and thirst. **Safety motives** involve the avoidance of dangerous and life-threatening situations. **Belongingness and love (affiliation) motives** have to do with our desire for affection from our parents, relatives, friends, or children. **Self-esteem motives** provide a means for the individual to improve his or her self-image in order to feel capable and gain respect from others. Finally, **competence motives** are related to our desire to improve the skills needed to successfully cope in the environment.

At the base of the pyramid are survival motives. Maslow suggested, for example, that a hungry and thirsty child, driven by a survival motive, is less likely to want to play with her friends, the latter behavior requiring an affiliation motive. Furthermore, when all the other motives are satisfied, an individual is then more likely to engage in an activity that will improve their competency skills. While other researchers have provided arguments against Maslow's hierarchy of motives (e.g., Dember, 1974), the model provides one way to identify or classify the various motives affecting behavior.

Factors Related to Participation in Physical Activity. Maslow's hierarchical model allows us to speculate about the above motive's contribution to our desire to participate in physical activity. The *type* of physical activity is likely to require certain motives. For example, a rampant car headed straight for you as you are walking on a sidewalk serves as a cue to invoke a survival motive. As a result, you immediately stop walking and run to safety. A belongingness motive is probably required for you to successfully engage in a team sport. Without it, you are less likely to want to participate in an activity that demands interactions with teammates. Self-esteem motives help direct you to physical activities that improve your self-image. You are more likely to choose physical activities that make you feel good about yourself. Finally, competence motives are required if you wish to try to improve your performance in a motor skill or learn a new skill. With this basic terminology covered, we are now ready to explore the various factors related to the participation in exercise, sport, and related physical activities.

Researchers in psychology and sport and exercise psychology have determined that there are many motives related to participation in sports, exercise, and rehabilitation programs. These motives may interact with each other and with other influences in peculiar and complex ways. Kenyon (1968 a, b) provided one of the first classifications of motives for why people engage in physical activity:

Physical activity as a social experience. Individuals participate in certain forms of physical activity, such as exercise and sports programs, for social interaction (e.g., to make new friends).

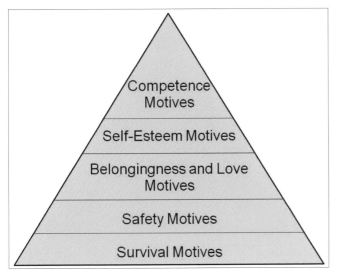

Figure 7.3

Maslow's hierarchy of motives, (adapted from Maslow, 1954).

Physical activity for health and fitness. People often choose to participate in exercise programs to improve their muscular strength or cardiovascular endurance, for example.

Physical activity for the thrill. Some individuals enjoy participating in activities that involve risk, danger, or high speed, such as bungee jumping, mountain climbing, or downhill skiing.

Physical activity as an aesthetic experience. Individuals who perform activities for their beauty, grace, or artistic quality, such as ballet, gymnastics, or synchronized swimming, do so for aesthetic reasons.

Physical activity as a catharsis. Some individuals participate in certain forms of physical activity to release tension.

Physical activity as an ascetic experience. Some individuals enjoy participating in long, strenuous, and painful activities such as marathons and triathlons.

Since Kenyon's classification, some researchers have found differences in motives for participating in physical activity among several groups of individuals, such as exercisers versus non-exercisers (Brunner, 1969), and males versus females (Biddle and Bailey, 1985). For example, Brunner (1969) found that motives to participate in physical activity for physically active men were to achieve physical fitness, feel better physically and mentally, and attain weight control. The inactive group listed relaxation, feel better, fun and enjoyment, and being outdoors as the motives to exercise. A study by Biddle and Bailey (1985) found that men were more motivated to participate in activities involving competition, whereas women favored aesthetic and cathartic activities. It appears from the literature that people have different motives for participation in physical activity.

Another major factor affecting one's desire to participate in physical activity is the **motivational orientation** of the individual (Weiss and Chaumeton, 1992). An individual may have either an extrinsic or intrinsic motivational orientation or disposition. Individuals with an *extrinsic motivational orientation* tend to want to participate in a physical activity to win (e.g., beat others), to gain rewards (e.g., money, trophies), or to receive social approval for their performance. These individuals are likely to choose competitive activities that are either easy or very difficult, such that they can increase the likelihood of winning (or beating an opponent) or gain social approval for their performance. Extrinsically oriented individuals tend to compare their performance outcomes to others. However, their perceived competence or beliefs about their own abilities may actually be quite poor. Additionally, they tend to have positive or pleasant feelings about their performance only by winning or beating an opponent. Sport psychologists generally feel that extrinsically oriented individuals have a stronger tendency to lose interest in or discontinue (drop out) a physical activity.

In contrast, individuals with an *intrinsic motivational orientation* tend to be interested in participating in a physical activity primarily to develop competence (e.g., improve skill), to have fun with others, or to improve fitness. Intrinsically motivated individuals will choose challenging but realistic activities. They are primarily interested in comparing their performance with previously set personal standards to improve their competence at tasks. These types of individuals generally have high perception of their competence with a strong sense of internal control. They tend to feel good about themselves regardless of whether they win or lose in a competitive situation. Sport psychologists generally feel that intrinsically motivated individuals are more likely to persist in a physical activity.

What factors predispose an individual for a particular motivational orientation? In other words, why do some individuals become extrinsically motivated, while others become intrinsically motivated to perform a particular task or physical activity? Researchers believe that the *motivational climate* of the individual plays an important role in shaping an individual's disposition for a particular motivational orientation (e.g., Roberts, 1993). The motivational climate, particularly of children, includes such influences as parents', teachers', and coaches' attitudes and corresponding reinforcement behaviors. For example, it is thought that if the parents, teachers, or coaches reinforce competitive goals such as winning or beating others, it is more likely the individual will adopt an extrinsic motivational orientation. Whereas, if the motivational climate predominantly emphasizes skill mastery, the child is likely to adopt an intrinsic motivational orientation (Ames, 1992). Duda (1993) has done extensive research examining goals from a social-cognitive perspective. Readers are encouraged to examine this work.

Finally, there is some evidence that the developmental age of the individual influences motivational orientation (Nicholls, 1978). Young children up to the age of approximately 12 years old apparently have difficulty distinguishing between the concept of effort and ability. This finding means young children believe that the successful completion of difficult tasks requires greater effort, and greater effort demonstrates greater ability. In essence, the child believes anyone can successfully complete a task if they simply try harder. The child has no concept that people can actually differ in ability. What this means is that up to the age of 12 or so, the child is more inclined to possess an intrinsic motivational orientation, one which predisposes the child to focus on effort and skill development rather than winning or beating a competitor. For reasons that are still unclear, the child becomes more susceptible to the extrinsic motivational orientation after the age of 12. It is not surprising that sport psychologists believe that the goal of the parent, teacher, and coach should be to develop a motivational climate that fosters the continued growth of an intrinsic motivational orientation within the child (e.g., Roberts, 1993; Weiss and Chaumeton, 1992). It is believed that this type of orientation will allow the child to persist longer at the physical activity, and adopt certain physical activities for the sake of skill improvement, fitness, and fun.

The Concept of "Flow"

You probably have heard the old adage about mountain climbing. When asking a climber why he wants to climb a mountain, the response is "Because it is there." From time to time, most of us have engaged in activity just for the sake of doing the activity. That is, we are not concerned about how others feel about what we are doing. We do not particularly care about the consequences of engaging in the activity. We are simply enjoying the moment. Sailors call this state of mind "in the groove," when everything seems to be going well, the sails are trimmed perfectly, the course is set, and the sea breeze is blowing comfortably in the sailor's face. The more scholarly term for this state of mind is called **flow**, a concept investigated by the psychologist Mihaly Csikszentmihalyi (pronounced *chick-saint-ma-hi*). In this section, we discuss the concept of flow and how flow might affect participation in physical activity.

Technology has increased leisure time and efficiency in the United States and other nations (see Chapter 9 for a further discussion). But the freeing up of more leisure time does not necessarily enhance enjoyment or the quality of life. Many Americans, young and old, use their leisure engaged in pleasureful activities. According to Csikszentmihalyi (1990), **pleasureful activities** are those which create feelings of contentment when biological and social conditions are met. Activities such as eating, resting, sex, and the use of drugs can bring on these feelings. But pleasureful activities do not ultimately bring happiness or personal growth. Often, leisure activities provide a relaxing reprise from the rigors of work but leisure activities usually consist of the passive absorption of information. In 1966 Americans watched an average of one hour and thirty minutes of television per day. By 1983 television viewing had increased to six hours and thirty minutes! By 2008, American households were watching TV over eight hours a day (*Los Angeles Times*, 2008)! Next to sleeping, television viewing occupies the greatest amount of children's leisure time (Dietz, 1990). It is safe to say that most television viewing involves only the passive intake of information accompanied by little personal growth.

According to Csikszentmihalyi, television viewing may bring pleasure but rarely enjoyment, with enjoyable activities being defined as those that 1) provide a sense of novelty and accomplishment in improving ability or skill; and 2) require mental energy. Other elements of enjoyable activities are the following:

- tasks that we have a chance of completing.
- tasks with clear goals.
- tasks that provide positive feedback.
- tasks in which the performer has a deep but effortless involvement in the absence of worry or frustration.
- tasks in which the performer has a sense of self-control over his or her actions.
- tasks in which the performer's concern for oneself disappears, and the performer feels mentally stronger after the activity.
- tasks in which the sense of time is lost.

When one engages in an activity or task under the above conditions, a mental state called "flow" is experienced, according to Csikszentmihalyi. In addition, virtually any activity can be enjoyed and flow can be experienced. But it is important to emphasize that flow and personal growth are intimately related. As one experiences flow in an activity, it is likely that the individual's skill level will increase. However, if the individual performs the task without new challenges, boredom may set in and possibly lead to a discontinuance of the task. On the other hand, if the individual's skill level does not improve as the task becomes more challenging, anxiety or frustration may emerge, also leading to a halt in personal growth (see HIGHLIGHT). It is only through experiencing flow that real enjoyment of the task materializes—a positive mental state that allows for personal growth without the requirement of extrinsic reward. If Csikszentmihalyi is correct, the lesson to be learned is that the benefits of participating in physical activities can be best achieved when one engages in them for enjoyment. Jackson and Csikszentmihalyi (1999) add some new interpretations to the concept of flow in sport that the reader is encouraged to examine.

HIGHLIGHT

Flow and Physical Performance

How is flow experienced as one is engaged in physical performance? Csikszentmihalyi has studied a wide variety of people who have had such experiences. These are some quotes from his book *Flow: The Psychology of Optimal Experience* (1990) that provide some insight into this phenomenon. From a mountain climber: "When you're [climbing] you're not aware of other problematic life situations. It becomes a world unto its own, significant only to itself. It's a concentration thing. Once you're into the situation, it's incredibly real, and you're very much in charge of it. It becomes your total world." From another: "It is as if my memory input has been cut off. All I can remember is the last thirty seconds, and all I can think ahead is the next five minutes." A dancer relates: "Your concentration is very complete. Your mind isn't wandering, you are not thinking about anything else; you are totally involved in what you are doing … Your energy is flowing very smoothly. You feel relaxed, comfortable, and energetic."

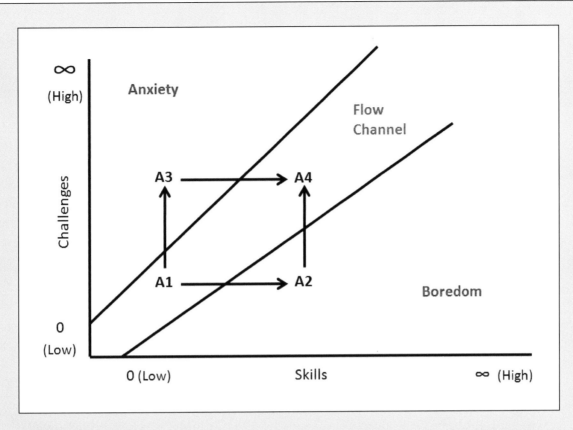

Figure 7.4

The relationship between the challenges inherent in a skill or activity and the skill level of the participant. Boredom results when skill level is greater that the task challenge, and anxiety results when the skill level fails to meet the task challenge. Note that "flow" exists only when the skill level balances task challenge (adapted from Csikszentmihalyi, 1990, figure on p. 74).

How is flow related to performance achievement? Figure 7.4 illustrates that only when an individual's skill level balances the challenges of the task can flow be experienced, which then leads the person to higher levels of personal growth and achievement. Let us say that an individual is engaged in learning a motor skill like tennis or golf. In condition A1, the person is at a relatively low level of skill, but the challenge of the task is low enough to allow the individual to experience flow, shown by the white portion of the graph. Because the individual is in the flow channel, the person is enjoying the task. If the person continues to practice, skill level will likely increase, but if the challenge of the task remains the same, boredom will set in (A2). To alleviate boredom, the person must increase the challenge in the task, such as by competing against a superior player. On the other hand, if at condition A1 the person attempts to perform at a more challenging level with the same skill, anxiety or frustration may set it. Both anxiety and boredom are not enjoyable states, and the person will either be motivated to return to the flow state or discontinue the activity. If it is decided to return to the flow state, the person must reach a new flow condition (A4) obviously at a different level than A1. As the person moves from one flow state to the other, personal growth is increased.

One of the interesting but as yet unanswered questions is how many elements of flow need to be met before flow can be realized (Kimiecik and Stein, 1992). Do *all* elements for flow mentioned previously have to be experienced, or will a smaller subset of elements do? Is there a relationship between the individual personality and the situation that best regulates the flow experience? Are some sports and physical activities more conducive to flow than others? How might sport and exercise psychologists assist the athlete to enhance the likelihood of flow experiences? More research is needed to answer these questions.

Table 7.1

Various Measures of Arousal

Psychological	Physiological	Biochemical
– Questionnaires	– EEG changes in brain wave patterns – Perspiration – Heart rate – Blood pressure – Muscle activity	– Hormonal changes in epinephrine,-norepineph-rine, and cortisol levels

Arousal and Motor Performance

Another major factor affecting performance is the arousal level of the individual. Arousal is viewed as an energizing function responsible for harnessing the body's resources for intense and vigorous activity (Sage, 1984). The arousal level of the individual can vary from that associated with deep sleep to intense excitement. Arousal is thought to represent the general activation of the nervous system and is controlled by several parts of the nervous system, including the cerebral cortex, the hypothalamus, and the reticular formation in the brain. These parts of the brain help to control the release of certain hormones into the bloodstream that either increase or decrease the activity of various physiological and psychological processes. As a result, the arousal level of the individual can be measured using both physiological and psychological instruments. We will first examine these measures and then attempt to answer the following questions. Is there an optimum level of arousal for any given task? Does the optimum level of arousal differ depending on the task?

Measures of Arousal. Before discussing the various measures of arousal, it is first necessary to distinguish arousal from two associated concepts: motivation and anxiety. We have previously defined motivation as the desire to direct an individual's behavior in a given direction, with some duration and intensity. Therefore, motivation can be thought of as a positive cognitive state in the sense of the individual's desire to want to perform a certain activity. On the other hand, anxiety is a negative cognitive state associated with the reluctance to perform a certain activity. For example, a basketball player may, in one situation, be motivated to make a foul shot during a game, while in another situation, may be anxious. In the first case, the player is looking forward to making the shot but in the second case, the player may be afraid or worried that the shot will be missed. The degree or intensity of the individual's motivation or anxiety is thought to represent the player's arousal level.

An analogy to an automobile's engine might be helpful in understanding the distinction between motivation, anxiety, and arousal (Martens, 1974). The speed of the automobile's engine can vary between slow and fast while it's in the idle position; however, the automobile does not go anywhere! It is only when the automobile is put into some gear that it begins to move. If put into drive, the car goes forward, and if put into reverse, the car goes backward. Thus, the car going forward is analogous to motivation, and the car going backward is analogous to anxiety. The speed with which the car goes in one of these directions is analogous to arousal. Thus,

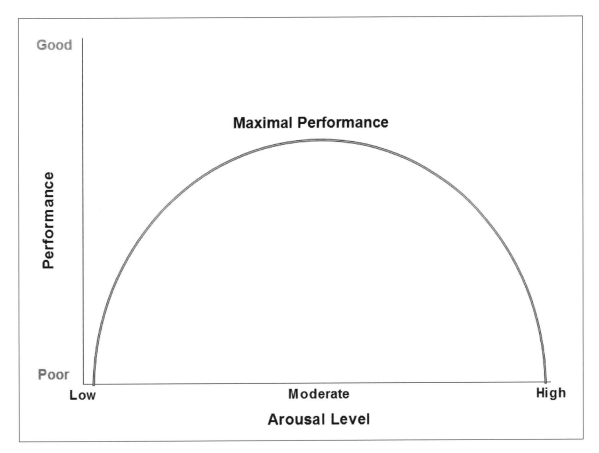

Figure 7.5

The "inverted-U" relationship between arousal and performance.

the engine's speed is similar to the overall level of activation of the nervous system or arousal. Given this analogy, it is easier to understand that the physiological effects of arousal can be similarly manifested in both motivating and anxious situations.

When one is aroused either in a motivated or anxious state, a variety of physiological responses may result that appear to vary across individuals (see Table 7.1). In one individual, an increase in motivation or anxiety might be seen in an elevated heart rate. In another, changes may be seen in muscular tension. For this reason, many researchers believe it is advisable to take multiple measures of arousal (e.g., Duffy, 1962). Biochemical measures can reveal hormonal changes as a result of arousal, but these tend to be expensive, time consuming to analyze. and invasive to subjects. Because of the lack of agreement among physiological measures and the impracticality of biochemical measures, some sport psychologists have turned to cognitive measures of arousal by using a variety of validated questionnaires that are completed by subjects and later analyzed by the researcher (see Landers and Boutcher, 1993, for a discussion). These methods have some advantages because they are relatively easy to administer and analyze. But they also may be susceptible to subject bias, such as the subject completing the questionnaire in a manner agreeable to the sport psychologist.

In summary, arousal is thought to represent the individual's general level of activation present during motivating or stressful situations. Arousal can be measured in a variety of ways, including physiological and psychological. Patterns of changes in arousal tend to vary from individual to individual.

The Relationship of Arousal to Performance. It is now appropriate to ask the following question: "How is performance affected by the level of arousal for any given task?" In attempting to answer this question, researchers have had to figure out ways to manipulate the level of arousal in the individual or to observe individuals in different

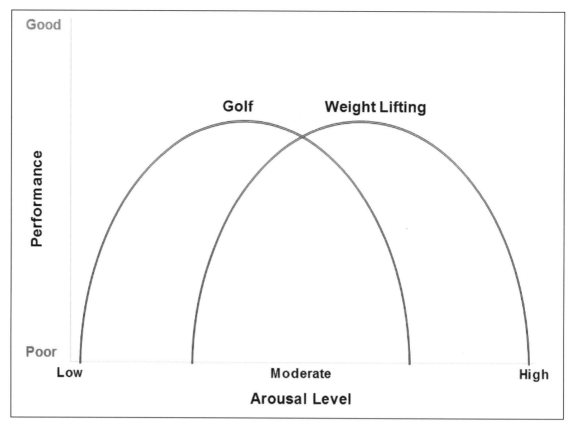

Figure 7.6

A hypothesized inverted-U relationship in two types of physical activities. Notice that the optimal level of arousal in weight lifting is higher compared to golf (adapted from Landers and Boutcher, 1993).

arousal settings. One way to manipulate arousal is to require individuals to exercise at different intensities (Levitt and Gutin, 1971). As the intensity of exercise increases, it is thought that the level of arousal also increases. Observing individuals in different arousal states has also been done, such as examining the relationship between arousal and performance during competitive athletic events (e.g., Gould, Petlichkoff, Simons, and Vevera, 1987; Burton, 1988; Klavora, 1979; Sonstroem and Bernardo, 1982).

Some of the research on arousal has shown what is known as an "inverted-U" relationship between arousal and performance. Figure 7.5 illustrates that performance is worst at low and high levels of arousal and best at medium or optimal levels of arousal. In other words, when arousal is too low, performance is poor; when arousal is too high, performance also suffers. Best performance is expected at more moderate or optimal levels of arousal. Results such as these have been found in both experimental (laboratory) situations (e.g., Martens and Landers, 1970; Levitt and Gutin, 1971) as well as in field settings or so-called "real-life" situations (e.g., Klavora, 1979; Sonstroem and Bernardo, 1982).

The optimal level of arousal may depend on the task (see Landers and Boutcher, 1993 for a discussion). For some tasks, such as putting in golf, the optimal level of arousal required for good performance is thought to be quite low. Whereas in other tasks such as weight lifting, the optimal level of arousal is thought to be much higher (see Figure 7.6). One explanation for this view is that tasks involving precise motor skill such as golf putting cannot tolerate unwanted muscle activity that can be brought on by high levels of arousal. However, tasks that involve intense activation of muscle contraction, such as weight lifting, require much higher levels of

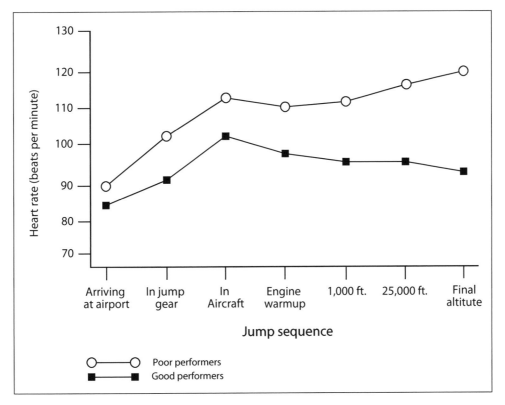

Figure 7.7

Changes in arousal (as measured by heart rate) in novice and experienced parachute jumpers. Notice that arousal increases in both skill types until entry into the aircraft. Subsequently, arousal continues to increase in novice jumpers but actually *decreases* in experienced jumpers as the aircraft ascends to its final altitude (adapted from Fenz and Jones, 1972).

arousal. There are a variety of psychological techniques that performers can use to modify their levels of arousal to fit the demands of different tasks.

Challenges to the Inverted-U Model. Is performance always related to anxiety in an "inverted-U" manner? Since the advent of the inverted-U model, much research has not shown an inverted-U relationship between anxiety and performance (see Hardy, 1996, for a review). One of the problems of the inverted-U model is that it does not take into account two important factors. One, the optimal level of arousal appears to be different for different individuals performing under different situations. Two, there are different components to anxiety, and each may have different effects on performance. Regarding the first factor, individuals may have particular "zones of optimal functioning" (Hanin, 1980) that characterize an individual's optimal level of anxiety related to peak performance. Some individuals may perform best at either low, moderate, or high levels of anxiety, contrary to the inverted-U model. Regarding the second factor, researchers have suggested that anxiety has at least two dimensions: cognitive and physiological anxiety. **Cognitive anxiety** refers to an individual's negative expectations and potential consequences about oneself in a given situation. **Physiological anxiety** is the physiological response of the individual (e.g., heart rate) in different performance situations. It is likely that the relationship between each of these types of anxiety to performance is different. Consequently, a number of theories have been proposed to account for the complex interaction between cognitive and physiological anxiety to performance that have been empirically studied. The reader is encouraged to examine Hardy (1996) for a discussion of this literature.

The Level of Skill and Arousal. There is interesting evidence that changes in arousal are affected by the skill level of the individual. Early studies by Fenz and coworkers on parachute jumping have shown that the pattern of arousal changes differs between novices and experts in anticipation of a jump. Figure 7.7 shows that arousal, as measured by changes in heart rate, similarly increased in novice and good jumpers from the point of arriving at the airport until the time of entering the plane. However, once in the plane, the heart rates of the novices continued to increase until the time of jump, whereas the heart rates of the good jumpers actually *decreased* to near-moderate levels, as predicted by the inverted-U theory. Other studies by Fenz showed that under stressful conditions, the heart rates of even good jumpers can approach those of novices. These results indicated that the skill level of the individual can influence the patterns of change in arousal.

SUMMARY

- There are several types of motives that serve to change or guide our behavior.
- There are several classifications of motives for participation in physical activity.
- The motivational orientation of the individual plays a major role in one's participation in physical activity.
- "Flow" is a positive mental condition accompanying participation in an activity in which the challenges of the skill are matched by the required skill level of the participant.
- The enjoyment of an activity is highest during a flow state.
- Adherence to an activity may be dependent on enjoyment.
- Arousal is the general activation of the nervous system.
- There are a variety of psychological and physiological measures of arousal.
- Motivation and stress provide the direction to behavior and arousal determines the intensity of the behavior.
- There is some research supporting an inverted-U relationship between arousal and performance. However, other relationships between arousal and performance have been shown.
- The optimal level of arousal is different depending on the individual and the task.
- Skill level appears to affect the pattern of arousal changes.

PERFORMANCE ENHANCEMENT USING COGNITIVE STRATEGIES

Improvements in motor or psychological skills can be accomplished in a variety of ways. As pointed out in Chapter 4, proper exercise training can facilitate improvements in cardiovascular endurance that allow one to run a faster 1500-meter race or lift a greater weight in strength training. Receiving the appropriate knowledge of results or knowledge of performance can facilitate the performance and learning of motor skills, as indicated in Chapter 6. Performance can also be enhanced through the use of cognitive strategies on the part of the performer. There are numerous psychological variables that have been shown to enhance performance. In this section, we will examine two major cognitive strategies that have been shown to influence performance: mental preparation and goal setting.

Mental Preparation

We have all probably watched elite athletes engaged in intense mental concentration prior to a performance. Athletes report that this type of mental concentration generally involves the use of different types of what may be called mental preparation. The cognitive activity of mentally preparing for a skill prior to execution is quite common in competitive athletics. For example, nearly 99 percent of the Canadian Olympic athletes participating in the 1984 Olympic games reported using mental imagery, one form of mental preparation

(Orlick and Partington, 1988). Nearly 70 percent of a sample of 587 elite track and field athletes reported the routine use of mental preparation techniques (Ungerleider, Golding, Porter, and Foster, 1989). In fact, many different kinds of elite athletes are known to have regularly used some form of mental practice in their training (see HIGHLIGHT).

HIGHLIGHT

Mental Imagery: A Common Form of Mental Preparation in Sport

What did Greg Louganis, Dwight Stones, Jack Nicklaus, Chris Evert, Fran Tarkenton, and Jean-Claude Killy have in common? They were all elite athletes who stated they regularly used some form of mental imagery to aid their performances. A good example of mental imagery is often observed with competitive divers. The following quote was taken from a Canadian Olympic diver:

"I did the dives in my head all the time. At night, before going to sleep, I always did my dives. Ten dives. I started with a front dive, the first one that I had to do at the Olympics, and I did everything as if I was actually there. I saw myself on the board with the same bathing suit. Everything was the same … If the dive was wrong, I went back and started over again. It takes a good hour to do perfect imagery of all my dives, but for me it was better than a workout. Sometimes I would take the weekend off and do imagery five times a day." (Orlick and Partington, 1988, p. 112).

One advantage of mental imagery training is that it can be done in several settings, not just on the playing field, court, or swimming pool. Jack Nicklaus, in his 1974 book *Golf My Way*, wrote that he used (and presumably still uses) mental imagery before each golf shot. Prior to each shot, Nicklaus would mentally picture his desired swing, the trajectory of the ball's flight to the target area, and the ball's desired destination. His intense mental concentration has been thought to be one of the important distinguishing features of his game. Research suggests that mental practice is better than no practice at all, and that a combination of mental and physical practice is best for improving performance (Murphy and Jowdy, 1992 [see text for details]).

Before discussing the effectiveness of the mental preparation for performance, it is necessary to distinguish between two commonly used terms: imagery and mental practice.

Imagery is the mental technique of imagining situations to improve performance (Block, 1981). The imagined situation may or may not be specific to the motor skill to be performed. For example, *nonspecific imagery*, such as imagining a peaceful scene (e.g., a waterfall, lying on a beach, etc.) can be used to help an individual relax prior to an actual performance. Nonspecific imagery can be used, for example, to reduce one's heart rate or relax the muscles prior to performance (Harris and Williams, 1992). *Specific imagery* is imagery of the actual motor skill. This type of imagery can be either internal or external (Mahoney and Avener, 1977). *Internal imagery* means that the individual visually imagines an object of attention (e.g., golf ball, ski slope, etc.) but not his or her own body. *External imagery* means the individual visualizes their own body performing the activity. Thus, imagery can be either specific or nonspecific to the motor skill to be performed. If it is specific to the motor skill, mental imagery can be either internal or external in nature.

Mental practice, on the other hand, is any type of cognitive rehearsal of the activity in advance of it actually being performed. Mental practice, may or may not involve imagery. For example, an individual might mentally rehearse the sequence of events of a tennis serve (e.g., correct body position, ball toss, backswing, follow-through) without actual internal or external imagery. This type of mental rehearsal is primarily through the use of overt (out loud) or covert (internalized) verbal behavior. Martina Navratilova, a great women's tennis player, could be often heard actually yelling at herself after a lost point, such as "Come on, follow through!" This type of overt verbal behavior could serve as a type of overt mental practice that helps her concentrate on a certain aspect of the skill to be subsequently executed. But mental practice can also involve the use of imagery such as imagining oneself successfully negotiating a ski slope or correctly hitting a golf ball (see HIGHLIGHT).

In sum, the two major types of mental preparation involve imagery and mental practice. It is possible to use imagery with or without mental practice, and it is also possible to use mental practice with or without imagery.

Evidence for Imagery and Mental Practice. Most of the research on mental preparation and its effects on performance has been done with participants using mental practice with some form of imagery. Using a meta-analysis technique, Feltz and Landers (1983) examined the results of 98 mental practice studies. One of their major conclusions corroborated by other comprehensive reviews (e.g., Suinn, 1993; Weinberg, 1981) was that mental practice is better than no practice at all. Thus, in the absence of actual physical practice, individuals can still expect benefits from mentally practicing a motor skill. Another conclusion reached by Feltz and Landers, and later by Feltz, Landers, and Becker (1988), was that mental practice better benefits those skills containing larger *cognitive* demands.

Theories of Imagery and Mental Practice. Why does imagery and mental practice work? Several theories have been constructed to help answer this question (see Suinn, 1993). Two of the earlier theories are the psychoneuromuscular theory and the symbolic learning theory. The **psychoneuromuscular theory** states that during imagery or mental practice, the muscles to be used in the actual performance are slightly innervated (or stimulated). Thus, this theory argues that mental practice is not completely mental! The idea is that the slight muscle innervation during imagery mimics the innervation during a subsequent motor performance. The muscle feedback created by this slight innervation might also be used in some way to help future performance. Some evidence suggests that during mental practice, small amounts of muscle contraction can be seen in the muscles used in the imagined movements with the help of electromyography (Jacobson, 1930; Harris and Robinson, 1986; Suinn, 1980; Wehner, Vogt, and Stadler, 1984).

The **symbolic learning theory** argues that mental practice allows the performer to prepare for the various elements or possible sequences of movements involved in the actual performance. During mental practice, the individual can mentally rehearse the sequential aspects of the task, task goals, and the movements to be performed in the actual performance. This theory is supported by evidence that shows more benefits of mental practice in tasks that require a large amount of cognitive activity (Ryan and Simons, 1981; Hird, Landers, Thomas, and Horan, 1991; Wrisberg and Ragsdale, 1979).

In sum, psychoneuromuscular theory, symbolic learning, and other theories of mental practice (see Murphy and Jowdy, 199,2 and Cox, 1994) have some empirical support. However, no single theory seems to provide a comprehensive explanation of the mental practice effect at this time. The reader is encouraged to examine other theories that have been proposed (Suinn, 1993).

Goal Setting

What might our objectives be in participating in an exercise, sport or rehabilitation program? Perhaps we want to improve our cardiovascular conditioning or muscular strength. We may wish to learn how to play golf or tennis. After a knee injury, we may want to rehabilitate the injured knee and restore the knee to normal function. All of these objectives might be better reached using a psychological strategy called **goal setting**. A formal definition of goal setting is "attaining a specific standard of proficiency on a task, usually within a specified time limit" (Locke, Shaw, Saari, and Latham, 1981, pg. 145). In this section, the purpose will be to identify different types of goals, to briefly overview research in goal setting, to examine some theoretical explanations of how goal setting may work, and finally, to describe some guidelines on how goal setting may be used in physical activity.

Types of Goals. When discussing goal setting, it is important to understand that there are different types of goals related to participation in physical activity. One distinction that has been made is the difference between general and specific objective goals (McClements, 1982). **General goals** are similar to those mentioned above, such as the desire to become more cardiovascularly fit or the desire to improve muscular strength. **Specific goals** involve more detail regarding behavioral objectives and the time period. A specific goal might be to improve one's best 1500-meter running time by 30 seconds within a six-month training period. Another way to characterize goals is in terms of goal orientation. Goals can be either outcome- or performance-oriented. An **outcome goal** involves the desire to achieve an outcome of a skill or behavior. An example of an outcome goal is to win a set in tennis against an opponent. A **performance goal** is related to the desire to execute a behavior or set of behaviors regardless of the overall outcome of the skill. An example of a performance goal is to get 80 percent of one's first tennis serves in during a set against an opponent. Additionally, performance goals are usually associated with comparing one's performance to one's *previous* level of performance. As such, performance goals are often related to intrinsic motivation.

Research Findings in Goal Setting. Much of the research on goal setting has been conducted by Locke and his co-investigators in the academic, business, and industrial settings (e.g., Locke et al., 1981). Some of the results have been applied to sports and physical activity settings, although much more research is needed. First of all, 90 percent of approximately 400 studies have shown positive or partially positive effects of goal setting. Generally, the studies indicate that goal setting is a very powerful technique to enhance performance. Research has also showed the benefits of goal setting for sport performance (e.g., Burton, 1989; Kyllo and Landers, 1995) even though some negative findings have also been shown (see Gould, 1993, and Cox, 1994, for a discussion). A statistical evaluation of 36 goal-setting studies in sport activities revealed strong support that goal setting is better than having no goals at all or performing under instructions to do one's best (Kyllo and Lander, 1995).

Explanations of Goal-Setting Effects. Several theoretical explanations of the relation between goal setting and performance have been offered (Gould, 1993). It should be pointed out that much more research is needed to verify and validate the following explanations:

1. *Goals may direct the performer's attention to important aspects of the task.* For example, a golfer might set a goal of eliminating a nasty slice and proceed to focus his or her attention on various methods to do so.
2. *Goals can help mobilize a performer's effort.* By setting a goal, the performer can direct his or her energies in a given direction.
3. *Goals can help increase one's persistence on a task.* Without a goal, the performer is more inclined to lose interest in an activity. Without interest, it is difficult or impossible to continue improving on a skill.
4. *Goals may facilitate the development of new learning strategies.* By having certain goals, the learner is likely to explore new strategies for learning.

Outcome goals may not be effective, because the outcome of a task may be beyond the performer's control. Having a goal such as "win the match" may not be effective, because much of the final outcome depends on the performance of the competitor and is beyond the complete control of the individual. Failure to achieve the outcome can thus lead to lower levels of confidence, cognitive anxiety, decreased effort, and possibly poor future performance.

Guidelines for Goal-Setting. It is difficult to provide accepted guidelines for goal setting because 1) goal-setting effects probably interact with many other variables such as the qualities of the individual performer (age, skill level, etc.) and the different environmental and task conditions; and 2) much more research is needed to tease out the complex interactions between goal setting and other variables. However, some guidelines suggested by Gould (1993) are presented, as well as the findings of a statistical review of 36 goal-setting studies in sport activities (Kyllo and Landers, 1995).

Prominent sport and exercise psychologist Daniel Gould offered some suggestions regarding the use of goal setting (Gould, 1993):

1. *Set specific goals in measurable and behavioral terms.* General goals are less effective than specific goals. It is helpful if the goals can be quantifiable, allowing the performer to more easily determine if the goals are met.

2. *Set difficult but realistic goals.* Goals that are difficult but achievable are more effective than easy goals, probably because they provide more of a challenge to the individual.

3. *Set short-term as well as long-term goals.* Short-term goals are important, because they allow the performer to experience early success, thereby enhancing motivation. Long-term goals are important for providing the overall direction to performance improvement.

4. *Set performance goals as opposed to outcome goals.* Performance goals are more under the performer's control. In addition, failure to achieve an outcome goal (e.g., winning every game) may result in failure to reset future goals and a lack of motivation to improve future performance.

5. *Set goals for practice and competition.* Performers (certainly athletes) spend most of their time practicing for some type of competition. Therefore, it makes sense to set goals for practice as well as for the competition. Some common practice goals are arriving to practice on time, maximizing effort on all drills, and paying attention to the coach or teacher.

6. *Set positive goals, not negative goals.* Try to identify the behaviors to be achieved, not the behaviors to be avoided, whenever possible. This approach allows the performer to focus on success, as opposed to failure.

7. *Identify target dates for attaining goals.* Stating goals with a clear target date helps to motivate the performer to meet objectives in a realistic time period.

8. *Identify goal achievement strategies.* It is just as important to have a *way* of achieving a goal as it is to have goal. Exactly how will the performer attempt to obtain a goal? Will it be by performing more drills, developing mental practice strategies, and so forth?

9. *Record goals once they have been identified.* Having written goals in a place that can be routinely observed (e.g., on the performer's bedroom door, inside a locker, etc.) constantly reminds the performer of what the goals are for the activity.

10. *Provide for goal evaluation.* Performance feedback from a coach or teacher can help the performer determine how close he or she is to obtaining the desired goal.

11. *Provide support for goals.* A goal-setting program will be facilitated if the significant others in the performer's life are involved, such as teacher, coach, or family. It is recommended that the significant others understand and support the performer's goals.

Some of these suggested guidelines have been statistically evaluated by
Kyllo and Lander (1995). Using a procedure called meta-analysis, which summarizes the effects of variables across many studies, Kyllo and Landers came to the following conclusions:

1. Goal setting is better for enhancing performance compared to no goal setting, or to situations where the performer is simply told to do his or her best.

2. There is no support for the view that setting difficult goals is better than setting either easy goals or no goals. Moderately difficult goals seem to lead to better performance.

3. Setting absolute goals, where all group members work toward the same goal, seems to be better than setting relative goals that are based on individual performance.

4. There is weak experimental support that performance goals are effective, even though many sport and exercise psychologists tend to favor performance goals over outcome goals.

5. Setting a *combination* of short- and long-term goals results in the best performance.

6. Sport performance is enhanced more if the performer is allowed to set his or her own goals, as opposed to being assigned goals from someone else.

Kyllo and Lander (1995) discuss several theoretical reasons for these results. The reader is encouraged to examine their paper for more details.

In summary, there appears to be experimental support for the benefits of goal setting to sport performance. However, because there are many factors influencing goal-setting behavior, much more research is needed in this interesting area. For example, Burton (1993) has conducted empirical research and extensively reviewed the literature on goal setting and sport. Burton suggests that the goal-setting effects found in the sport literature may not parallel those found in the industrial literature because of individual difference variables. These individual difference variables may mask some goal-setting effects, especially when individuals are performing difficult tasks that are near their maximum ability levels.

SUMMARY

- Two major types of cognitive strategies for performance enhancement are mental practice and goal setting.
- Mental imagery and mental practice are common forms of mental preparation among elite performers.
- Mental imagery and mental practice are particularly effective in tasks composed of a large cognitive element.
- The two major theories of mental practice are psychoneuromuscular theory and symbolic learning theory.
- There are different types of goals such as general, specific, performance, and outcome goals.
- Goal setting has been shown to enhance performance.
- Several theoretical explanations have been given for the relationship between goal setting and performance.
- A variety of guidelines have been established for the development of a goal setting program.

THE EXERCISE PSYCHOLOGY AREA

The third subfield of sport and exercise psychology is exercise psychology. The exercise psychology subfield is a broad one, encompassing several research issues such as the importance of exercise on mental health, factors associated with adherence to exercise and physical activity programs, and the relationship of psychological stress to physical injury. In this section, we will focus on problems associated with participation in exercise and physical activity programs and the relationship of psychological stress to injury.

The Problem of Participation in Exercise and Physical Activity Programs

In Chapter 4 we overviewed the benefits of exercise to physiological function. These benefits can include improvements in the various physiological systems (e.g., cardiovascular, muscular, etc.) in the body, depending on the type of exercise. However, there is additional evidence that the maintenance of a chronic (or long-term) exercise program can have positive effects on psychological and emotional functioning of the individual as well. A position statement by the International Society of Sport Psychology, based on available evidence, summarizes the following:

1. Exercise can be associated with a reduced state (or short-term) anxiety.
2. Exercise can be associated with a decreased level of mild to moderate depression.
3. Long-term exercise is usually associated with reductions in neuroticism and anxiety.
4. Exercise may be an adjunct to the professional treatment of severe depression.
5. Exercise can result in the reduction of various stress indices.
6. Exercise can have beneficial emotional effects across all ages and both genders.

In spite of the documented benefits of exercise to both psychological and physiological functioning, there is a severe problem related to motivating people to exercise on a routine basis as well as to maintain an exercise program once it is started. LaFontaine et al. (1992) indicated that nearly 60 percent of the U.S. adult population is sedentary, up to 25 percent suffer from mild anxiety to moderate depression, and 50 percent of individuals who begin an exercise program drop out after the first six months.

Before attempting to identify factors related to adherence to exercise programs, it is helpful to understand the time-course of events related to the participation in exercise programs. Figure 7.8 shows major phases in behavior related to exercise programs (Sallis and Hovell, 1990). *Phase 1* is the sedentary stage, signifying a lifestyle without any significant exercise. *Phase 2* is the adoption phase when the individual decides to begin an exercise program. *Phase 3* is a critical phase in which the individual makes a decision to either maintain (or adhere) to an exercise program or to drop out and discontinue the program. If the individual drops out of the program. he or she may revert back to the sedentary stage or decide, once again, to resume the program (*Phase 4*). This model of exercise participation is important in identifying the major phases an individual experiences as he or she decides whether or not to participate in an exercise program.

The model also indicates that there are major transitional states between each of the phases. One transition is between the sedentary and adoption phase. Surprisingly, very little is known about the factors contributing to an individual's decision to become physically active on a routine basis. *Social-cultural influences* are likely to have a major bearing on this decision, such as parental encouragement for participation in sports (see next chapter). The degree of social-cultural influences in a given society could depend on the race, social class, and ethnicity of the individual (Weiss and Chaumeton, 1992). *Personal influences* may be related to the individual's confidence in succeeding in the exercise program and having knowledge about the importance of exercise and physical activity for a healthy lifestyle (Sallis, Haskell, Fortmann, Vranizan, Taylor, and Solomon, 1986). Another transition is between adoption to either maintenance or to dropping out of the exercise program. This transition from

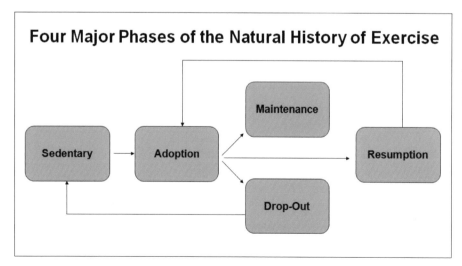

Four Major Phases of the Natural History of Exercise

Figure 7.8

The major phases of participation in an exercise program (from Sallis and Hovell, 1990).

adoption to maintenance or dropping out is related to the problem of exercise adherence and has garnered the most attention by researchers. Some of the factors identified to be related to exercise adherence are:

1. Intrinsic motivation.
2. Support from the individual's spouse or significant other.
3. Time availability to exercise.
4. Easy access to the exercise facility.
5. Recognition of the importance of exercise to good health.
6. Social interaction (such as meeting new friends, having conversation).

In fact, some research has indicated that social interaction is a major contributor to one's adherence to an exercise program (Wankel, 1985). Factors identified to be related to *non-adherence* to an exercise program are the following:

1. Type of profession, such as "blue collar" workers.
2. Being overweight.
3. Lack of enjoyment.
4. Individuals who smoke.
5. Individuals suffering from tension or depression.

Ironically, the very individuals who could benefit from participation in an exercise program such as smokers, overweight, and depressed people, tend to drop out of an exercise program in a short time! Increasing the adherence to a physically active lifestyle is still a very important issue facing society today.

The remaining transition is from dropping out to exercise resumption. Very little is known about this transition. As pointed out by Cox (1994), many interesting questions remain to be answered about this transition. For example, is it harder to resume an exercise program after dropping out compared to initially starting one? Are resumers of an exercise program good adherers? What are the factors associated with resuming an exercise program?

As we can see, the engagement and maintenance to a physically active lifestyle is a complicated issue. However, it can be expected that sport psychologists in the field of kinesiology will play a major role in improving our understanding of this complex and important phenomenon.

The Relationship of Psychological Stress to Injury

As we all know, physical injuries can occur as a result of participating in sports or exercise programs. In fact, of the over 70 million injuries requiring medical attention each year, between 3 and 5 million occur within sports or recreation settings (Boyce and Sobolewski, 1989). Physical and psychological trauma, reduction in physical activity, and financial cost can be a result of these types of injuries. Thus, isolating the causes of these injuries is important in helping to prevent them. As mentioned in Chapter 4, *over*training can result in injury, for example. Accidents (e.g., a collision with another skier), equipment failure (e.g., sudden flat tire while cycling), or poor environmental conditions (e.g., slipping on wet trail while running) are major contributors to injuries while engaging in an exercise program. There is growing evidence however, that *psychological* factors may also influence one's predisposition to injury. Just how might one's mental state play a role in increasing the likelihood of injury while participating in some type of physical activity? In addition, can one's mental state affect the recovery or rehabilitation process *following* an injury? These are the questions to be addressed in this section.

One psychological factor identified to be related to injury is stress. Recall that earlier in the chapter, stress was defined as a negative cognitive state associated with the reluctance to perform a certain activity. There are a variety of incidents in daily living (i.e., life events, daily hassles) that can elevate the stress level (or life stress) of an individual. Research has shown a relationship between life stress and injury. Williams and Roepke (1993) reviewed several studies showing a positive relationship between life stress of an individual and athletic injuries in a number of different sports. The strength of relationship varied considerably across these studies. There were some studies showing no relationship between life stress and injury, but in general, athletes with high life stress were between two and five times as likely to be injured compared to athletes with low life stress.

How might the life stress and injury relationship be explained? One theoretical model proposed by Andersen and Williams (1988) focused on the elicitation of what they termed "**the stress response**" and its effects on an individual's behavior. According to Andersen and Williams, if the athlete or performer perceives inadequate resources to meet the demands of the situation, then a stress response increases in magnitude. The stress response is hypothesized to be manifested as an increase in muscle tension, narrowing of the visual field, and increased distractibility. As a result, the individual is more inclined to be susceptible to injury. For example, before a race, cross country runners typically do some jogging near the starting line. Often, runners can be seen actually

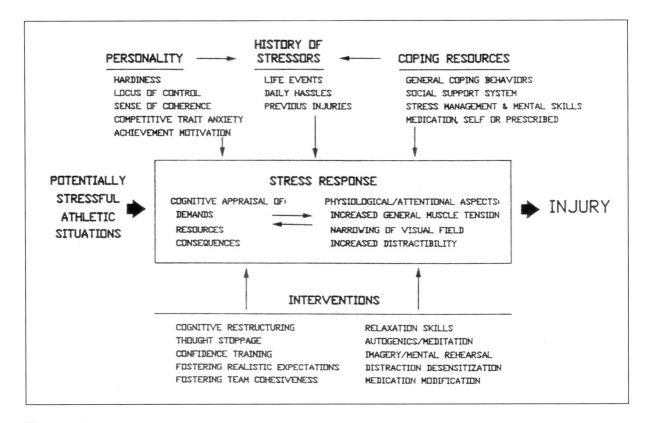

Figure 7.9

A model of stress and injury. Personality, history of stressors, and the individual's coping resources contribute to changes in the stress response to potentially stressful situations. The individual's cognitive appraisal of the situational demands relative to the individual's resources determines the magnitude of the stress response. The stress response can affect the person's muscle tension, narrowing of the visual field, and distractibility, which can lead to physical injury. A variety of intervention techniques can also be used to control the stress response (from Andersen and Williams, 1988).

colliding with one another and sometimes causing an injury! It is likely that because of stress associated with the impending race, the runners' visual field becomes so narrow that they fail to see a nearby competitor until it's too late. Figure 7.9 provides a schematic representation of the Andersen and Williams (1988) model of stress and athletic injury. There are three major factors that can contribute to a change in the stress response:

1. Personality factors—highly anxious individuals; less intrinsically motivated.
2. History of stressors—more life stress and daily hassles; previous injury.
3. Coping resources—social support from family, coach, teacher, friends, relatives; stress management techniques such as relaxation and imagery techniques.

The basic idea behind the model is that in a demanding situation, the individual makes an appraisal of the task demands relative to the available resources to meet those demands. If the individual perceives that the resources are inadequate, the stress response increases. As the stress response increases, muscles become tenser, attention is narrowed, and/or the individual becomes more easily distracted, which increases the likelihood of an injury. There is substantial support for the model, but more research is needed to verify the relative importance of all the components within the model (see Williams and Roepke, 1993; Williams and Andersen, 1998).

Psychological Factors in Injury Rehabilitation

After an injury occurs, it has been found that individuals can experience mood disturbances such as anxiety, depression, irritability, etc. (e.g., Smith, Scott, O'Fallon, and Young, 1990). It has also been shown that fast recovery is associated with the mental state of the individual. For example, Ievleva and Orlick (1991) found that individuals who engaged in goal setting, positive self-talk, and even mental imagery experienced faster rehabilitation. Thus, the rehabilitation process might be facilitated by incorporating a variety of mental strategy techniques such as stress management to help reduce mood disturbances accompanying an injury. Stress management during injury rehabilitation is becoming an important research area in the field of sport psychology. Recently, Wiese-Bjornstal, Smith, and Shaffer (1998) proposed an integrated model involving both psychological and sociological factors to explain the complex response to athletic injury. Coaches and physical therapists might find this material helpful when dealing with injured performers or athletes.

SUMMARY
- There are a number of benefits to mental health by participating in a regular exercise or physical activity program.
- However, a majority of adult Americans are sedentary and nearly half of those who begin an exercise program drop out after the first six months.
- The four major phases to participation in an exercise program are the sedentary, adoption, adherence or drop out, and resumption phases.
- A number of factors have been identified as influencing one's adherence to an exercise program.
- A significant number of injuries occur as a result of participating in an exercise, sport, or physical activity program.
- The psychological state of the individual has been shown to be related to injury and safety.
- Life stress has been associated with changes in what sport psychologists call the stress response, which can lead to increased muscle tension, attentional narrowing, and increased distractibility.
- Personality factors, history of stressors, and coping resources can contribute to changes in the stress response.

- Mood disturbances often accompany a physical injury.
- Faster injury rehabilitation can be associated with a positive mental state.
- Mental training techniques such as stress management can be used to facilitate a positive mental state during rehabilitation.

CHAPTER SUMMARY

- The three major areas of sport and exercise psychology are social psychological, performance enhancement, and health psychology.
- There are a number of motives that serve to change and guide behavior.
- The motives for participation in physical activity are numerous.
- Individuals may possess extrinsic or intrinsic motivational orientations.
- Flow is a positive mental state that is a relationship between the skill level of the performer and the task difficulty.
- The arousal level of the individual can vary both within and between individuals.
- Changes in arousal may be manifested both physiologically and psychologically within an individual.
- The relationship between arousal and performance is complex.
- Changes in arousal appear to be affected by the skill level of the individual.
- Mental preparation is widely used among elite performers.
- Two common types of mental preparation are mental imagery and mental practice.
- There are several theories to account for the relationship between performance and mental imagery/practice.
- Goal setting is an important performance enhancement technique shown to influence performance in a number of settings.
- Sport psychologists have developed a number of guidelines for the use of goal setting.
- In spite of documented benefits, there is a serious problem motivating people to begin and maintain a physically active lifestyle.
- Several factors have been identified that influence one's adherence to an exercise program.
- There appears to be a relatively strong relationship between life stress and physical injury.
- Several factors contribute to changes in one's stress response.
- Psychological techniques can be used to aid the rehabilitation process following a physical injury.

IMPORTANT TERMS

Arousal – the level of activation of the nervous system

Belongingness and love (affiliation) motives – a motive in Maslow's hierarchy

Cognitive anxiety - an individual's negative expectations and potential consequences about oneself in a given situation

Competence motives - related to our desire to improve the skills needed to successfully cope in the environment

Drive - a physiological need that leads to an aroused state but with no directionality

Drive cues – stimuli that evoke a drive

Extrinsic motivation – an orientation to want to participate in a physical activity to win (e.g., beat others), to gain rewards (e.g., money, trophies), or to receive social approval for their performance.

Flow – a psychological state that optimizes personal growth

General goals – vague goals that are probably not very effective

Goal setting – a process of setting future expectations

Intrinsic motivation – the desire to participate in the activity for its own sake without the expectation for external rewards

Inverted-U relationship – the relationship between arousal level and performance

Measures of arousal – arousal can measured with both psychological and physiological techniques

Mental imagery – imagining a future motor performance

Mental practice – a cognitive technique to improve performance

Mental preparation – a variety of cognitive techniques to improve performance

Motivation - the desire to direct behavior in a given direction, with some duration and intensity

Motivation orientation – the tendency to be either intrinsically or extrinsically motivated

Motivational climate – an environment that reinforces either intrinsic or extrinsic motivation

Motive - the arousal of an individual to strive for some goal

Outcome goals – the desire to achieve some result from a physical performance

Performance enhancement area – a major subdivision in sport and exercise psychology focusing on psychological techniques to improve performance

Performance goals – the desire to improve performance or the execution of a skill without an emphasis on outcome

Physiological anxiety – physiological manifestations of stress

Pleasureful activities - According to Csikszentmihalyi (1990), activities that create feelings of contentment when biological and social conditions are met

Psychoneuromuscular theory – a theory that states that during imagery or mental practice, the muscles to be used in the actual performance are slightly innervated (or stimulated)

Safety motives - involve the avoidance of dangerous and life-threatening situations

Self-esteem motives - provide a means for the individual to improve his or her self-image in order to feel capable and gain respect from others

Specific goals – goals that contain more detail regarding behavioral objectives and the time period

Stress and injury - a relationship showing that increased psychological stress can lead to physical injury

Stress response – an increase in muscle tension, narrowing of the visual field, and increased distractibility as a result of psychological stress

Survival motives – motive that relate to satisfying basic physiological needs, such as hunger and thirst

Symbolic learning theory – a theory that argues that mental practice allows the performer to prepare for the various elements or possible sequences of movements involved in the actual performance

Zones of optimal functioning - characterize an individual's optimal level of anxiety related to peak performance

INTEGRATING KINESIOLOGY: PUTTING IT ALL TOGETHER

1. Identify three physical activities you enjoy participating in. Describe the motives for why you participate in these activities. Are the motives the same for each activity?

2. What is your motivational orientation for each of the activities mentioned above?

3. What type of motivational climate existed in your family that may have influenced the type of physical activities you participate in now?

4. Have you ever had a "flow" experience while participating in a physical activity? Can you think of ways to help repeat this experience?

5. Have you ever used any type of mental preparation for participating in a physical activity? Is so, do you feel it was beneficial to your future performance?

6. If you decided to start a personal fitness program (Chapter 4), what types of goals could you use to enhance the benefits of the programs and to improve your adherence to the program?

7. If you injured yourself while participating in some physical activity, what psychological strategies would you use to enhance the rehabilitation process?

KINESIOLOGY ON THE WEB

- www.issponline.org/documents/physactstatement.pdf—This website is the International Society of Sport Psychology position statement on physical activity and psychological benefits.
- http://appliedsportpsych.org/home—This is the website for the Association for Applied Sport Psychology. Click on "Resource Center" to obtain valuable information on the subfield of sport and exercise psychology.
- http://www.naspspa.org—This is the website for North American Society for Sport Psychology and Physical Activity
- http://apa47.org/—This is the website for Division 47, Exercise and Sport Psychology, in the American Psychology Association

REFERENCES

Abernethy, B., Summers, J. J., and Ford, S. (1998). Issues in the measurement of attention. *Advances in Sport and Exercise Psychology Measurement.* Morgantown, WV: Fitness Information Technology, pp. 173–93.

Ames, C. (1992). The relationship of achievement goals to student motivation in classroom settings. In G. C. Roberts (ed.), *Motivation in Sport and Exercise.* Champaign, IL: Human Kinetics.

Andersen, M. B., and Williams, J. M. (1988). A model of stress and athletic injury: prediction and prevention. *Journal of Sport and Exercise Psychology, 10,* 294–306.

Biddle, S., and Bailey, C. (1985). Motives for participation and attitudes toward physical activity of adult participants in fitness programs. *Perceptual and Motor Skills, 61,* 831–34.

Block, N. (1981). *Imagery.* Cambridge, MA: MIT Press.

Boyce, W. T., and Sobolweski, S. (1989). Recurrent injuries in schoolchildren. *American Journal of the Disabled Child, 143,* 338–42.

Brunner, B. C. (1969). Personality and motivating factors influencing adult participation in vigorous physical activity. *Research Quarterly, 3,* 464–69.

Brustad, R. J. (1998). Developmental considerations in sport and exercise

psychology measurement. In J. L. Duda (ed.), *Advances in Sport and Exercise Psychology Measurement.* Morgantown, WV: Fitness Information Technology, pp. 461–70.

Burton, D. (1988). Do anxious swimmers swim slower? Reexamining the elusive anxiety-performance relationship. *Journal of Sport and Exercise Psychology, 10,* 45–61.

Burton, D. (1989). Winning isn't everything: examining the impact of performance goals on collegiate swimmers' cognitions and performance. *The Sport Psychologist, 3,* 105–32.

Burton, D. (1992). The Jekyll/Hyde nature of goals: reconceptualizing goal setting in sport. In T. S. Horn (ed.), *Advances in Sport Psychology.* Champaign, IL: Human Kinetics.

Burton, D. (1993). Goal setting in sport. In R. N. Singer, M. Murphey, and L. K. Tennant (eds.). *Handbook of Research on Sport Psychology.* New York: Macmillan, pp. 467–91.

Cacioppo, J. T., and Berntson, G. G. (1992). Social psychological contributions to the decade of the brain: doctrine of multilevel analysis. *American Psychologist, 47,* 1019–28.

Cox, R. H. (1994). *Sport Psychology: Concepts and Applications.* Madison, WI:Brown and Benchmark.

Csikszentmihalyi, M. (1990). *Flow: The psychology of Optimal Experience.* NY: Harper and Row.

Dember, W. N. (1974). Motivation and the cognitive revolution. *American Psychologist, 29,* 161–68.

Dietz, W. H. (1990). Children and television. In M. Green and R. J. Hagerty (eds.), *Ambulatory Pediatrics IV*, Philadelphia: W. B. Saunders, 39–41.

Dishman, R. K. (1994). Biological psychology, exercise, and stress. *Quest, 46*, 28–59.

Dishman, R. K. (1983). Identity crises in North American sport psychology: academics in professional issues. *Journal of Sport Psychology, 5*, 123–34.

Duda, J. L. (1993). Goals: A social-cognitive approach to the study of motivation. In R. N. Singer, M. Murphey, and L. K. Tennant (eds.). *Handbook of Research on Sport Psychology*. NY: Macmillan, pp. 421–36.

Duda, J. L., and Hayashi, C. T. (1998). Measurement issues in cross-cultural research within sport and exercise psychology. In J. L. Duda (ed.), *Advances in Sport and Exercise Psychology Measurement*. Morgantown, WV: Fitness Information Technology, pp. 471–83.

Duffy, E. (1962). *Activation and Behavior*. New York: Wiley.

Feltz, D. L. (1992). The nature of sport psychology. In T. S. Horn (ed.), *Advances in Sport Psychology*. Champaign, IL: Human Kinetics.

Feltz, D. L., and Landers, D. M. (1983). The effects of mental practice on motor skill learning and performance: a meta-analysis. *Journal of Sport Psychology, 5*, 25–57.

Feltz, D. L., Landers, D. M., and Becker, B. J. (1988). A revised meta-analysis of the mental practice literature on motor skill learning. In D. Druckman and J. Swets (eds.), *Enhancing Human Performance: Issues, Theories and Techniques*. Washington, DC: National Academy Press, pp. 1–65.

Fenz, W. D., and Jones, G. B. (1974). Cardiac conditioning in a reaction time task and heart rate control during real-life stress. *Journal of Psychosomatic Research, 18*, 199–203.

Gould, D. (1993). Goal setting for peak performance. In J. M. Williams (ed.). *Applied Sport Psychology: Personal Growth to Peak Performance*. Palo Alto, CA : Mayfield.

Gould, D., Petlichkoff, L., Simons, J., and Vevera, M. (1987). Relationship between Competitive State Anxiety Inventory-2 scores and pistol shooting performance. *Journal of Sport Psychology, 9*, 33–42.

Hannin, Y. L. (1980). A study of anxiety in sports. In W. F. Straub (ed.), *Sport Psychology: An Analysis of Athletic Behavior* (pp. 236–49). Ithaca, NY: Mouvement.

Hardy, L. (1996). Testing the predictions of the cusp catastrophe model of anxiety and performance. *The Sport Psychologist, 10*, 140–56.

Harris, D. V., and Williams, J. M. (1993). Relaxation and energizing techniques for regulation of arousal. In J. M. Williams (ed.). *Applied Sport Psychology: Personal Growth to Peak Performance*. Palo Alto, CA : Mayfield.

Harris, D. V., and Robinson, W. J. (1986). The effects of skill level on EMG activity during internal and external imagery. *Journal of Sport Psychology, 8*, 105–11.

Hird, J. S., Landers, D. M., Thomas, J. R., and Horan, J. J. (1991). Physical practice is superior to mental practice in enhancing cognitive and motor task performance. *Journal of Sport and Exercise Psychology, 13*, 281–93.

Ievleva, L., and Orlick, T. (1991). Mental links to enhancing healing: An exploratory study. *The Sport Psychologist, 5*, 25–40.

International Society of Sport Psychology (1992). Physical activity and psychological benefits: A position statement from the international society of sport psychology. *Journal of Applied Sport Psychology, 4*, 94–98.

Jackson, S. A., and Csikszentmihalyi, M. (1999). *Flow in Sports*. Champaign, IL: Human Kinetics.

Jacobson, E. (1930). Electrical measures of neuromuscular states during mental activities (Part 1). *American Journal of Physiology, 91*, 567–608.

Kenyon, G. S. (1968a). A conceptual model for characterizing physical activity. *Research Quarterly, 39*, 96–105.

Kenyon, G. S. (1968b). Six scales for assessing attitudes toward physical activity. *Research Quarterly, 39*, 566–74.

Kimiecik, J. C., and Blissmer, B. (1998). Applied exercise psychology: measurement issues. Advances in Sport and Exercise Psychology Measurement. Morgantown, WV: Fitness Information Technology, pp. 447–60.

Kimiecik, J. C., and Stein, J. C. (1992). Examining flow experiences in sport contexts: conceptual issues and methodological concerns. *Journal of Applied Sport Psychology, 4*, 144–60.

Klavora, P. (1979). An attempt to derive inverted-U curves based on the relationship between anxiety and athletic performance. In D. M. Landers and R. W. Christina (eds.), *Psychology of Motor Behavior and Sport*. Champaign, IL: Human Kinetics.

Kyllo, L. B., and Landers, D. M. (1995). Goal setting in sport and exercise: a research synthesis to resolve the controversy. *Journal of Sport and Exercise Psychology, 17*, 117–37.

LaFontaine, T. P., DiLorenzo, T. M., French, P. A., Stucky-Ropp, R. C., Bargman, E. P., and McDonald, D. G. (1992). Aerobic exercise and mood: a brief review, 1985–1990. *Sports Medicine, 13*, 160–70.

Landers, D. M., and Boutcher, S. H. (1993). Arousal-performance relationships. In J. M. Williams (ed.). *Applied Sport Psychology: Personal Growth to Peak Performance*. Palo Alto: CA: Mayfield.

Levitt, S., and Gutin, B. (1971). Multiple-choice reaction time and movement time during physical exertion. *Research Quarterly, 42*, 405–10.

Locke, E. A., Shaw, K. M., Saari, L. M., and Latham, G. P. (1981). Goal setting and task performance: 1969–1980. *Psychological Bulletin, 90*, 125–52.

Los Angeles Times (2008). http://latimesblogs.latimes.com/technology/2008/11/americans-now-w.html.

Mahoney, M. J., and Avener, M. (1977). Psychology of the elite athlete: an exploratory study. *Cognitive Therapy and Research, 1*, 135–41.

Martens, R. (1974). Arousal and motor performance. In J. H. Wilmore (ed.). *Exercise and Sport Sciences Reviews*. New York: Academic.

Martens, R., and Landers, D. M. (1970). Motor performance under stress: a test of the inverted-U hypothesis. *Journal of Personality and Social Psychology, 16*, 29–37.

Maslow, A. H. (1954). *Motivation and Personality*. New York: Harper.

McClements, J. (1982). Goal setting and planning for mental preparations. In L. Wankel and R. B. Wilberg (eds.)., Psychology of sport and motor behavior: research and practice. *Proceedings of the Annual Conference of the Canadian Society for Psychomotor Learning and Sport Psychology*. Edmonton, Canada.: University of Alberta.

McKeachie, W. J., Doyle, C. L., and Moffett, M. M. (1976). *Psychology*. Addison-Wesley. (3rd ed.)

Murphy, S. M., and Jowdy, D. P. (1992). Imagery and mental practice. In T. S. Horn (ed.), Advances in Sport Psychology. Champaign, IL: Human Kinetics.

Nicholls, J. (1978). The development of the concepts of effort and ability, perception of attainment, and the understanding that difficult tasks require more ability. *Child Development, 49*, 800–14.

Nicklaus, J. (1974). Golf My Way. New York: Simon & Schuster.

Orlick, T., and Partington, J. (1988). Mental links to excellence. *The Sport Psychologist, 2,* 105–30.

Roberts, G. (1993). Motivation in sport: Understanding and enhancing the motivation and achievement of children. In R. N. Singer, M. Murphey, and L. K. Tennant (eds.). Handbook of Research on Sport Psychology. New York: Macmillan, pp 405–20.

Ryan, E. D., and Simons, J. (1981). Cognitive demand imagery and frequency of mental practice as factors influencing the acquisition of mental skills. *Journal of Sport Psychology, 4*, 35–45.

Sage, G. (1984). *Motor Learning and Control: A Neuropsychological Approach*. Dubuque, IA : Wm. C. Brown.

Sallis, J. F,. and Hovell, M. F. (1990). Determinants of exercise behavior. In K. B. Pandolf, and J. O. Holloszy (eds.), *Exercise and Sport Science Reviews, Vol. 18*, 307–30. Baltimore: Williams and Wilkins.

Sallis, J. F., Haskell, W. L., Fortmann, S. P., Vranizan, K. M., Taylor, C. B., and Solomon, D. S. (1986). Predictors of adoption and maintenance

of physical activity in a community sample. *Preventive Medicine, 15*, 331–41.

Smith, A. M., Scott, S. G., O'Fallon, W. M., and Young, M. L. (1990). Emotional responses of athletes to injury. *Mayo Clinic Proceedings, 65,* 38–50.

Sonstroem, R. J., and Bernardo, P. (1982). Intraindividual pregame state anxiety and basketball performance: A reexamination of the inverted-U curve. *Journal of Sport Psychology, 4,* 235–45.

Suinn, R. M. (1980). Psychology and sports performance: principles and applications. In R. Suinn (ed.), *Psychology in Sports: Methods and Applications* (pp. 26–36), Minneapolis: Burgess.

Suinn, R. M. (1993). Imagery. In R. N. Singer, M. Murphey, and L. K. Tennant (eds.). *Handbook of Research on Sport Psychology.* New York: Macmillan, pp. 492–510.

Tolman, E. C. (1951). *Collected Papers in Psychology.* Berkeley:University of California Press.

Triplett, N. (1897). The dynamogenic factors in pacemaking and competition. *American Journal of Psychology, 9,* 507–33.

Ungerleider, S., Golding, J. M., Porter, K., and Foster, J. (1989). An exploratory examination of cognitive strategies used by master track and field athletes. *The Sport Psychologist, 3,* 245–53.

Wankel, L. M. (1985). Personal and situational factors affecting exercise involvement: The importance of enjoyment. *Research Quarterly for Exercise and Sport, 56,* 275–82.

Wehner, T., Vogt, S., and Stadler, M. (1984). Task-specific EMG-characteristics during mental training. *Psychological Research, 46,* 389–401.

Weinberg, R. S. (1981). The relationship between mental preparation strategies and motor performance: A review and critique. *Quest, 33,* 195–213.

Weinberg, R. S., Bruya, L. D., Jackson, A., and Garland, H. (1986). Goal difficulty and endurance performance: a challenge to the goal attainability assumption. *Journal of Sport Behavior, 10,* 82–92.

Weiss, M. R., and Chaumeton, N. (1992). Motivational orientations in sport. In T .S. Horn (ed.), *Advances in Sport Psychology.* Champaign, IL: Human Kinetics.

Wiese-Bjornstal, D. M., Smith, A. M., and Shaffer, S. M. (1998). An integrated model of response to sport injury: psychological and sociological dynamics. *Journal of Applied Sport Psychology, 10,* 46–69.

Wiese-Bjornstal, D. M., and Weiss, M. R. (1992). Modeling effects on children's form kinematics, performance outcome, and cognitive recognition of a sport skill: an integrated perspective. *Research Quarterly for Exercise and Sport, 63,* 67–75.

Williams, J. M., and Andersen, M. B. (1998). Psychosocial antecedents of athletic injuring: review and critique of the stress and injury model. *Journal of Applied Sport Psychology, 10,* 5–21.

Williams, J. M., and Roepke, N. (1993). Psychology of injury and injury rehabilitation. In R. N. Singer, M. Murphey, and L. K. Tennant (eds.). *Handbook of Research on Sport Psychology.* New York: Macmillan, pp. 405–20.

Wrisberg, C. A., and Ragsdale, M. R. (1979). Cognitive demand and practice level: factors in the mental practice of motor skills. *Journal of Human Movement Studies, 5,* 201–08.

Chapter Eight
Developmental Foundations

CHAPTER SUMMARY

Important terms

Integrating Kinesiology: Putting It All Together

Kinesiology on the Web

References

CHAPTER EIGHT

Developmental Foundations

All the world's a stage,
And all the men and women merely players.
They have their exits and their entrances,
And one man in his time plays many parts,
His acts being seven ages …

— William Shakespeare, "As You Like It" (Act II, Scene VII)

STUDENT OBJECTIVES

1. To understand the important scientific terms related to development.
2. To know the various developmental periods.
3. To understand the difference between cross-sectional and longitudinal research designs.
4. To be able to describe changes in growth with age in the skeletal, muscular, and nervous systems.
5. To know the various types of infantile reflexes.
6. To describe the various phases related to the development of locomotion and prehension.
7. To understand gender differences in performance of skills during childhood and beyond.
8. To be able to describe the differences between various aging concepts such as life expectancy and the average and maximum life span of a population.
9. To appreciate the different theories of aging.
10. To describe the various physiological and performance decrements accompanying the aging process.
11. To explain the positive effects of exercise and physical activity on the aging process and health.

DEVELOPMENTAL CONCEPTS AND TERMS

Developmental Kinesiology—A Life Span Approach

In the previous chapters, we have examined a number of subfields within kinesiology. With a few exceptions, our examination has primarily focused on the structure and function of the mature or adult human being. But as we know, all life forms have distinct beginnings, a period of growth and maturation, and finally, a decline in function prior to the end of life. The discipline of kinesiology has a keen interest in the study of movement and physical activity across the *entire* life span, from the developing child to the older adult. Infants come into this world possessing a variety of reflexes and rudimentary movements that have some functional purposes. These behaviors then become dominated by the emergence of *voluntary* skilled behavior, whose proficiency usually takes many years of practice. It is also during this period that an individual begins to adopt a certain lifestyle involving physical activity habits. As mid-life approaches, motor skills and physical performance typically begin to decline. Much of this decline appears to be due to changes in anatomy and physiological functioning. However, there is evidence that the adoption of an active lifestyle throughout one's life can influence the aging process by minimizing the rate of decline of functional capacity even into later adulthood. There are many interesting issues and questions related to the aging process we examine in this chapter. Before doing so, let us begin with an understanding of some important terms used in development.

Important Terms: Growth, Maturation, Learning, Heredity, Environmental Influences, Development

Developmental scientists have defined important terms that often are inappropriately interchanged or confused with one another. The term **growth** means a quantitative increase in the size of a certain anatomical structure. For example, the size of child's hand increases as a function of age, reflecting a change in growth. Changes in the height of an individual is another good example of growth changes. **Maturation** is the increase in the functional capacity of an individual (or of a body part) without necessarily an increase in growth. Maturation usually depends on the functioning of the cells of the body's organs. It is possible for an organ to reach a particular level of growth without maturation because the cells have not become fully functional. **Learning** is defined as a relatively permanent change in performance as a function of practice. It is possible for an individual to be physically mature without achieving a high level of performance. To reach higher levels of performance beyond the state of full maturation requires additional learning through practice. **Heredity** refers to a set of qualities that are relatively fixed at birth, which tend to set limits of certain characteristics of the individual. The genetic makeup of the individual determines whether the person is a male or female, what the hair color is, and to a large extent, how tall he or she will be as an adult, as just a few examples. **Environmental influences** refer to a large set of variables such as parental and family interactions, dietary factors, and educational opportunities. **Development** is thought to be the changes in the capacities and skill level of the individual with age. The individual's developmental state is believed to be influenced by all the other factors mentioned above.

Stages (or Phases) of Development

During the late Middle Ages in Europe, society viewed individuals as being in one of two age-related states: either infancy or adulthood. Infants were thought to range from birth to seven years of age and adults from seven years of age to death. Thus, at seven years of age, individuals were expected to assume the various roles of an adult, such as working in factories and even fighting in wars! As a result of much research and changing societal viewpoints, a number of developmental periods or stages have been identified (see Table 8.1).

Table 8.1

Chronological Ages for Various Developmental Periods.
(Payne and Isaacs, 2008)

Developmental Period	Approximate Chronological Age
Prenatal	
Embryo	2 weeks to 8 weeks
Fetus	8 weeks to birth
Neonate	Birth to 4 weeks
Infancy	Birth to 1 year
Childhood	
Toddlerhood	1 to 4 years
Early childhood	4 to 7 years
Middle childhood	7 to 9 years
Late childhood	
(preadolescence)	9 to 12 years
Adolescence	
Girls	11 to 19 years
Boys	13 to 21 years
Adulthood	
Early adulthood	20 to 40 years
Middle adulthood	40 to 60 years
Older adulthood	60 years and over

These age-related periods should only be considered as guideposts, because there is much variation between individuals.

Research Methodology

How do researchers determine the effects of age on motor development? While there are a number of ways to conduct developmental research, the two predominant research designs are the cross-sectional design and the longitudinal design. The **cross-sectional design** involves the use of at least two separate groups of subjects differing in age. For example, a researcher may want to determine whether there are motor skill differences between the ages of six and ten years. Using a cross-sectional design, the researcher would assign an equal number of subjects to the 6-year-old and 10-year-old groups and measure performance on the same motor skill. The researcher would conclude that age is a contributing factor to motor skill development if a significant difference is found between the groups. The major advantage of the cross-sectional design is that a large amount of data can be collected in a short period of time. The major disadvantage of this design is that changes in a given *individual* as a function of age cannot be examined.

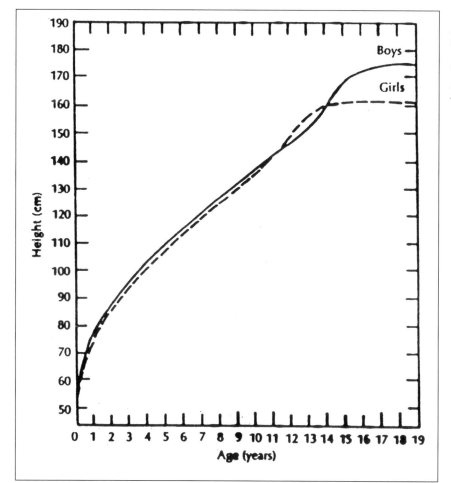

Figure 8.1

Typical individual height-attained curves for boys and girls. From Tanner, Whitehouse, and Takaishi (1966).

To examine individual changes as a function of age requires the use of the **longitudinal design**. In this design, the researcher examines an individual (or group of individuals) over an extended period. For example, a group of six year olds are studied periodically between the ages of six and 10 years. The major advantage of the longitudinal design is that large amounts of data can be gathered on individual subjects. Some disadvantages are that 1) these types of designs are generally very expensive, given the time and effort it takes to collect the data (four years in the above example); and 2) there is a greater chance for subjects dropping out of the study (e.g., moving away, illness, etc.). Despite these limitations, researchers believe the longitudinal design can provide important data on the effects of age on motor performance and other measures.

SUMMARY

- Developmental kinesiology is the study of age-related changes of the human body and its functioning during physical activity.
- Important developmental concepts that are sometimes misunderstood are growth, maturation, heredity, environmental influences, and development.
- There are a number of stages (or phases) of development: prenatal, neonate, infancy, childhood, adolescence, and adulthood.
- The two major research designs in studying development are the cross-sectional and the longitudinal design.

ANATOMICAL AND PHYSIOLOGICAL DEVELOPMENTAL CHANGES

In this section, we examine the various anatomical and physiological changes that accompany the aging process. We first examine gender differences in height and weight changes followed by a comparison in relative body proportions of infants and adults. Then, we explore the unique effects of aging on a number of physiological systems in the body such as the skeletal, muscular, nervous, and cardiovascular systems. An attempt is made to understand how these changes might affect motor skill and functioning within the developing individual.

Growth Changes with Age

When my daughter, Makaila, was born, I decided to chart her growth changes in height over the span of several years. I was also interested in how her growth might change compared to established norms. Before sharing these data, let us first examine two types of growth curves for boys and girls that reflect the average changes in height from birth to about 19 years of age. The first curve, shown in Figure 8.1, is the average change in height for boys and girls. Up to the ages of approximately 11 or 12, growth appears to increase rather consistently, for both boys and girls. At the beginning of the teenage years, girls show a greater increase in growth than boys for a short period of time. This increase occurs during a period called **puberty**, when the sexual organs become fully mature. Puberty for boys, on average, begins later than for girls, and the associated increase in growth also occurs later in boys. The increase in growth for boys during puberty is large enough on average to eventually surpass the girls. Growth in height for boys and girls reaches its maximum following puberty, with boys remaining taller than girls. After the ages of 25–29 in males and 16–20 in females, height begins to gradually decrease primarily due to compression of the vertebrae from bone mineral loss (Frisancho, 1990; Galloway, Stini, Fox, and Stein, 1990).

The next curve shown in Figure 8.2 illustrates the height gain (or change) of growth in boys and girls computed from the data in Figure 8.1. There are a number of observations that can be made from this graph. First, growth change is greatest from birth to one year. Notice that even though height gain is decreasing during this time period, the absolute values (from 24 to 14 cm/yr) are still the highest of all ages. This type of change would be analogous to a car slowing down from 55 to 30 miles per hour—the car is slowing down, but it is still moving at a high rate of speed! During childhood, height gain is still occurring (between approximately 5 and 10 cm/year), but is gradually slowing down. During puberty, height gain dramatically increases (called the pubescent spurt), after which it reduces to zero (around 16 for girls and 18 for boys). The pattern of weight changes for boys and girls into young adulthood are similar, but not identical, to those for height. Maximum weight for males and females is reached by around 40 and 50 years, respectively, before weight begins a slow gradual decline (Frisancho, 1990).

Illustrated in Figure 8.4 is a percentile growth curve for girls from 2 to 18 years of age. (Note that the vertical axis is in inches, whereas the previous two graphs were in centimeters.) The dark line in the middle indicates that 50 percent of the population of girls is above and below the line, representing a type of average growth curve. The line at the top indicates that 95 percent of the population is below this line. Individuals following this line end up being much taller than 95 percent of the population One interpretation of this graph is that individuals should always follow a given line from ages 2 to 20 years. However, it should be noted that there is variation among individuals, and that people can change from one percentile line to another.

Returning to my daughter as an example, I have plotted her growth curve shown by the solid circles. Between the ages of 2 and 10, she closely followed the 50th percentile curve. From ages 11 to 13, she fell below the 50th percentile curve and then suddenly jumped to above the 75th percentile curve during her puberty! These kinds of changes suggest that it is often difficult to precisely predict the maximal height (and probably the weight) of a fully grown individual based on childhood data.

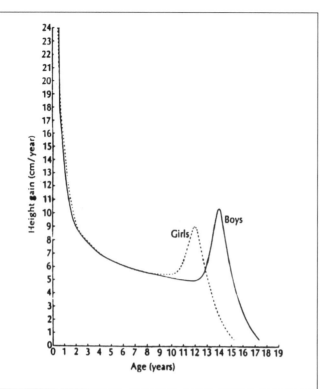

Figure 8.2

Typical individual velocity curves for supine length or height in boys and girls. From Tanner, Whitehouse, and Takaishi (1966).

HIGHLIGHT

Is Growth a Continuous Process?

After examining the curve in Figure 8.1, one might conclude that growth proceeds in a rather continuous fashion. Continuous growth implies that changes in growth (from birth to just after puberty) occur without any periods of rest (or no growth, called *stasis*). Discontinuous growth is suggested when "spurts" of growth (called *saltation*) are followed by periods of stasis. Actually, there is considerable debate about whether growth is a continuous process or not (Lampl, Veldhuis, and Johnson, 1992; Lampl; Cameron, Veldhuis, and Johnson, 1995). Sparking the debate was a recent study by Lampl, Veldhuis, and Johnson (1992), who measured body length changes in infants during their first 21 months. Lampl et al. took measures of the infants frequently during this time period, sometimes every day. The results from one of the infants in the study are shown in Figure 8.4. During a period from 90 to 218 days, they found approximately 13 separate "spurts" of growth separated by short periods of stasis. Lampl et al. concluded that growth is a discontinuous process. In addition, they argued that stasis is a normal, rather than a pathological, process as some researchers have suggested. Other researchers have questioned Lampl et al.'s results, but there does appear to be supportive evidence in favor of the discontinuous growth hypothesis (e.g., Gibson and Wales, 1994). It is unknown what underlying processes or mechanisms are responsible for discontinuous growth, but periodic releases of certain hormones are suspected. In addition, it is unknown whether discontinuous growth is the norm in humans from infancy to adolescence. More research is clearly needed to further understand this complex and interesting issue in human development.

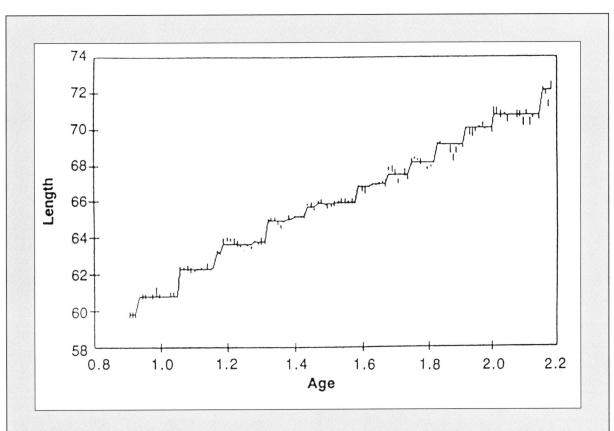

Figure 8.3

Daily length measurements of a male infant from 90 to 218 days of age. Data plotted by length (centimeters) and age (days x 102). Notice the periods of growth (saltation) or rapid increases in length and the periods of no growth (stasis). From Lampl, Veldhuis, and Johnson (1992, Figure 1).

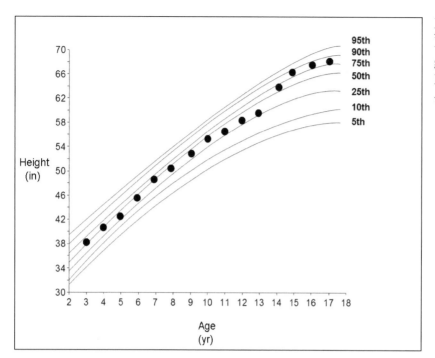

Figure 8.4
Percentile height curves for girls. (My daughter Makaila's height curve is represented by the dark dots).

The Skeletal System and Development

The skeletal system of humans during the embryonic stage (see Table 8.1) is primarily made of **cartilage**, a tough but flexible tissue. During the fetal stage, bones begin to ossify, or harden. Ossification begins in **primary ossification centers**, located within the mid-portions of long bones such as the femur and humerus (see Chapter 3). After birth, bone growth occurs at **secondary ossification centers** near the end of the shaft, called the **epiphyseal (or growth) plates**. Virtually all growth within the growth plates is complete by the ages of 18–19 years.

It is important for the growth plates to maintain a continuous blood supply and remain free from injury as the child grows. While the bones themselves typically heal rather quickly following injury, damage to the growth plates during childhood can cause early cessation of cell growth in the affected bone. Damage can occur, for example, as a result of overuse or stress caused by physical contact. It has been suggested that young children avoid participation in certain contact sports that could result in injury to the growth plates (Haywood and Getchell, 2009).

The Muscular System and Development

Individual muscles increase in size by either the production of more muscle cells or by an increase in the size of the muscle cells. Prior to and shortly after birth, muscle mass increases primarily by an increase in number of muscle cells. During infancy and childhood, increases in the size of the muscle cells are largely responsible for increases in muscle mass. During childhood, there is little difference in changes in the muscular systems of boys and girls. However, following adolescence, muscle mass accounts for over 50 percent of total body weight in boys, but only 45 percent of total body weight in girls (Malina, 1978).

The percentage of slow twitch fibers in an individual appears to be set by the end of the first year of life. But the percentage of fast twitch fibers is much greater in adults than in infants. How muscle fiber type changes as a function of age is still a very open question (Baldwin, 1984).

The Nervous System and Development

At birth, the spinal cord and lower brain centers are much more advanced than the brain. Important basic physiological functions such as breathing and the control of heart rate, largely controlled by these structures of the nervous system, are quite mature at birth. In addition, a large number of important reflexes exist at birth that are supported by the spinal cord and lower brain centers (see Motor Development section below). These basic physiological functions and reflexes are critical for the survival of the infant.

Full maturity of the brain is reached several years after birth. Changes in brain weight following birth is due to an increase in the size of the brain cells, the amount of branching of the axons and dendrites, and the increase of myelin and glial cells, which improve nerve conduction and nourish the brain cells, respectively.

SUMMARY

- Growth changes in boys and girls are similar up to puberty.
- There are debates as to whether growth is a continuous or discontinuous process.
- Growth of the skeletal system is typically completed by the ages of 18–19 years.
- Bone growth occurs in the epiphyseal (or growth) plates.
- Muscle growth is due to increases in more muscle cells and in the size of existing muscle cells.
- At birth, the spinal cord and lower brain centers are more advancedthan the brain.

MOTOR DEVELOPMENT

The normal period of **gestation** of the human fetus (defined as the time from conception to birth) is approximately 40 weeks. During this time period, the fetus can produce arm, leg, and head movements. Within only a few months following conception, fetal movements can be felt by the mother. In fact, pregnant mothers can often differentiate the type of movements made by the fetus! The movements produced by the fetus, particularly in the last months of gestation, are reflexive and spontaneous, as opposed to voluntary and well planned. In this section, the concept of a reflex is defined and a number of infant reflexes are described. Next, voluntary motor skill development is described by focusing on the development of two major skills: prehension and locomotion. Finally, changes in physical activity patterns in boys and girls are discussed, and factors that may influence gender differences in physical activity patterns are provided.

Reflexes

A **reflex** is defined as an involuntary response to a specific stimulus, such as a touch, sound, odor, light, or muscle stretch. Reflexes provide us with a mechanism to quickly respond to certain stimuli without the delays usually associated with conscious, willful activity. The knee-jerk stretch reflex was described in Chapter 6 as a contraction of a muscle in response to a stretch of that same muscle. The first muscle response in the knee jerk reflex can be elicited within 30–60 msec, illustrating the rapidity of the neuromuscular system to respond to stimuli. Sneezing is also a good example of a rapid, as well as a functional, response to certain odors or irritants.

Fetuses and infants possess a multitude of reflexes. Many of the reflexes are quite complex, because their responses may involve the use of many muscles and limbs. These reflexes may be categorized into three groups: primitive reflexes, postural reflexes, and locomotor reflexes (Peiper, 1963; Haywood and Getchell, 2009).

Primitive reflexes are those primarily under the control of the lower brain centers and can be elicited in utero (within the womb) or in premature infants. It is thought that these reflexes allow the fetus to move around within the womb and put itself in the proper position for birth. In addition, some of these primitive reflexes are important for the infant's survival. The palmar grasping reflex is elicited by touching the fetus's or infant's palm with an object or a finger. The response to this stimulus is the rapid closure of the hand around the object. The palmar grasping reflex is the precursor to voluntary and willful grasping behavior. The sucking reflex is initiated by touching the infant's face above or below the lips. The response is a sucking motion that enables the newborn to feed (Milani-Comparetti, 1981). There are at least a dozen primitive reflexes (see Haywood and Getchell, 2009, for a description). Most of these reflexes cease between 2 weeks to a year after birth, depending on the reflex.

Postural reflexes assist the infant in maintaining posture (i.e., head or body position) in a changing environment. For example, the labyrinthine righting reflex is invoked when the infant, held by someone, is tilted. The response is a head movement to keep in an upright position, allowing the breathing passages to remain open. There are at least a half dozen different types or variations of postural reflexes that can be actively invoked from between birth and one year, but can emerge under specific situations even in adulthood.

Locomotor reflexes are a complicated series of movements that resemble certain voluntary ambulatory motions such as crawling, walking, or even swimming. The walking reflex can be elicited by placing an upright infant on the flat surface. The response is a series of alternating leg movements that resembles walking. In addition, a swimming reflex can be invoked by holding an infant in the prone position (face down) in or just over water. This stimulus will produce complicated leg and arm movements resembling swimming. All of the locomotor type reflexes are thought to be inhibited by approximately five months following birth.

The Role of Reflexes

It is generally agreed that most reflexes have some functional purpose for the fetus or infant. Some reflexes assist the infant in eating and drinking. Other reflexes assist in the fetus's and infant's body orientation and posture in their respective environments. Reflexes provide the infant a means to learn about and interact with the environment. Reflexes provide opportunities to strengthen muscles, later helpful in the development of voluntary movement. In addition, reflexes are an indicator of the maturity and soundness of the infant's nervous system. If reflexes cannot be invoked by a clinician or medical doctor at a time when they should be normally active, problems with the development of infant's nervous system are presumed. Furthermore, the persistence of a reflex past the point in time when it is normally inhibited also sends a warning to the physician or clinician. Finally, there is some belief that the neural connections responsible for infant reflexes are incorporated into later voluntary movement (Pieper, 1963; Easton, 1972). Others insist that most infant reflexes must eventually give way to the development of future motor skills (Molnar, 1978). Thus, there is much debate regarding the transitional process from reflexive to voluntary behavior. However, there is no debate about the eventual emergence of voluntary motor skills, a topic we turn to next.

Spontaneous Movements

In contrast to reflexes, infants also produce movements that occur without any apparent stimulation. These behaviors are called **spontaneous movements**. Infants can be observed producing these types of movements with their arms and legs. Spontaneous movements have been shown to be coordinated, and there is speculation that they may serve as building blocks for later voluntary, functional movements (Thelen, 1979, 1981, Jensen, Thelen, Ulrich, Schneider, and Zernicke, 199.5)

Voluntary Motor Skill Development

Beginning within the first year of the infant's life, there is a gradual increase in *voluntary* motor skill development. The repertoire of the infant's movements changes from predominantly reflexive behaviors and spontaneous movements to more willful and purposeful behavior. Beginning in infancy, the number of different types of voluntary motor skills increases as a result of the maturation of the nervous system, increases in physical growth, and experience or practice. Two of the most important motor skills that develop are locomotion and prehension. In this section, we briefly discuss how the development of these two motor skills unwinds over time by focusing on the different stages or phases of each.

Locomotion

The adage "You have to learn to crawl, before you can walk" implies that performance of certain tasks in life depends on the successful completion of previous tasks. The adage also originates from the basic observation in motor development that the development of voluntary walking may depend on the successful completion of so-called "pre-walking behaviors." In fact, there are a number of "motor milestones" prior to actual walking, which may need to be successfully completed in a progressive fashion before the infant can walk unassisted (Bayley, 1935; Shirley, 1963; Haywood and Getchell, 2009).

During the first few months after birth, babies are rather immobile. It is not possible for them to easily change their body positions or move around in the environment. Later on, they become able to change their body position, move across the ground, walk with assistance, and then finally, walk unassisted. Each of the first three milestones may have to occur before the infant can walk unassisted. The motor milestones prior to actual walking (i.e., changing body position and moving across the ground) may be categorized as **pre-walking behaviors** and are seen in Figure 8.5. The behaviors related to changing body position concern the infant's ability to produce body turns, such as side to back, back to side, and rolling from back to stomach. Pre-walking progression behaviors include *crawling* or moving on hands and knees with the abdomen on the floor and *creeping* or moving on hands and knees with the abdomen *off* the floor. Following the pre-walking behavior milestones, there are several **walking behavior** milestones.

Somewhere between 6 and 12 months, the infant is able to perform stepping movements using some assistance, either from another person or by holding on to stable fixtures in the environment, such as a piece of furniture. Later, the infant has the ability to walk with assistance. Finally, sometime typically between 9 and 12 months, the infant is able to walk unassisted. This is about the time parents start to move valuable items in the house to higher shelves! Also notice in Figure 8.5, within all the above milestones, there is considerable variability between infants of the onset and completion of these milestones. It should be pointed out that in some cultures, crawling is not permitted. Crawling is perceived to be hazardous and primitive by Jamaican parents (Hopkins and Westra, 1989) and demeaning to Balinese parents (Mead and Macgregor, 1951). Cintas (1995) concluded that the necessity of crawling for subsequent developmental skills must be questioned based on these types of sociocultural differences. Thus, the points about crawling being a necessary milestone for subsequent walking may very well be based on a more "Euro-American" bias.

Prehension

Locomotion provides the means for the infant to move and interact with the environment. But another important motor skill, prehension, allows the infant to *literally* make life a hands-on experience. The human hand is a rather unique organ among animals. One of its significant features is the anatomical position of the thumb relative to the rest of the fingers. Compared to our nearest relatives, the apes, our thumbs allow us to

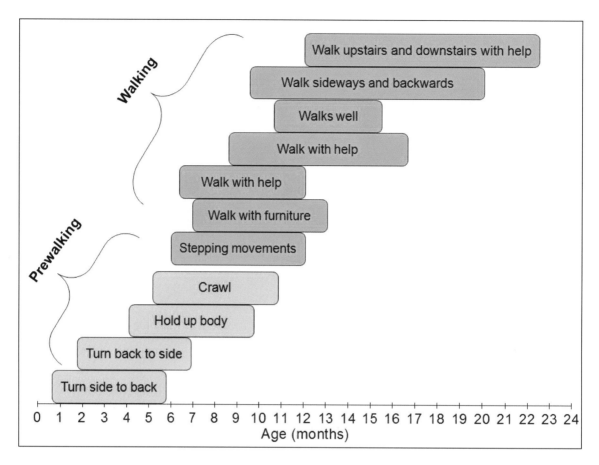

Figure 8.5

Motor milestones associated with the development of walking, (adapted from Keogh and Sugden (1985).

grasp objects much better because of the characteristic called "finger-thumb opposition." This characteristic allows the tip of our thumb to make contact with the tips of each finger, improving the manual dexterity required for manipulating, grasping, and moving objects. As adults, we take for granted this dexterity because like most of the other motor skills we possess, the use of our hands has become an almost automatic process. However, from a developmental point of view, the efficient use of our hands is a gradual and time-dependent process, requiring many years of growth, maturation, and practice.

Evidence suggests that the developmental process regarding the use of our hands involves a series of phases and associated motor milestones. A classic paper by Halverson (1931) described the developmental process of prehension. Figure 8.6 illustrates this developmental process by showing the relationship between the types of grasping behaviors and the age of the infant (and child). Halverson found that up to the age of 16 weeks, infants did not have the ability to voluntarily grasp an object (an one-inch cube was used). The infant could move the hand in the vicinity of the object and occasionally make contact with it. Primitive squeezes on the object occurred at roughly 20 weeks of age. At a later age, the infant could reach out and use a whole-hand grasp or a palm grasp on the object. It was at approximately 10 months of age that the infant could voluntarily perform a smooth and directed reach and grasp of the object with one hand.

Some researchers have criticized Halverson's work because only one type of object (a one-inch cube) and environmental condition (voluntary reaching and grasping) were tested. If different types of objects and environmental conditions are used, infants have been shown to exhibit quite an array of different reaching behaviors.

Since Halverson's work, developmental researchers have observed three types of reaching behaviors in infants: prereaching, visually guided reaching. and visually elicited reaching (Haywood and Getchell, 2009).

Prereaching behavior involves the reaching for and occasional contact with objects. The actual grasping of an object is rare but contact is more likely if the object is moving and the infant is looking directly at it (von Hofsten, 1982). This behavior is usually observed from birth to four months of age.

As the name suggests, in **visually guided reaching**, infants use vision to guide their hands to the object. Infants are able to adjust the hands during the reach, depending on the location, shape, and orientation of the object. This behavior begins around four months of age, predominates at seven months, and then gradually gives way to visually elicited behavior.

In **visually elicited behavior**, the beginning of the reach is so accurate that infants do not have to watch their hands. The movement is made quite rapidly toward the target, and vision is not required until the hand is close to the object. Visually elicited behavior begins around nine months of age and continues to improve until adultlike reaching and grasping behavior is achieved. Adultlike reaching is characterized by a great deal of anticipatory behavior; that is, the hand is *pre-shaped* during the reach for an object, depending on the speed or movement time of the reach and the object's shape and orientation (e.g., Jeannerod, 1984; Wallace and Weeks, 1988).

Type of Grasp	Weeks of Age
No Contact	16
Contact Only	20
Primitive Squeeze	20
Squeeze Grasp	24
Hand Grasp	28
Palm Grasp	28
Superior-Palm Grasp	32
Inferior-Palm Grasp	36
Forefinger Grasp	52
Superior-Forefinger Grasp	52

Figure 8.6

Motor milestones associated with unimanual prehension, (adapted from Halverson, 1931 and Haywood, 1993, Figure 3.6).

Another type of prehension, called **bimanual prehension**, involves the use of two hands in reaching and grasping for objects. There are many tasks that require bimanual prehension, such as hammering a nail or twisting the lid off of a container. The development of bimanual prehension also involves a series of phases. At approximately two months of age, infants are able to extend and raise both arms simultaneously. At four and a half months, objects are often reached for with both hands. Around the fifth month, the prevalence of bimanual reaching declines, and infants tend to reach and grasp objects with one hand. By seven months, infants can use either unimanual or bimanual reaches, depending on the task. After eight months, infants begin to use the two hands in a cooperative manner. However, the full cooperation of the hands in manipulating objects does not typically occur until the end of the second year (Bruner, 1970).

Gender Differences in Physical Activity Patterns

Between the ages from birth until puberty, there are few differences in anatomy and physiological function between boys and girls (Adams, 1991; Haywood and Getchell, 2009). In early childhood, boys are only slightly taller and heavier, but as indicated previously, girls tend to mature faster than boys and begin puberty earlier. Muscle mass differences between boys and girls tend to be minimal during childhood until puberty. Yet in spite of anatomical and physiological similarities during childhood, there are marked differences between boys and girls in motor skill performance during this time period, between the age of five and puberty.

A comprehensive review of over 60 research studies examining differences between boys and girls during childhood was done by Thomas and French (1985). This review confirmed that boys, on average, performed either slightly or much better than girls in running, throwing, long jumping, sit-up performance, and in grip strength. However, in balance and flexibility task, and tasks involving eye-hand coordination in fine motor skill, girls were generally better. What can explain these gender differences? One hypothesis suggests that there are significant upbringing differences between boys and girls that result in more involvement and participation in certain physical activities than in others. There are a number of studies showing that boys are more encouraged by their parents to participate in more vigorous activities than girls. In contrast, parents have been shown to discourage and even punish girls for participating in vigorous activities and sports (DiPietro, 1981; Eaton and Enns, 1986). As a result, many girls choose not to continue participation in these types of activities (Greendorfer, 1983) and may instead migrate toward less vigorous activities or those activities considered to be more girl-like in the eyes of society, such as ballet and gymnastics. If true, this hypothesis could explain why girls generally perform poorer than boys in the standardized sports skills test mentioned above, but do relatively better in fine motor skill or flexibility and balance tests.

Another related factor in determining the level of participation of girls in more vigorous sports is the degree of available opportunity for participation. As mentioned in Chapter 2 (The History of Kinesiology), one significant factor in increasing opportunities for girls and women in sports was the passage of Title IX of the Educational Amendment Act in 1972. Title IX mandated equal treatment for women and men in accessing educational programs or activities that receive federal assistance. In 1988 the Civil Rights Restoration Act extended this mandate of equal opportunity for men and women. It is generally agreed that these mandates have had a pronounced effect on the increase of female participation in sports and other physical activities (e.g., Adams, 1991; Wuest and Bucher, 1995).

Many of the performance differences between boys and girls that exist prior to puberty, continue to increase into adulthood (see Haywood and Getchell, 2009, for a review). After puberty, significant changes in body composition occur that contribute to these differences. Specifically, following puberty, boys tend to have more muscle mass, which probably allows them to perform many physical activities with greater strength and speed relative to females.

It should be emphasized that much of the performance differences between males and females after puberty are partially accounted for by greater training opportunities for males. Following Title IX, however, opportunities for participation in sports and vigorous physical activities have increased. As predicted by the hypothesis above, performance differences between males and females in several sports and physical activities have reduced over the last several years. There is even some suggestion that female performance in running events may surpass males' early into the 21st century (Whipp and Ward, 1992, see HIGHLIGHT)! Whether this hypothesis materializes remains to be seen.

SUMMARY

- The three types of reflexes are primitive, postural, and locomotor reflexes.
- Spontaneous movements by infants may serve as building blocks for later voluntary, functional movement.
- There are several motor milestones for locomotion and prehension.
- In spite of anatomical and physiological similarities, boys generally perform better in running, throwing, jumping, and strength-related activities, whereas girls are often better in flexibility, balance, and fine eye-hand coordination tasks.
- Upbringing differences can be cited as a major factor in the above performance differences between boys and girls.
- However, high-level performance differences between males and females have continued to reduce over the last century.

HIGHLIGHT

Will Women Win the Battle of the Sexes?

An investigation by Whipp and Ward (1992) examined the world records in running events set by men and women throughout the 20th century. Figure 8.7 illustrates the results of their investigation by plotting mean running speed in running events from 200 meters to the marathon (26.2 miles). For both men (a) and women (b), the mean running speed has systematically increased over the century (all the lines have a positive slope). In all events, men have run faster than women throughout this time period until 1992, the last year included in their investigation. However, the slopes of these lines for women were greater than the men, indicating that women have been improving their speed at a faster rate compared to men. What would happen if men and women continue to improve their running speeds at the present rate suggested by the slopes of the lines? When Whipp and Ward extrapolated the results into the future (see dotted lines in c), they found an interesting result. If men and women continue their running speed improvements at the present rate, the data suggest that women will surpass men somewhere early in the next century (noted by the intersection of the dotted lines for men in women (in Figure 8.7c)! In general, it is estimated that women will surpass men sooner in the longer running events. Do you believe these predictions? What assumptions must be met if these results are to materialize? In addition, can you speculate as to why the improvements by women over the last century have been greater compared to men?

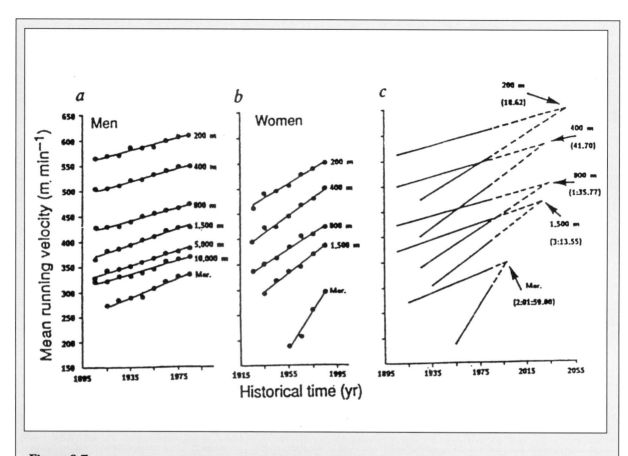

Figure 8.7

World record progression, expressed as mean running velocity versus historical time, for men (a) and women (b). In (c), the regression lines for the common events for men and women (solid lines) are extrapolated (dashed lines) to their points of intersection; the predicted world record times at these intersection points are shown in parentheses (h:min:s). From Whipp and Ward (1992).

AGING

Demographics of Aging

As living organisms, humans follow a well defined aging process from conception to death. The average number of years of life remaining for a population of people, usually expressed from birth is called **life expectancy** (Spirduso, 1995). The life expectancy for the average American is approximately 75.1 years and 80.2 years for males and females, respectively (Heron, Hoyert, Murphy, Xu, Kochanek, and Tejada-Vera, 2009). It is interesting to note that the average life expectancy of humans in various countries has increased over the last several hundred years. Figure 8.8 shows this increase in life expectancy from birth over the last several centuries, a pattern called the **secular trend of life expectancy**. It is generally recognized that this increase is not due to changes in any type of inheritable longevity. Rather it is thought to be due to rapid increases in medical knowledge and improvements in socioeconomic conditions (Espenschade and Eckert, 1980). But there are large differences in socioeconomic conditions today across various countries and as a result, there are dramatic differences in life expectancies. As illustrated in Figure 8.9, the average life expectancy in some African countries is drastically less than in Scandinavia and other countries with better socioeconomic conditions. In these latter countries, including the United States and Canada, females are expected to live longer than males, which has been the case throughout history (Holden, 1987). Why females tend to outlive their male counterparts is a subject of much debate (Spirduso, 1995). Furthermore, it is wrong to assume that the secular trend in most countries *always* continues to increase. A number of factors, such as reductions in socioeconomic conditions experienced in the early 1990s in Russia (State Committee on Statistics, 1994) and the sudden rise of infectious diseases, such as AIDS, have been shown to abruptly reduce life expectancy in certain countries (U.S. Census Bureau, 1994).

In the United States, the proportion of older individuals (65 years and over), as well as very old individuals over 85 years of age, has sharply risen over the last several decades. Figure 8.10 illustrates the actual numbers of individuals (in millions) on the vertical axis and the year on the horizontal axis. Taeuber and Rosenwaike (1992) estimated that there were about 31 million people 65 years and older representing 12 percent of the population. By the year 2040, the projections estimate that this number will increase to 60 million people or 20 percent of the population. The number of very old individuals age 85 years and older is expected to increase to between 10 and 18 million. The fact that our population is aging is expected to have a major impact on socioeconomic and health care conditions in the United States (DiPietro and Seals, 1995; Spirduso, 1995).

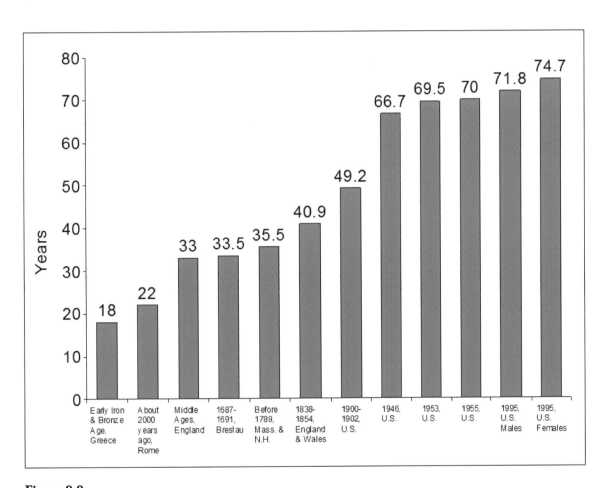

Figure 8.8

Average length of life from ancient Greece to modern times, (adapted from von Mering and Weniger, 1959 and Espenschade and Eckert, 1980, Figure 10-13).

Largely because of the realization that older individuals are likely to play a more prominent role in the United States and elsewhere into the foreseeable future, a relatively new area of the study of aging, called **gerontology**, has developed. The field of gerontology investigates several questions such as the causes of aging, the physiological decrements of aging, and the effects of physical activity on the aging process. A brief examination into what researchers in kinesiology and other disciplines have learned about these issues will now be provided.

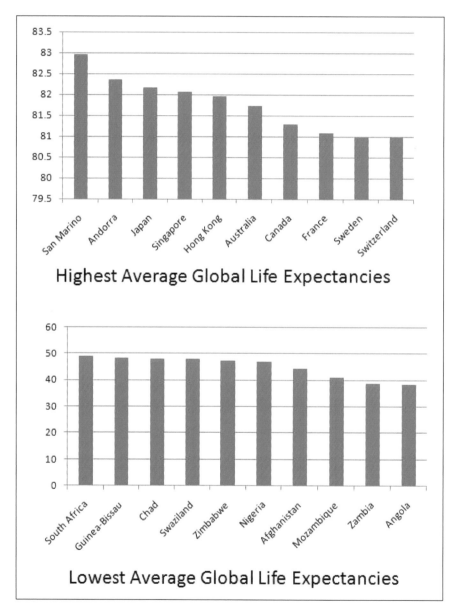

Figure 8.9

Life expectancy at birth for selected countries. (adapted from Central Intelligence Agency website: https://www.cia.gov/library/publications/ the-world-factbook/rankorder/2102rank.html)

Theories of Aging

What causes aging? There is no simple answer to this question because researchers believe there may be any number of causes of the aging process. In general, there are three general classes of theories that attempt to explain the causes of aging: microscopic (cellular) theories, macroscopic (organ/system) theories, and multiple factor theories (Chodzko-Zajko, 1995). **Microscopic theories** state that the basis for the aging process is at the cellular level of the organism (Dice, 1993). One of these theories suggests that the growth of the organism is limited to the number of cell divisions that may be "programmed" into the cell's genetic code, called the **Hayflick Limit** (Hayflick, 1976). Once this limit is reached, the cell dies. Other microscopic theories argue that the genetic makeup of cells produces mutations that eventually alter the cell's function, eventually causing disruption and finally, cell death. Still other microscopic theories propose the cause of aging does not necessarily lie in the genetic material of the cell, but is due to the accumulation of by-products of the cell division process that cannot be dealt with adequately by the cell. These by-products impede normal cell function, and eventually lead to cell death. **Macroscopic theories**, on the other hand, propose that aging is the result of deterioration of any number of physiological systems in the body such as the cardiovascular, immune, or nervous systems. For example, a deterioration of the immune system would render the body incapable of warding off disease or infection, which would eventually cause the death of the organism. Finally, **multiple factor theories** state that there is no single cause of the aging process. It may very well be the case that aging is caused by *both* microscopic and macroscopic factors. For example, even in the absence of disease or injury, the cells, organs, and various physiological systems undergo an inevitable decline in function prior to death (DiPietro and Seals, 1995),

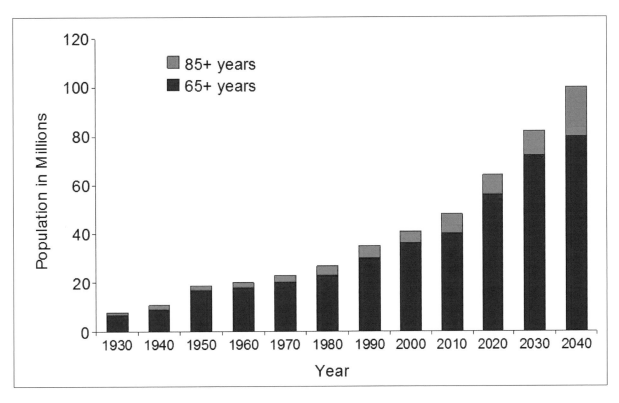

Figure 8.10

Population estimates and projections for persons aged 65 years and older and those aged 85 years and older who lived in the United States between the years 1930 and 2040. (adapted from Taeuber and Rosenwaike, 1992 and DiPietro and Seals, 1995, Figure 1-1)

a process referred to as **primary aging** (Spirduso, 1995). It is likely that microscopic factors are significant contributors to the normal aging process. However, outside factors such as poor environmental conditions, diseases, and inadequate nutritional and physical activity patterns also contribute to the effects of aging, called **secondary aging** (Spirduso, 1995).

It is unknown how long humans could live if the microscopic and macroscopic factors mentioned above were eliminated or at least controlled in some way. But some researchers believe it is possible to increase the maximum life span of humans to 115 to 120 years (Rothenberg, Lentzner, and Parker, 1991). Vladimir Chebotaryov, a gerontology researcher from the former Soviet Union, knew one individual who lived to be 158 years old (Turner and Helms, 1983)! But the oldest known individual actually documented was 120 years (Olshansky, Carnes, and Casel, 1990). Interestingly, in spite of increases in the life expectancy in developed countries (the secular trend of life expectancy), the maximum life span of humans has been relatively unchanged. In fact, the only strategy for increasing life span, verified only in mice, is food restriction (Walford and Crew, 1989; also see Spirduso, 1995, for a discussion). Mice fed only two-thirds of their normal food intake have been shown to live longer. These findings, while robust, remain controversial, and have not been verified in humans.

Physiological and Performance Decrements with Aging

Whatever the causes of aging, its effects on physiological functioning and performance are beginning to be better understood. In this section, a variety of physiological and performance decrements associated with aging are briefly described. Some of the major physiological systems affected by the aging process such as the skeletal, muscular, nervous, and cardiovascular systems will be examined. In addition, the effects of aging on reaction and movement time and balance and posture will be discussed.

The Skeletal System

As mentioned in Chapter 3, with increasing age, there is a decline in bone mineral mass, typically after the age of 30 in both males and females. The loss of bone mineral mass, usually associated with calcium loss, results in the bones becoming more brittle and more susceptible to fracture or breakage. In addition, older adults, particularly women, may suffer from a disease called **osteoporosis**, a condition that results from a lack of the mineral calcium, which leaves bones even more susceptible to injury. The maintenance of ample levels of calcium is affected by weight-bearing physical activity. If weight-bearing activity is greatly reduced, calcium can be lost and the bones become more fragile. Individuals bedridden for extended time periods and astronauts in space under weightless conditions have been shown to lose significant amounts of calcium from their skeletal system (Arnaud, Schneider, and Morey-Holton, 1986; Schneider and McDonald, 1984). Thus, it is recommended that appropriate levels of physical activity are required throughout one's life to maintain healthy bones. In addition, it is important to maintain an adequate intake of calcium as one grows older, particularly for women.

The Muscular System

Both the number and size of muscle cells, as well as the number of motor units, are reduced in older adults. Seventy-year-olds have been shown to lose up to 25 percent of their muscle mass when compared to 20-year-olds (Young, Stokes, and Crowe, 1985). This loss in muscle mass effectively reduces the amount of contractile force the muscle can produce. Therefore, muscular strength declines with aging. Figure 8.11 illustrates the results of a study on changes with grip strength in individuals from 20 to 100 years of age (Kallman, Plato, and Tobin, 1990). As one can see, there is considerable variation in grip strength loss; however, the overall trend is visible. The research is unclear as to whether aging affects the loss of fast and slow twitch muscle fibers differently (see Spirduso, 1995, for a discussion).

The loss of flexibility in joints is also associated with the aging process (Chapman, deVries, and Swezey, 1972). Flexibility around a joint is reduced because the muscles that surround it shorten due to infrequent use. In addition, the stretching of muscles, an activity that improves joint flexibility, is practiced less and less as we grow older. The reduction of flexibility due to less frequent joint movement as we age effectively reduces the full range of joint motion, affecting joint movement speed.

The Nervous System

As we grow older, the number of neurons in the nervous system declines. Due to the reduction in neural connections, the amount of interference among neurons increases and negatively impacts sensory and motor function. As a result, older individuals experience losses in visual acuity that can affect their balance (Tobis, Reinsch, Swanson, Byrd, and Scharf, 1985). Older individuals also have poorer depth perception and peripheral vision than their younger counterparts, affecting their ability to perform many perceptual motor tasks such as walking over obstacles (Chen, Ashton-Miller, Alexander, and Schultz, 1991) and driving an automobile (Kline, Kline, Fozard, Kosnik, Schieber, and Sekuler, 1992).

The Cardiovascular and Pulmonary Systems

The cardiovascular and pulmonary systems of the older individual also are impaired relative to his or her younger counterpart. The major blood vessels of older adults become more rigid (arteriosclerosis) and lipid deposits within the blood vessels increase (atherosclerosis). It is not clear whether these effects are due to primary or secondary aging but a sedentary lifestyle certainly plays a role. Increases in arteriosclerosis and atherosclerosis

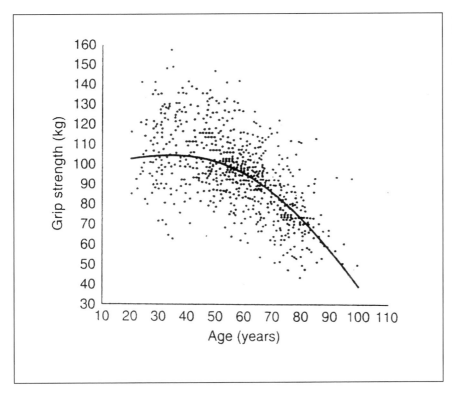

Figure 8.11
Changes in grip strength for 847 males from 20 to 100 years of age. From Spirduso (1995, Figure 5.2) and Kallman, Plato, and Tobin (1990, pg. M83).

accompanying the aging process usually lead to high blood pressure (hypertension). Roughly 40 percent of individuals in the United States over 65 have hypertension (Vokonas, Kannel, and Cupples (1988). Maximum heart rate decreases with aging. A rough approximation of maximum heart rate can be obtained by subtracting one's age from 220 (beats per minute). For example, a 65-year-old's maximum heart rate would be expected to be approximately 155 beats per minute (220 - 65 = 155). Stroke volume of the heart also reduces in older adults, possibly because the heart muscle loses contractile strength and becomes stiffer (Klausner and Schwartz, 1985). Because maximum heart rate and stroke volume tend to decrease in older individuals, total cardiac output also declines (see Chapter 4, Exercise Physiology Foundations). Because cardiac output is a major determinant of oxygen delivery and nutrients to the body, reductions in cardiac output with aging hamper cardiovascular performance. However, these reductions are seen, to a larger degree, in *less* healthy or sedentary individuals, compared to physically active older people (see below). In the absence of disease, pulmonary (lung) function in the elderly is maintained quite well during rest and moderate exercise. Thus, the ability to retrieve oxygen from the atmosphere is not typically a problem. However, because cardiac output is less in the elderly (due to reductions in maximum heart rate and possibly stroke volume), maximum oxygen uptake (VO2 max) tends to decline in older individuals. It has been shown that VO2 max in 70-year-olds can be 40 percent less than in 30-year-olds (Astrand and Rodahl, 1986). The generally accepted decline in VO2 max is approximately 10 percent per decade, with the most significant reductions occurring after the age of 60 (Buskirk and Hodgson, 1987). Reductions in VO2 max limit cardiovascular performance or aerobic capacity in the elderly.

Response Speed and Motor Skills

The ability to react and respond to stimuli in the environment is generally reduced as a function of aging. This overall slowness of response affects many aspects of the older individual's life. It takes older people longer to complete most tasks, from doing household chores to performing work-related duties. Research has confirmed that both reaction time, movement time and overall response time are significantly slower in older individuals. The slowness of response is thought to be due to a combination of physiological decrements mentioned above, although a number of theories have been proposed (see Spirduso, 1995, for a review). In addition, it has been shown that movement skill and coordination decline with age in tasks such as aiming movements, visual tracking, handwriting, throwing skills, and automobile driving. However, in spite of this decline, older individuals use compensation strategies such as trading speed for accuracy. Contrary to popular belief, even though older individuals perform more poorly on perceptual-motor tasks related to driving skills, senior citizens are *least* likely to be involved in traffic accidents (Evans, 1988). It is likely that older drivers take fewer risks and drive with more caution as a result of experience.

Aging and the Effects of Physical Activity

In this last section, we explore the effects of exercise and physical activity on aging. A large body of research has been done in kinesiology and other related disciplines on the influence of exercise and an active lifestyle on the aging process. How do the various physiological systems (e.g., skeletal, muscular, nervous, cardiovascular) respond to exercise with advancing age? Does an active lifestyle affect life expectancy? In other words, can exercise extend one's life? Does exercise influence primary or secondary aging?

Skeletal System Changes

As noted previously, there is significant bone mineral loss as we grow older, particularly for women. However, there is growing evidence that individuals living a sedentary lifestyle lose more bone mineral content than individuals who regularly exercise (Dalsky, Stocke, Ehsani, Slatopolsky, Waldon, and Stanlexy, 1988; Dilson, Berker, Ordl, and Varan, 1989). Apparently, even very old women (over 80 years old) can increase bone mineral mass by exercising regularly (three times per week) over a three-year period (Smith, 1982). These results suggest that maintaining a regular exercise program throughout one's life is important in limiting bone loss in older age. However, much more work needs to be done to verify this assertion (LeBlanc and Schneider, 1991). In addition, it would appear that the *type* of exercise required to maintain bone mineral content must be weight bearing. Weight-bearing exercise (against gravity) stresses bone cells into producing calcium (Smith, 1982).

Muscular System Changes

As pointed out earlier in the chapter, muscular strength decreases with advancing age, largely due to the loss of muscle mass. We know that resistance training such as weight lifting, for example, improves muscular strength in young adults (see Chapter 4, Exercise Physiology Foundations). Can physical activity improve muscular strength in the elderly? It has been well documented that older adults who stay physically active are stronger than their sedentary counterparts (Laforest, St-Pierre, Cyr, and Gayton, 1990; Viljanen, Viitasalo, and Kujala, 1991). In addition, resistive exercise can improve strength at any age in adults (Frontera, Meredith, O'Reilly, Knuttgen, and Evans, 1988). Even very old and quite frail individuals over 80 years of age have been shown to dramatically increase muscular strength as a result of an eight-week progressive resistance leg exercise program (Fiatarone, Marks, Ryan, Meredith, Lipsitz, and Evans, 1990). Not only did the individuals increase muscular strength, they also exhibited increased mobility and function as a result of the exercise program such as being able to rise from a chair unassisted and eliminate the use of a cane to walk. In spite of these and other research findings, many physicians and exercise physiologists have expressed caution and have even discouraged progressive resistance training in older adults (ACSM, 1991, p. 166). There is some belief that heavy resistance training may put undue pressure within the chest cavity, increasing the likelihood of cardiovascular accidents. However, there is limited research supporting the view that heavy resistance training is dangerous for the elderly. It is probably safe to say that under the supervision of a physician, clinician, or trained exercise physiologist, older adults can expect significant improvements in muscular strength and even mobility by engaging in progressive resistance exercise programs. However, individuals with cardiac impairments or disease should avoid heavy resistance training.

As noted previously, joint flexibility significantly decreases as we grow older, largely due to disuse. However, engaging in a longer-term exercise and physical activity program can offset much of this limitation. Research has shown that physical activity, such as muscle stretching and progressive resistance training, can improve flexibility (increase joint range of motion) in the elderly (Brown and Holloszy, 1991; Chapman, deVries, and Swezey, 1972).

Nervous System Changes

As we grow older, there is a reduction in the number of neurons in the nervous system and corresponding neuronal interconnections. Nerve conduction velocity has also been shown to reduce with advancing age. Retzlaff and Fontaine (1965) demonstrated that exercising rats have faster nerve conduction velocities than non-exercising rats. The actual diameters of neurons have also been shown to be larger in mice who have participated in lifelong exercise, compared to sedentary mice. Thus, there could be a relationship between activation of muscle in exercise and neuron function. In addition, physical practice (repetition) of a motor skill has been shown to increase the number of, and interactions among, neuronal cells in the brains of animals (Floeter and Greenough, 1979; Pysh and Weiss, 1979). There is some belief that older animals also are capable of neuronal regeneration with physical practice (Spirduso, 1995). However, more research is needed on older humans to verify the results of animal studies.

Cardiovascular and Pulmonary Changes

With advancing age, we have discussed accompanying decrements in a variety of cardiovascular and pulmonary parameters that influence performance. Does exercise help to minimize these decrements in older adults? In regard to aerobic capacity, individuals who have sustained a lifelong habit of vigorous exercise can expect 50 percent less decrement in VO2 max with age (Dempsey and Seals, 1995, for a review). However, aerobic capacity can decline even in endurance athletes who have stopped or significantly reduced their activity level (Buskirk and Hodgson, 1987; Marti and Howald, 1990). In general, the aerobic capacity with advancing age appears to be due to several factors such as physical activity level, gender, and disease factors (secondary aging), as well as the process of aging itself (primary aging). Figure 8.12 illustrates how VO2 max (aerobic capacity) is expected to decline with age, depending on the type of individual.

As shown in Figure 8.12, the decline in aerobic capacity depends on the type of lifestyle and physical condition of the individual. Endurance athletes who maintain their conditioning throughout life have the best maintenance of aerobic capacity. In stark contrast, non-athletes exhibiting a sedentary lifestyle and eventually developing some form of cardiovascular disease have the poorest maintenance of aerobic capacity over the years. It should be emphasized that there is great variability among people, even within a given lifestyle condition. The decline in aerobic capacity is likely due to some combination of primary and secondary aging. One of the difficult tasks facing gerontology researchers is to determine the separate contributions of primary and secondary aging to changes in aerobic capacity with advancing age. There is some evidence that older individuals who are healthy and fit have better pulmonary (lung) function than their sedentary counterparts, allowing better ventilation and gas exchange. However, it is not clear whether extended exercise training over the years *causes* these advantages, or whether those individuals with more age-resistant lung function chose to stay active longer (Dempsey and Seals, 1995).

Exercise and Longevity

Let us address the issue of whether exercise and the adoption of a more active lifestyle can extend one's life. To address this issue, we must again make the distinction between the concepts of life expectancy, average life span, and maximum life span. *Life expectancy* is the average number of years of life remaining for a population of people, usually expressed from birth. As stated earlier, the life expectancy of Americans is roughly 72 and 75 years of age for males and females, respectively. The *average life span* is the average age that all but a few individuals decease. The average life span of most Americans is around 85 years of age, and has not significantly changed for centuries (Spirduso, 1995). *Maximum life span* is the maximum age any member of a species or race of people can live, presumably set by biological limits. The maximum life span of humans is thought to be 120 years of age. How does regular exercise and certain health practices affect each of these aging parameters? There is actually very little research on this topic, particularly with humans, but it is generally agreed that a physically active lifestyle does *not* increase maximum life span. Individuals who are physically active, have good nutritional habits, and are otherwise free from disease do not exceed the 120-year maximum life span limit. However, there is considerable evidence, particularly for animals, that exercise increases the *average* life span and the average life expectancy (see Holloszy, 1988, for a review). Individuals who are physically active throughout their lives are expected to live longer than sedentary individuals. A large study on nearly 17,000 Harvard graduates, ages 35 to 74, confirmed that those individuals who were more physically active lived longer (Paffenbarger, Hyde, Wing, and Hsieh, 1986). The amount of physical activity required to show this effect was roughly equivalent to jogging four hours per week.

When an individual initiates or begins an exercise program and whether they maintain the exercise program appears to be critical to resulting life expectancy. The research generally suggests that the sooner in life an individual starts and maintains an exercise program, the better. Furthermore, having a physically active lifestyle early in one's life does not ensure a longer life expectancy if the individual reverts to a sedentary lifestyle. For example, studies show that being a former athlete provides no advantages to longevity if a sedentary lifestyle is subsequently adopted by the individual (Montoye, Van Huss, Olson, Pierson, and Hudec, 1957; Paffenbarger, et al., 1986). Maintaining a physically active lifestyle *throughout* one's life is the key to longevity (Spirduso, 1995).

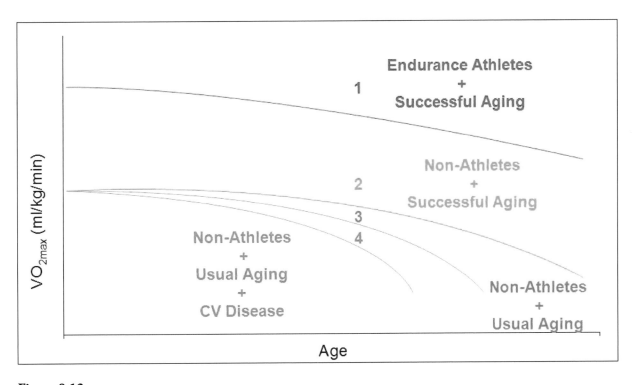

Figure 8.12

Age-related declines in VO2 max with advancing age in different populations of humans: 1) endurance athletes who remain highly active and otherwise undergo successful aging, i.e., they gain minimal body fat, are free of disease, and have genetic resistance to adverse effects of aging (Rowe and Kahn, 1987); 2) non-athletes who remain somewhat active and otherwise undergo successful aging; 3) non-athletes who become progressively less active and undergo usual aging (i.e., gradual increase in body fat, minor disease states, less aging-resistance genetic makeup, etc.); and 4) non-athletes who become progressively less active and develop serious cardiopulmonary disease (e.g., coronary heart disease with left ventricular dysfunction, (adapted from Dempsey and Seals, 1995, Figure 6-1).

Exercise Intensity

Does the *intensity* of the exercise program provide additional benefits to longevity? There is considerable debate about this issue among exercise physiologists, as well as gerontologists. A study by Blair, Kohl, Paffenbarger, Clark, Cooper, and Gibbons (1989) on over 13,000 men and women suggested that the intensity of exercise does not increase longevity. Blair et al. evaluated five separate groups of individuals who participated in exercise programs over an eight-year period. They found that those individuals in the lowest fit group died at 3.5 to 4.5 times the rate of the fit men and women, respectively. The low fit individuals had the highest rates of cancer and cardiovascular disease. However, low to moderate levels of fitness were as effective in increasing longevity as high fitness. The view that vigorous exercise does not promote increased longevity has been recently disputed (Lee, Hsieh, and Paffenbarger, 1995). In a study of 22,000 men who graduated from Harvard from 1924 to 1954, it was found that hard and regular workouts provided up to a 25 percent reduction in death risk, compared to moderate exercise. It is still too early to make conclusions regarding the relationship between exercise intensity and life expectancy. However, most researchers would agree that maintaining some form of an active lifestyle is critical to a long and healthy life.

Exercise and Health

In this regard, is there a relationship between physical activity and physical health? **Physical health** depends on at least three major elements of the individual: the physical condition, the functional condition, and the subjective health status of the individual (Spirduso, 1995). The *physical condition* is related to the number of health problems experienced by the individual. For example, well over three-quarters of people over 70 years of age experience one or more chronic diseases. We have shown that exercise, particularly the lifelong variety, helps to reduce the incidence of cardiovascular disease. The *functional condition* is related to the ability of the individual to perform so-called activities of daily living (ADL), such as bathing, dressing, climbing stairs, rising from a chair, and so forth. While there is not as much specific research on the relationship between physical activity and the functional condition of the aging individual (specifically ADL tasks), it is likely that exercise and physical practice are beneficial.

Some researchers have introduced the concept of **active life expectancy** to better evaluate the influence of exercise and physical activity on longevity. Active life expectancy was defined by Katz, Branch, Branson, Papsidero, Beck, and Greer (1983) as the number of years free of disability in one's life. If given a choice, one would wonder whether living a longer life would be desirable if it was not free of disability or if it required dependence on others for survival. It has been determined that active life expectancy is always less than life expectancy for everyone (Branch, Guralnik, Foley, Kohout, Wetle, Ostfeld, and Katz, 1991). However, there is little research on whether long-term exercise programs increase active life expectancy, an issue that deserves more attention. Finally, the *subjective health status* of the individual is related to the individual's own perceptions and feelings about his or her health. Often, the feelings an individual has about one's own physical condition can influence his or her behavior. Spirduso (1995) concluded that research generally shows exercise can have both short-term and long-term effects on one's feelings of well-being, life satisfaction, and other psychological states. But there is little research on how exercise affects the feelings and perceptions of the elderly. Clearly, more research is needed in this area.

In conclusion, exercise has been shown to improve the physical condition of the elderly, particularly if a lifelong or relatively long-term exercise program is maintained. The functional condition of the elderly is likely to benefit from improvements in physical health, but more specific research is needed to verify this expectation, particularly in ADL tasks. Finally, while exercise has been shown to contribute to the enhancement of a variety of mental states in younger adults, its effects, both short- and long-term, need to be determined in older individuals.

SUMMARY

- Life expectancy has increased in the United States over the last several centuries, a phenomenon called the secular trend of life expectancy.
- Life expectancy in any country is influenced by many factors.
- The elderly population in the United States is rapidly increasing.
- Gerontology is the study of aging.
- There are primary and secondary factors associated with aging.
- The three general theories of aging are microscopic, macroscopic, and multiple factor theories.
- There are several physiological and performance decrements associated with the aging process.
- Exercise and physical activity can influence life expectancy and average life span but not maximum life span.
- The maximum life span of a human appears to be limited to around 120 years of age.
- The physical health of an individual depends on his or her physical condition, functional condition, and the subjective health status of the individual.

CHAPTER SUMMARY

- The factors contributing to the development of the individual are growth, maturation, learning, heredity, and environmental influences.
- There are several stages (or phases) of development.
- The two major research designs used to study development are the cross- sectional and longitudinal designs.
- Boys and girls show similar growth patterns until puberty.
- Physical growth of various physiological systems (e.g., skeletal, muscular, and nervous) continues until after the completion of puberty.
- Infants possess a number of reflexes that have some functional purpose.
- Locomotion and prehension are two major voluntary motor skills that progressively develop in phases, called "motor milestones."
- Probably due to upbringing differences, boys typically perform better in running, throwing, jumping, and strength-related activities, and girls are often better in balance, flexibility, and tasks requiring fine eye-hand coordination.
- Performance differences between male and female elite athletes have consistently reduced over the last century.
- Life expectancy in many countries, including the United States, has increased over the last several centuries, a phenomenon called the secular trend of life expectancy.
- The percentage of people 65 years of age and older has sharply increased over the last several decades.
- The study of aging is called gerontology.
- There are several theories of aging.
- The maximum life span of humans appears to be around 120 years of age.
- Aging is accompanied by several physiological and performance decrements.
- However, exercise and physical activity, particularly if maintained over one's life, improves functional capacity and health with advancing age.

IMPORTANT TERMS

Active life expectancy - the number of years free of disability in one's life

Average life span - is the average age that all but a few individuals decease

Bimanual prehension – the use of two hands in manipulating an object

Cartilage - a tough but flexible tissue

Cross-sectional design - use of at least two separate groups of subjects differing in age

Development - the changes in the capacities and skill level of the individual with age

Environmental influences - a large set of variables such as parental and family interactions, dietary factors, and educational opportunities

Epiphyseal plates – the location of bone growth

Functional condition - the ability of the individual to perform so-called activities of daily living (ADL), such as bathing, dressing, climbing stairs, rising from a chair, and so forth

Gender differences- differences between males and females

Gerontology – the study of the aging process

Gestation – the period of growth of the human fetus

Growth - a quantitative increase in the size of a certain anatomical structure

Hayflick Limit - the number of cell divisions that may be "programmed" into the cell's genetic code

Hereditary - a set of qualities that are relatively fixed at birth, which tend to set limits of certain characteristics of the individual

Learning – the relatively permanent change in performance as a function of practice

Life expectancy - the average number of years of life remaining for a population of people, usually expressed from birth

Locomotion – ambulation of the whole body

Locomotor reflexes - a complicated series of movements that resemble certain voluntary ambulatory motions such as crawling, walking, or even swimming

Longitudinal design - examines an individual (or group of individuals) over an extended period

Macroscopic theories - propose that aging is the result of deterioration of any number of physiological systems in the body such as the cardiovascular, immune, or nervous systems

Maturation - the increase in the functional capacity of an individual (or of a body part) without necessarily an increase in growth

Maximum life span - the maximum age any member of a species or race of people can live, presumably set by biological limits

Microscopic theories - state that the basis for the aging process is at the cellular level of the organism

Multiple factor theories - state that there is no single cause of the aging process

Osteoporosis - a condition that results from a lack of the mineral calcium, which leaves bones more susceptible to injury

Physical condition - the number of health problems experienced by the individual

Physical health - depends on at least three major elements of the individual: the physical condition, the functional condition, and the subjective health status of the individual

Postural reflexes - assist the infant in maintaining posture (i.e., head or body position) in a changing environment

Prehension – the voluntary use of the hands in manipulating objects

Prereaching - involves the reaching for and occasional contact with objects

Pre-walking behaviors – the motor milestones of crawling and creeping

Primary aging – the decline in function prior to death in the absence of disease or injury, of the cells, organs, and various physiological systems

Primary ossification centers –located within the mid-portions of long bones such as the femur and humerus where the beginning of ossification occurs

Primitive reflexes - are those primarily under the control of the lower brain centers and can be elicited in utero (within the womb) or in premature infants

Puberty - a process of physical changes in which a child's body matures into an adult body capable of sexual reproduction to enable fertilisation

Reflex – an involuntary response to a stimulus

Saltation – periods of growth spurts

Secondary aging - outside factors such as poor environmental conditions, diseases, and inadequate nutritional and physical activity patterns that contribute to the effects of aging

Secondary ossification centers - areas of bone growth after birth occurring near the end of the shaft

Secular trend in life expectancy - the increase in life expectancy from birth in a population of people over the last several centuries

Stages of development – significant age related changes in aging

Stasis – periods of no growth

Subjective health status - related to the individual's own perceptions and feelings about his or her health

Visually elicited behavior – rapid infant reaching movements where vision is not required until the hand is close to the object

Visually guided reaching - the infant's use vision to guide their hands to the object

Walking behaviors milestones – significant advances in locomotion that lead to adult walking behavior

INTEGRATING KINESIOLOGY: PUTTING IT ALL TOGETHER

1. Based on your knowledge of the development of the skeletal system, what physical activities should be discouraged for a growing child?

2. After receiving permission by the parents, try to identify some of the primitive, postural, and locomotor reflexes in an infant (avoid testing for the swimming reflex for obvious reasons!). If you are able to examine a young child less than two years old, examine evidence for prereaching, visually guided reaching, or visually elicited behavior.

3. Using the Web site below, determine the current world records held by men and women in various track and field events. Plot the differences in the records between men and women on a graph. Are these the kinds of trends that would be predicted by the Whipp and Ward (1992) study?

4. In doing the previous assignment, you may have noticed that men and women from the United States have tended to dominate the short-distance running events over the last 15–20 years. However, this domination is not the case at the longer running events. Can you think of any explanations for these performance differences?

5. If you are planning to develop a fitness program for yourself (Chapter 4), what developmental issues should you consider?

6. What is your definition of a "healthy" lifestyle? Take into account the concepts of life expectancy, average life span, maximum life span, physical health, and active life expectancy.

KINESIOLOGY ON THE WEB

- www.ehow.com/motor-development—This is an interesting Web site, covering many aspects of motor development.
- www.trinity.edu/~mkearl/geron.html—This is a Web site containing varied information on aging, such as the biology and psychology of aging.
- These two Web sites contain current U.S. and Canadian records for men and women in track and field:
  http://en.wikipedia.org/wiki/United_States_records_in_track_and_field html
  http://en.wikipedia.org/wiki/Canadian_records_in_track_and_field

REFERENCES

Adams, W. C. (1991). *Foundations of Physical Education, Exercise and Sport Sciences.* Philadelphia: Lea and Febiger.

American College of Sports Medicine. (1991). *Guidelines for Exercise Testing and Prescription.* Philadelphia: Lea and Febiger.

Arnaud, S. B., Schneider, V. S., and Morey-Holton, E. (1986). Effects of inactivity on bone and calcium metabolism. In H. Sandler and J. Vernikos (eds.), *Inactivity: Physiological Effects.* Orlando: Academic.

Astrand, P.-O., and Rodahl, K. (1986). *Textbook of Work Physiology: Physiological Bases of Exercise.* New York: McGraw-Hill, 1986, 3rd ed., p. 409.

Baldwin, K. M. (1984). Muscle development: neonatal to adult. In R. L. Terjung (ed.), *Exercise and Sport Science Reviews. Vol. 12,* pp. 1–19. Lexington, MA: Collamore.

Bayley, N. (1935). The development of motor abilities during the first three years. *Society for Research in Child Development Monograph, 1,* 1.

Blair, S., Kohl, H. W., Paffenbarger, R. S., Clark, D. G., Cooper, K. H., and Gibbons, L. W. (1989). Physical fitness and all-cause mortality: a prospective study of healthy men and women. *Journal of the American Medical Association, 262,* 2395–401.

Branch, L. G., Guralnik, J. M., Foley, D. J., Kohout, F. J., Wetle, T.T., Ostfeld, A., and Katz, S. (1991). Active life expectancy for 10,000 Caucasian men and women in three communities. *Journal of Gerontology: Medical Science, 46,* M145–M150.

Brown, M., and Holloszy, J. O. (1991). Effects of a low-intensity exercise program on selected physical performance characteristics of 60- to 71-year-olds. *Aging, 3,* 129–39.

Bruner, J. S. (1970). The growth and structure of skill. In K. J. Connolly (ed.), *Mechanisms of Motor Skill Development* (pp. 63–94). London: Academic.

Buskirk, E. R., and Hodgson (1987). Age and aerobic power: the rate of change in men and women. *Federation Proceedings, 46,* 1824–29.

Chapman, E. A., deVries, H. A., and Swezey, R. (1972). Joint stiffness: effects of exercise on young and old men. *Journal of Gerontology, 27,* 218–21.

Chen, H., Ashton-Miller, J. A., Alexander, N. B., and Schultz, A. B. (1991). Stepping over obstacles: gait patterns of healthy young and old adults. *Journal of Gerontology: Medical Sciences, 46,* M196–203.

Chodzko-Zajko, W. (1995). The aging process: structural changes and functional consequences. A paper presented at the American Academy of Kinesiology and Physical Education, Vail, Colorado, Oct. 7–10.

Cintas, H. L. (1995). Cross-cultural similarities and differences in development and the impact of parental expectations on motor behavior. *Pediatric Physical Therapy, 7,* 103–11.

Dalsky, G. P., Stocke, K., Ehsani, A. A., Slatopolsky, E., Waldon, C. L., and Stanlexy, J. B. (1988). Weight-bearing exercise training and lumbar bone mineral content in postmenopausal women. *Annals of Internal Medicine, 108,* 824–28.

Dempsey, J. A., and Seals, D. R. (1995). Aging, exercise, and cardiopulmonary function. In D. R. Lamb, C. V. Gisolfi, and E. Nadel (eds.). *Perspectives in Exercise Science and Sports Medicine: Volume 8, Exercise in Older Adults.* Carmel, IN: Cooper, pp. 237–304.

Dice, J. F. (1993). Cellular and molecular mechanisms of aging. *Physiological Review, 73,* 149–59.

Dilson, G., Berker, C., Ordt, A., and Varan, G. (1989). The role of physical exercise in prevention and management of osteoporosis. *Clinical Rheumatology, 8,* 70–75.

DiPietro, J. A. (1981). Rough and tumble play: a function of gender. *Developmental Psychology, 17,* 50–58.

DiPietro, L., and Seals, D. R. (1995). Introduction to exercise in older adults. In D. R. Lamb, C. V. Gisolfi, and E. Nadel (eds.). *Perspectives in Exercise Science and Sports Medicine: Vol. 8, Exercise in Older Adults.* Carmel, IN: Cooper, pp. 1–10.

Easton, T. A. (1972). On the normal use of reflexes. *Scientific American, 60,* 591–99.

Eaton, W. O., and Enns, L. R. (1986). Sex differences in human motor activity levels. *Psychological Bulletin, 100,* 19–28.

Espenschade, A. S., and Eckert, H. D. (1980). *Motor Development* (2nd ed.). Columbus, OH: Merrill.

Evans, L. (1988). Older driver involvement in fatal and severe traffic crashes. *Journal of Gerontology: Social Sciences, 43,* S186–S193.

Fiatarone, M. A., Marks, E. C., Ryan, N. D., Meredith, C., Lipsitz, L. A., and Evans, W. J. (1990). High-intensity strength training in nonagenarians. *Journal of the American Medical Association, 263,* 3029–34.

Floeter, M., and Greenough, W. T. (1979). Cerebellar plasticity: modification of Purkinje cell structure by differential rearing in monkeys. *Science, 206,* 227–29.

Frisancho, A. R. (1990). *Anthropometric Standards from the Assessment of Growth and Nutritional Status.* Ann Arbor: University of Michigan Press.

Frontera, W. R., Meredith, C. N., O'Reilly, K. P., Knuttgen, H. G., and Evans, W. J. (1988). Strength conditioning in older men: skeleton muscle hypertrophy and improved function. *Journal of Applied Physiology, 64,* 1038–44.

Galloway, A., Stini, W. A., Fox, S. C., and Stein, P. (1990). Stature loss among an older United States population and its relation to bone mineral status. *American Journal of Physical Anthropology, 83,* 467–76.

Gibson, A. T., and Wales, J. K. H. (1994). *Human Biology: Budapest, 24,* 34.

Greendorfer, S. L. (1983). Shaping the female athlete: the impact of the family. In M. A. Boutilier and L. Sangiovanni (eds.), *The Sporting Woman* (pp. 135–55). Champaign, IL: Human Kinetics.

Halverson, H. M. (1931). An experimental study of prehension in infants by means of systematic cinema records. *Genetic Psychology Monographs, 10,* 107–286.

Hayflick, L. (1976). The cell biology of human aging. *New England Journal of Medicine, 195,* 1302–1308.

Haywood, K. M., and Getchell, N. (2009). *Life Span Motor Development.* Champaign, IL: Human Kinetics. 5th ed.

Heron, M., Hoyert, D. L., Murphy, S. L., Xu, J. Kochanek, K. D., and Tejada-Vera, B. (2009). *National Vital Statistics Reports, 57,* 1–136.

Holden, C. (1987). Why do women live longer than men? *Science, 238,* 158–60.

Holloszy, J. (1988). Exercise and longevity: studies on rats. *Journal of Gerontology: Biological Sciences, 43,* B149–B151.

Hopkins, B., and Westra, T. (1989). Maternal expectations of their infants' development: some cultural differences. *Developmental Medicine and Child Neurology, 31,* 384–90.

Jeannerod, M. (1984). The timing of natural prehension movements. *Journal of Motor Behavior, 16,* 235–54.

Jensen, J. L., Thelen, E., Ulrich, B. B., Schneider, K., and Zernicke, R. F. (1995). Adaptive dynamics of the leg movement patterns of human infants: III. Age-related differences in limb control. *Journal of Motor Behavior, 27,* 366–74.

Kallman, D. A., Plato, C. C., and Tobin, J. D. (1990). The role of muscle loss in the age-related decline of grip strength: cross-sectional and longitudinal perspectives. *Journal of Gerontology: Medical Science, 45,* M82–M88.

Katz, S., Branch, L. G., Branson, M. H., Papsidero, J. A., Beck, J. C., and Greer, D. S. (1983). Active life expectancy. *New England Journal of Medicine, 309,* 1218–24.

Keogh, J., and Sugden, D. (1985). Movement skill development. New York: Macmillan.

Klausner, S. C., and Schwartz, A. B. (1985). The aging heart. *Clinical Geriatric Medicine, 1,* 119–41.

Kline, D. W., Kline, T. J. B., Fozard, J. L., Kosnik, W., Schieber, F., and Sekular, R. (1992). Vision, aging, and driving: the problems of older drivers. *Journal of Gerontology: Psychological Sciences, 47,* P27–P34.

Laforest, S., St-Pierre, D. M. M., Cyr, J., and Gayton, D. (1990). Effects of age and regular exercise on muscle strength and endurance. *European Journal of Applied Physiology, 60,* 104–11.

Lampl, M., Cameron, N., Veldhuis, J. D., and Johnson, M. L. (1995). A response in *Science, 268,* 445–47, to Heinrichs, C., Munson, P. J., Counts, D. R., Cutler, G. B, and Baron, J. (1995). Patterns of growth, *Science, 268,* 442–45.

Lampl, M., Veldhuis, J. D., and Johnson, M. L. (1992). Saltation and stasis: a model of human growth. *Science, 258,* 801–803.

LeBlanc, A., and Schneider, V. (1991). Can the adult skeleton recover lost bone? *Experimental Gerontology, 26,* 189–201.

Lee, I. M., Hsieh, C. C, Paffenbarger, R. S. Jr. (1995). Exercise intensity and longevity in men: the Harvard alumni health study. *Journal of the American Medical Association, 273,* 1179–84.

Malina, R. M. (1978). Growth of muscle tissue and muscle mass. In F. Falkner and J. M. Tanner (eds.), *Human Growth: Vol. 2. Postnatal Growth*. New York: Plenum.

Marti, B., and Howald, H. (1990). Long-term effects of physical training on aerobic capacity: controlled study of former elite athletes. *Journal of Applied Physiology, 69*, 1451–59.

Mead, M. and MacGregor, F. C. (1951). Growth and Culture: A Photographic Study of Balinese Childhood. New York: G. P. Putnam.

Mering, O. von, and Weniger, F. L. (1959). Social-cultural background of the aging individual. In J. E. Birren (ed.), *Handbook of Aging and the Individual*. Chicago: University of Chicago Press.

Milani-Comparetti, A. (1981). The neurophysiological and clinical implications of studies on fetal motor behavior. *Seminars in Perinatology, 5*, 183–89.

Molnar, G. (1978). Analysis of motor disorder in retarded infants and young children. *American Journal of Mental Deficiency, 83*, 213–22.

Montoye, H. J., Van Huss, W. D., Olson, H. W., Pierson W. R., and Hudec, A. J. (1957). *The Longevity and Morbidity of College Athletes*.

Indianapolis: Phi Epsilon Kappa Fraternity. National Center for Health Statistics (1993). Monthly vital statistics report. *Advance Report of Final Mortality Statistics*, 1990, 41 (7). Washington, DC: U.S. Government Printing Office.

Olshansky, S. J., Carnes, B. A., and Cassel, C. (1990). In search of Methuselah: estimating the upper limits to human longevity. *Science, 250*, 634–40.

Paffenbarger, R. S., Jr., Hyde, R. T., Wing, A. L., and Hsieh, C. C. (1986). Physical activity, all-cause mortality, and longevity of college alumni. *New England Journal of Medicine, 314*, 603–13.

Payne, V. G., and Isaacs, L. D. (2008). *Human Motor Development: A Lifespan Approach*. New York: McGraw-Hill, 7th ed.

Peiper, A. (1963). *Cerebral Function in Infancy and Childhood*. NY: Consultants Bureau.

Pysh, J. J., and Weiss, G. M. (1979). Exercise during development induces an increase in Purkinje cell dendritic tree size. *Science, 206*, 230–31.

Retzlaff, E., and Fontaine, J. (1965). Functional and structural changes in motor neurons with age. In A. T. Welford and J. E. Birren (eds.),

Behavior, Aging, and the Nervous Systems (pp. 340–52). Springfield, IL: Charles C. Thomas.

Rothenberg, R., Lentzner, H. R., and Parker, R. A. (1991). Population aging patterns: the expansion of mortality. *Journal of Gerontology: Social Sciences, 46*, S66–S70.

Rowe, J. W., and Kahn, R. L. (1987). Human aging: usual and successful. *Science, 237*, 143–49.

Schneider, V. S., and McDonald, J. (1984). Skeletal calcium homeostasis and countermeasures to prevent disuse osteoporosis. *Calcified Tissue International, 36*, S151–S154.

Shirley, M. M. (1963). The motor sequence. In D. Wayne (ed.). *Readings in Child Psychology*. Englewood Cliffs, NJ: Prentice Hall.

Smith, E. L. (1982). Exercise for the prevention of osteoporosis: a review. *Physician and Sports Medicine, 3*, 72–80.

Spirduso, W. W. (1995). *Physical Dimensions of Aging*. Champaign, IL: Human Kinetics.

State Committee on Statistics (1994). Reported in the *Daily Camera*, Boulder, CO.

Taeuber, C. M., and Rosenwaike, I. (1992). A demographic portrait of America's oldest old. In: R. M. Suzman, D. P. Willis, and K. G. Manton (eds.) *The Oldest Old*. Oxford, England: Oxford University Press, pp. 17–49.

Tanner, J. M., Whitehouse, R. H., and Taikishi, M. (1966). Standards from birth to maturity for height, weight, velocity , and weight velocity: British children, 1965–1. *Archives of Disease in Childhood, 41*, 454–71.

Thelen, E. (1979). Rhythmical stereotypes in normal human infants. *Animal Behaviour, 27*, 699–715.

Thelen, E. (1981). Kicking, rocking, and waving: Contextual analysis of rhythmical stereotypes in normal human infants. *Animal Behaviour, 29*, 3–11.

Thomas, J. R., and French, K. E. (1985). Gender differences across age in motor performance: A meta-analysis. *Psychological Bulletin, 98*, 260–82.

Tobis, J. S., Reinsch, S., Swanson, J. M., Byrd, M., and Scharf, T. (1985). Visual perception dominance of fallers among community-residing older adults. *Journal of the American Geriatrics Society, 33*, 330–33.

Turner, J. S., and Helms, D. B. (1979). *Life Span Development.* Philadelphia: W. B. Saunders.

United Nations Monthly Bulletin of Statistics, April 1971.

U.S. Census Bureau (1994). Reported in *Daily Camera,* Boulder, CO, April 29, 1994.

Viljanen, T., Viitasalo, J. T., and Kujala, U. M. (1991). Strength characteristics of a healthy urban adult population. *European Journal of Applied Physiology, 63*, 43–47.

Vokanus, P. S., Kannel, W. B., and Cupples, L. A. (1988). Epidemiology and risk of hypertension in the elderly: the Framingham study. *Journal of Hypertension, 6 (Suppl. 1)*, S3–S9.

Von Hofsten, C. (1982). Eye-hand coordination in the newborn. *Developmental Psychology, 18*, 450–61.

Walford, R. L., and Crew, M. (1989). How dietary restrictions retard aging: an integrative hypothesis. *Growth, Development and Aging.* Winter, 139–40.

Wallace, S. A., and Weeks, D. L. (1988). Temporal constraints in the control of prehensile movements. *Journal of Motor Behavior, 20*, 81–105.

Whipp, B. J., and Ward, S. A.(1992). Will women soon outrun men? *Nature, 255*, 25.

Wuest, D. A., and Bucher, C. A. (1995). *Foundations of Physical Education and Sport.* St. Louis: Mosby.

Young, A., Stokes, M., and Crowe, M. (1985). The size and strength of the quadriceps muscles of old and young men. *Clinical Physiology, 5*, 145–54.

Chapter Nine
Sociocultural Foundations

Sociocultural Factors in Participation Patterns of Physical Activity

The Process of Socialization

Importance of the Family, Peer Groups, and the School

Gender Differences in Sport and Physical Activity Participation

Racial Factors in Sport and Physical Activity

Adult Patterns of Physical Activity

Improving Participation in Physical Activity from a Sociocultural Perspective

Viewing the Human Body from a Sociocultural Perspective

Summary

CHAPTER SUMMARY

Important Terms

Integrating Kinesiology: Putting It All Together

Kinesiology on the Web

References

CHAPTER NINE

Sociocultural Foundations

"Human beings learn to be social beings."

– (Loy and Ingham, 1973)

STUDENT OBJECTIVES

1. To understand the important terms related to sociology and culture.
2. To appreciate the contribution of sociocultural kinesiology to the study of physical activity.
3. To become familiar with different sociological theories and how they relate to the study of kinesiology.
4. To identify the various factors influencing the socialization into physical activity.
5. To understand how participation in physical activity and sport is influenced by gender and race.
6. To identify the various sociocultural strategies for improving society's interest and participation in physical activity.

In this chapter, we describe the final subfield in kinesiology dealing with sociological and cultural factors in the study of physical activity. In the first section, definitions of sociology and culture are provided, followed by a rationale for why this approach to the study of human behavior is relevant to kinesiology. The subfield of sociocultural factors related to kinesiology is briefly described, along with some of the issues that are studied within this subfield.

THE RELEVANCE OF SOCIOCULTURAL FACTORS TO KINESIOLOGY

Sociology and Culture

Much of what we know about how society and culture affect our participation in physical activity has been greatly influenced by another scientific field of study called sociology. **Sociology** is the scientific study of human behavior in group (or social) settings (Brezina, Selengut, and Meyer, 1990). Sociology stresses the importance of a society (or a social system) in shaping our thoughts, feelings, and behavior. Sociology seeks to explain the order as well as the disorder in society (Spenser, 1976).

A **society** is a large, relatively permanent group of people such as the group of people within a country, who share common values and beliefs. Within a society, various social systems may exist. A **social system** within a society is a group, large or small, of individuals who interact. Some examples of a social system are a group of neighborhood children, a sports team, or the convicts within a given prison. Sociologists study group behavior within a society as a whole or within the social systems of a society, and try to determine how the society or the social system affects individual thoughts, feelings, and/or behaviors.

There are a number of questions that sociologists address, such as why societies and social systems are the way they are, and what factors led them to develop in a certain way. Sociologists often deal with issues that affect society as a whole such as cultural differences between the races, the nature of social deviancy (behaving against established norms), and social factors affecting the cognitive and emotional development of girls and boys.

The collective behavior within a society or its social systems changes dynamically over time. The behavior of the members within a society or social system can be cooperative or competitive. **Cooperative behaviors** are those behaviors that help or assist others for the purpose of achieving some goal. **Competitive behaviors** are those behaviors that attempt to outperform or defeat another individual. Different reward systems may exist within a society or its social systems to promote either cooperative or competitive behavior among its members (Coakley, 1998).

Culture refers to the fundamental values and beliefs of a given society or social system as well as the norms followed and the material goods created by its members (Giddens, 1991). There are many elements of culture, called **cultural universals**, shared by all cultures in some way. For example, all cultures possess some type of language and writing system that allows its members to communicate with one another. All cultures have some form of family system and some type of institution of marriage. Other cultural universals have been identified such as the norms regulating art, dance, body adornment and personal hygiene, and sexual activity (Murdock, 1945).

There are marked cultural differences among the various societies throughout the world. Cultural differences between societies may be larger than their geographic separation. Many Americans who visit Canada for the first time notice how similar it is to the United States in terms of the services, businesses, modes of travel, language, and other elements of its society. But other cultural elements between Canada and the United States have been noted, such as attitude differences between Canadians and Americans (Malcolm, 1990).

Cultural differences between the United States and Mexico are more obvious. An obvious cultural difference is that Mexicans speak Spanish, while Americans primarily converse in English. The cultures of Canada, the United States, and Mexico, with all their differences, probably are more similar (having all been influenced by European culture) than most countries in the Middle East, for example. There is often a tendency among peoples of a given culture to judge other cultures relative to their own, which is known as **ethnocentrism**. Sociologists, as scientists, attempt to avoid ethnocentrism in their studies of other cultures, an approach called **cultural relativism**. In addition, sociologists believe that society cannot exist without culture, and culture cannot exist without society. In this way, society and culture are thought to be inseparable. We might view society as the structural foundation of a group of people, and culture as its functional foundation.

Socialization is an important process of study within the field of sociology. By definition, socialization is a learning process through which a person acquires the necessary values, norms, roles, and skills necessary to function in a society (Spenser, 1976). What are the major factors contributing to the socialization process? How do we learn our respective roles in society? How do these roles change over time? These are just some of the many important questions addressed by sociologists pertaining to socialization.

Sociocultural Kinesiology

How does the study of sociology and culture relate to kinesiology? In general, the type of culture we live in greatly influences how we view ourselves, our bodies, and the importance of physical activity. Socialization into society begins at birth. We shall see that even as very young children, the extent and type of physical activity in which we participate is related to the roles, norms, and values we learn. We shall see that males and females are often socialized quite differently regarding participation in physical activity, a topic we previously touched on (Chapters 7 and 8).

Our ethnic backgrounds certainly influence our participation in physical activity. In addition, our participation and experiences in organized sports, particularly as children, have had a significant impact on our values about physical activity. Much of the research on the sociocultural influences of physical activity has been done in sports or sports-related settings, an area of study traditionally called **sport sociology**. As I discussed in the first chapter, there are many types of physical activity, with sport being one. **Sociocultural kinesiology** can be defined as the study of social and cultural factors influencing physical activity in many environments, such as play, recreation, games, and exercise as well as sport settings. While this term is not commonly used, I believe that it better represents the sociocultural factors that influence participation in all types of physical activity, including sport.

Sport sociologists Kenyon and Loy warned in 1965 that it should not be assumed that all physical activity is necessarily good (Loy and Kenyon, 1965). According to these scholars, the study of sport and physical activity should be "value-free" if it is to be objectively understood. In addition, they emphasized that the social scientist studying physical activity should not be an "evangelist for exercise." It should pointed out that the notion of "value-free" research has been the subject of debate among social scientists over the years (see Greendorfer, 1981, for a discussion). While the benefits of physical activity and exercise have been emphasized throughout this book, there are also negative consequences. In this chapter, we will explore some of the "darker sides" of physical activity. Thus, in addressing Kenyon and Loy's concern, both the positive *and* the negative consequences of physical activity should be studied and better understood by kinesiologists.

SUMMARY
- Sociology is the scientific study of human behavior in group (or social) settings.
- A society is a large, relatively permanent group of people, such as the group of people within a country, who share common values and beliefs.
- A social system within a society is a group of individuals, large or small, that engages in interaction.
- Cooperative and competitive behaviors are two of the major behaviors expressed by people within a society or social system.
- Culture refers to the fundamental values and beliefs of a given society or social system, as well as the norms followed and the materials and goods created by its members.

- Elements of culture common to most societies such as language are called cultural universals.
- The tendency among peoples of a given culture to judge other cultures relative to their own is called **ethnocentrism**.
- Socialization is the process through which a person acquires the necessary values, norms, roles, and skills necessary to function in society.
- Sociocultural kinesiology is the study of social and cultural factors in all types of physical activity settings.

General Theoretical Approaches

In the field of sociology, a number of theories have been developed that attempt to explain why social life is organized in particular ways. In this section, brief descriptions of functionalism, conflict theory, critical theory, and symbolic interactionism are provided, along with some discussion as to how each of the theories may be applied to social and cultural issues in physical activity.

Functionalism

Starting with the writings of Auguste Comte (1789–1857), **functionalism** emerged as a theory that focused on how elements of a society contribute to the continuation of the society as a whole. One of the key features of functionalism is that to understand how a society functions, it is important to determine how one element of a society is related to another. A good analogy might be that to understand how the heart works, it must be shown how the heart affects the functioning of other parts of the body such as the lungs (Giddens, 1991). In a similar way, to understand how religion functions as an element in society, it must be determined how religion impacts other elements of society such as politics, for example. Another aspect of functionalism is an underlying view that all elements of society work together for the continuation and maintenance of the society.

One application of functionalism related to kinesiology has been to the study of sport. When a functionalist approach is used to study sport, one key question is how does sport contribute to the functioning of society as a whole. Coakley (1998), a leading sport sociologist, has summarized how functionalism has been applied to sport. Most studies of functionalism applied to sport address the following issues:

1. Does participation in sports help the individual learn the rules and values of society?
2. Do sports help bring people together and create the unity needed to maintain order and efficiency in society?
3. Do sports help teach people to be committed to social progress and do sports help people become more self-disciplined?
4. Do sports teach people the skills required to contribute to the survival of the society? (Coakley, 1998)

These are interesting and important questions, and the reader is encouraged to explore Coakley's (1998) examination of the research on these topics. In general, he argues that people often use functionalist theory to demonstrate that participation in sports is important for the overall health of the society.

A functionalist approach would also help us understand attitudes and beliefs toward other types of physical activities such as exercise. Since the 1960s, there has been an increasingly popular belief that participation in exercise improves the psychological and physiological state of the individual. A number of professional organizations (e.g., American College of Sports Medicine, American Psychological Association) and agencies (President's Council on Physical Fitness and Sports, National Institutes of Health, etc.) have acknowledged the psychological and physiological benefits of participation in physical activity and the contribution of physical activity to disease prevention. Yet, as we have pointed out in previous chapters, the levels of participation in physical activity are low in the United States and many other societies. One would think that with all this increased awareness of the benefits of physical activity and exercise, more people would adopt an active lifestyle. A functionalist approach has a difficult time explaining why most people choose not to engage in adequate levels of physical activity. This approach has difficulty explaining why a physically active lifestyle is adopted by some people and not others.

Conflict Theory

Conflict theory has its origins in the writings of Karl Marx (1818–1920), and a number of sociologists who followed him have provided a number of reinterpretations of his views (Giddens, 1991). Marx believed that, particularly in countries based on market (or capitalistic) economies, it is inevitable that certain groups of people will achieve power and wealth at the expense of others. Aspiration to achieve power and wealth will result in a stratification of people into different classes. Conflict among the "haves" and the "have-nots" in capitalistic societies will inevitably arise. According to Coakley (1998), workers in such societies require some type of escape to deal with tensions produced by such conflicts. One way to deal with these types of tensions is to search for activities such as sport, which provide escape and entertainment. Thus, one application of conflict theory to kinesiology is that participation in certain types of physical activity such as sports is viewed as an escape from the tensions and conflicts experienced in the workplace. In addition, activities like sport, particularly in market economies, help perpetuate the power of the elite through economic exploitation. Sports are not viewed by conflict theorists as inspirational, but are considered the "opiate" of the masses (Coakley, 1998). Conflict theory has been used to illustrate the negative side of participation in sports, such as how sports promote racism (addressed later in the chapter) and sexism, and how it is used to understand the relationship between sports and economic exploitation. Much more empirical support is needed to support the application of conflict theory to participation in physical activity.

Critical Theory

A number of sociologists since Karl Marx have argued that social life is not just dominated by class struggles due to economic exploitation. **Critical theory** was formulated to take into account many other factors that can create struggles among people in a society. Among these factors are power struggles in social systems, deviancy, and female and male relationships, to name just a few. Critical theorists also believe that these struggles change over time and that both competition and agreement cooperation among people coexist to help determine a given social climate. Critical theorists also seek to understand the potential of a society, given its social and material resources, and to determine what the society should do to maximize its potential.

Critical theory relates to kinesiology in several ways. In relation to sport, critical theory attempts to understand how sports have evolved and changed in our society. In other words, what are the factors that have determined how sports fit into our society and how sports have impacted social life. In a similar manner, critical theory can be used to ask questions about how the importance of physical activity and exercise has changed over the decades in our society as well as others. In addition, critical theory can ask what can be done by the society and its social groups to improve participation in physical activity.

Symbolic Interactionism

The last major theoretical approach used by sociologists that we will discuss is **symbolic interactionism**, originally developed by George Herbert Mead (1863–1931). Mead argued that much of human existence is driven by symbolism. A symbol is something that represents something else. For example, the words in a language symbolize the actual objects, people, places and events in nature. Once we have mastered a language, we are able to contemplate the objects in nature even when they are not visible or present by using symbols. Mead also believed that much of the interaction between individuals is an exchange of symbols. The clothes that we wear often signify the type of person we are, or think we are, and they may send these impressions to others. For example, if we wear a suit to an interview, this sends a message that we are serious about the job. The tone of our voice represents our emotional state at any particular time. The tone of our voice tells others how we feel about the topic of conversation. In general, the visual and verbal "symbols" we use while interacting with other people have much to do with our intentions and our behavior in a social setting. In addition, the symbolic interactionist approach argues that our *interpretations* of the symbols that are displayed by others can affect how we interact and behave in a social setting. Furthermore, the same event can be interpreted differently, depending on the social context.

There is a variety of symbols that are exchanged among people in our society when it comes to participation in sports, exercise, and physical activity in general. The wearing of a team logo T-shirt typically sends a message to others that we support that team. But the social context where this T-shirt is worn can influence how this type of symbol is interpreted by others. If a Colorado sports fan wears a University of Colorado football team T-shirt at a home game and sits with other home town fans, he or she is viewed as just another supportive fan. If this same fan wears the Colorado T-shirt at an away game in Nebraska and sits near Cornhusker fans, he or she might be viewed as a troublemaker. It certainly might be a wise choice to wear a *different* T-shirt in downtown Lincoln after the game!

The symbolic interactionist approach can be used to understand behavior in other physical activity settings. A sociocultural kinesiologist might be interested in determining whether Nike commercials featuring Michael Jordan have any influence on increasing physical activity patterns of African-American children. One aspect of this research may be to identify the important symbols in these commercials and how these symbols are interpreted by children.

There are some limitations of the symbolic interactionist approach. It tends to focus more on an *individual's* behavior and decision making, rather than the behaviors of social groups. In this way, the symbolic interactionist approach is as much a psychological theory as it is a sociological one. However, the symbolic interactionist approach has advanced our knowledge of sociological phenomenon through many empirical studies both within and outside physical activity settings (Coakley, 1998).

SUMMARY

- Functionalism is a sociological theory that attempts to understand how elements of a social system are functionally related to one another for the maintenance and continuation of the society.
- Conflict theory postulates that class struggles among the "haves" and "have-nots" dominate those societies with primarily market economies.
- Critical theory argues that there are many factors within society that create struggles among groups of people. Both cooperation and competition among people within a society or social group influence how the social group evolves over time.
- Symbolic interactionism focuses on the use of symbols by people in a social group. How symbols are used, as well as interpreted by others, is a major aspect of this theory.
- All of these theoretical approaches have been used to understand sociocultural influences in the participation in sport, exercise, and physical activity in general.

RATIONALIZATION AND SOCIETY

As we have seen in the last section, there are a number of theoretical perspectives developed by sociologists that provide insight into the nature of society, and why people interact as they do in social groups. These theories also provide us with ways to understand the factors that influence participation in physical activity, as well as how exercise and sport are valued in our society. There is another perspective, related to some of the notions in conflict theory, that deserves an examination within the context of our discussion. This perspective is called **rationalization**, defined as a society's increasing reliance on the scientific method and technology (e.g., Schlucter, 1979). Rationalization will be seen as providing many benefits for society. However, the negative aspects of rationalization will also be discussed, particularly with reference to participation in physical activity.

Elements of Rationalization

When we think of a modern society, like the United States, Canada, or Japan, what comes to mind? Other than the physical or geographical features, we could perhaps describe modern (or developed) societies on the basis of the technological innovations they possess. Communication technologies like the telephone, television, e-mail, and the Internet have greatly enhanced our ability to interact with one another. Transportation breakthroughs associated with automobile, rail, and air travel allow us to get to places with greater speed and efficiency. Significant advances in medicine and health care have reduced mortality rates and increased life expectancy in most modern societies. We have been able to create a whole host of time- and effort-saving devices that have made many of our jobs less physically demanding and have contributed to increases in leisure time. All of these achievements have come as a result of scientific advancements. We generally view these scientific and technological achievements in a positive way. Yet, some social scientists have argued that there has been a downside to these changes (e.g., Mitchell, 1982; Schlucter, 1979; Tenner, 1996). With these changes, a trend has developed for increasing the management of our lives. As we have strived to make our lives more efficient and practical, modern society has tended to reject the impractical and spontaneous in favor of the measured and predictable. This trend, which sociologists call "rationalization," has invaded all aspects of our lives, both in work, leisure and other physical activities. With rationalization comes a need to quantify everything we do, to increase the predictability of events in our lives and to rid ourselves of behaviors that are deemed inefficient or impractical.

Progressive Differentiation. At first glance, these so-called negative aspects of rationalization may seem trivial. However, many sociologists believe that rationalization has led to the stifling of the quality of life. How has rationalization negatively affected modern life? One effect of rationalization is that it has led to a compartmentalization of life's activities, a process called **progressive differentiation** (Mitchell, 1992). Progressive differentiation is a modern term for the concept of division of labor originally conceived by Adam Smith (1776). With progressive differentiation, new (and presumably more efficient) units of organization are developed in society that provide specific functions. For example, in ancient times, virtually all aspects of one's life were dominated by the family. With rationalization, other institutions were developed, such as schools, churches and factories, to provide specific functions originally performed by the family. But, progressive differentiation has also resulted in the division between work and leisure. We have come to view work and leisure as separate activities. Since work is supposed to be efficient and productive, behaviors we commonly reserve for leisure activities, such as laughing, singing, and other playful acts are generally discouraged in work environments. Play, for most working adults, is restricted to designated time slots called vacations and holidays. Work has become, for many individuals, routine, predictable, and boring.

Disassociation. Another downside is that rationalization has led to a disassociation between the benefits of technological advancements and a real understanding of our way of living (Mitchell, 1982). We all acknowledge the benefits of modern air travel, but few of us really have any idea of how an airplane flies, and some of us could care less. We buy medication at a drug store with the expectation that it will be effective but with little understanding of its contents. In contrast, so-called "primitives" living in the bush have a much better appreciation of the nature of their environment, how and why their tools work, and the contents of the food they eat (Freund, 1969).

Revenge Effects. Finally, some social scientists have argued that technological advancements are a type of double-edged sword, in that the benefits from them are offset by new unanticipated problems, called **revenge effects** (Tenner, 1996). One revenge effect brought on by technology is environmental damage caused by the development of the automobile, not to mention the injuries and deaths in automobile accidents and heightened stress levels during rush hour traffic. The increased reliance on the computer has forced ten million Americans each year to seek professional help due to eye strain (U.S. National Institute for Occupational Safety and Health, 1991). Carpal tunnel syndrome (CTS) sufferers experience severe hand pain and dysfunction due to a compression of the median nerve by inflammation of the surrounding tendons. While debated, the syndrome is associated with computer keyboard typing, yet evidence of CTS in standard typewriter users is nonexistent. One explanation is that the computer keyboard allows for more keystrokes per second than either standard or electric typewriters. This type of overuse injury suggests that our hands were not designed to perform small, rapid, and repeated motions for hours on end (Tenner, 1996).

As a final example, we briefly turn to the revenge effects due to advances in medical technology. No one can doubt that improvements in medical technology have had a profound effect on health care. The polio vaccine, developed by Jonas Salk (1914–1995) in the 1950s has nearly eradicated poliomyelitis. Cardiac bypass surgery has extended the lives of thousands of heart patients every year. A variety of medications provides relief from headaches, upset stomachs, and sore throats, to name just a few ailments. But advances in medical technology have not come without revenge effects. Survivors of many acute diseases eventually develop more chronic or long-term problems. It is estimated that nearly a fourth of the quarter million survivors of the polio epidemic in the 1950s will eventually develop post-polio muscular atrophy, which affects muscular strength (Moody, 1988; Munsat, 1991). Laboratory tests that accompany routine physical examinations cost the nation $30 million a year, but rarely reveal any pathological problems. Patients apparently expect these tests and doctors order them primarily to protect themselves from malpractice suits (Grady, 1996). Following successful cardiac surgery, most heart patients undergo years of rehabilitation that often require close supervision of highly trained health professionals, a process that escalates the rising costs of health care. While medicine has had success dealing with many acute health problems, such as trauma from accidents, relief from pain, and some viral and bacterial infections, it has done less well in dealing with chronic health problems. Chronic health problems are often the result of a multitude of complex, interacting factors, cost the nation $425 billion, and affect 100 million Americans every year. Much of this cost is associated with hospital stays (Recer, 1996). In addition, for modern medical technology to properly work, a high degree of craftsmanship on the part of medical personnel and highly reliable, costly, and sophisticated surgical, diagnostic, and therapeutic equipment are required, both of which have a tendency to fail at an alarming rate. It has been estimated that nearly 1 out of every 25 hospital patients will suffer some unintended adverse effect during their stay. One out of every 100 patients will suffer from medical negligence and 1 in 400 will needlessly die (Brennan, et al., 1991; Leape, 1991). Instead of relieving medical personnel from constant attention to patients and equipment, modern medical technology demands more from the practitioner (Tenner, 1996).

In sum, today's rationalized society is predictable and measurable, characteristics that allow us to function efficiently. We have come to define the quality of life, in part, by the so-called "improvements" brought by science and technology. However, with these improvements may come revenge effects of technology that do not necessarily enhance the quality of life in modern society. Similar arguments can be made with reference to rationalization and its effect on our participation in physical activity.

Influence of Rationalization on Sport and Physical Activity

According to some social scientists, rationalization has influenced most elements of our society including leisure, sport, and other forms of physical activity (Mitchell, 1982; Tenner, 1996). As mentioned above, progressive differentiation, an element of rationalization, has compartmentalized our lives into work and play. We are supposed to restrict playful activities to times outside of the working environment. Many of us use our leisure time to engage in recreational activities, sports, exercise, and other physical activities. How has rationalization affected these activities?

Quantification of Physical Activity. One major element of rationalization has been the increasing emphasis on the quantification of life's activities and the need to predict and control the events in our world. This emphasis is thought to come at the expense of the joys associated with spontaneous and perhaps "impractical" behavior. Rationalization has also affected our participation in physical activity. No clearer evidence of this emphasis can be presented than the influence of rationalization on sports. Sports historian Allen Guttman pointed out that the ancient Greeks and Romans admired their athletes not for setting records but for their bravery and demonstrated skill (Guttman, 1982). Of course, it could be argued that they lacked the technology for detailed record keeping.

Today, the emphasis on technology in sport is widespread. We have so many detailed records on athletic performance it is even possible to play simulated baseball based on the statistics of each player in the league! Indeed, technology has impacted virtually every sport. First, we should consider the apparent success stories of technology applied to sport. The recent developments of the bicycling helmet has dramatically reduced death and injury rates in those states that require their use (Rivara, 1994; Health News, 1990). Technological improvements in the luge, a sport in which occupants of the sled fly down frozen tracks at 80 miles an hour, have reduced injuries so substantially, that it is once again an Olympic sport after being banned in the 1960s (Tenner, 1996).

Revenge Effects. In many sports, technological improvements have been accompanied by revenge effects. In football, artificial turf, once thought of as a panacea for better traction and improved maintenance, has taken its toll on the player's ligaments, tendons, and joints. Evidence suggests that knee sprains have a higher rate of occurrence on artificial turf than on natural grass surfaces (Powell and Schootman, 1992). In spite of improvements in the football helmet and padding, two out of three professional football players required surgery in the 1970s and 1980s, compared to one in three in the 1950s. In addition, two thirds of the National Football League's players retired with a chronic injury in the last two decades (Couzens, 1992). Furthermore, there is some evidence that face masks used in hockey may actually enhance injuries, not reduce them (Chambers, 1996). According to Colorado College coach Don Lucia, "Since we've put face masks on [1980–1981 NCAA season], people have become gladiators. We've traded stitches on chins for broken necks and separated shoulders. The game would be better and safer if we take off the face masks." Tenner (1996) has suggested that improvements in the design of these pieces of equipment have allowed, perhaps encouraged, players to play the game with more intensity. Furthermore, it has become commonplace for players to use drugs, painkillers, and steroids to prolong their careers. But the toll taken on their bodies has been, in many cases, a high price to pay for the glory of the game. For example, Joe Namath, once a star quarterback in the 1970s, has had both of his knees replaced with artificial joints.

Other sports and recreational activities have also witnessed the negative consequences of technological improvements. Boxing, an ancient sport, has had its share of technological improvements. Prior to the 1880s, opponents would fight only with their bare fists. The sport was considered a brutal activity. But with the advent of boxing gloves and electric lighting, interestingly enough, the sport has become more glorified with the likes of Mohammed Ali, Sugar Ray Leonard, and other champions who have garnered millions of dollars in contests sometimes lasting less than a minute (Gorn, 1986). But with the technological changes came severe revenge effects: cumulative and chronic injuries. According to Tenner (1996), boxing gloves have allowed fighters to deliver more repeated blows to the body and head than was ever the case with bare-fisted fighting. Fighters today are much more likely to suffer from eye and nerve damage than their non-boxing counterparts. Muhammad Ali, in describing his boxing style, used to chant the he would "float like a butterfly and sting like a bee" in the boxing ring. But he may have accumulated many more blows to his head than even he realized. Today, Ali, considered by some the greatest athlete of all time, suffers from incurable Parkinson's disease, possibly as a result of repeated blows rather than knock-down punches. It should be noted that protective head gear used in amateur boxing reduces the likelihood of injury (Schmid, et al., 1968; Lubell, 1989). But its appeal has not reached the professional ranks.

There are many more examples of revenge effects in physical activities such as in skiing (Anders, 1993; Underwood, 1995), running and the influence of the modern running shoes (Edell Health Letter, 1991; Hoeberigs, 1992; Jones, Cowan, Knapik, 1994; Rolf, 1995; Thomas, 1994; Van Mechelen, 1992), mountain climbing (Robinson, 1993; Tenner 1996) and tennis (Arthur, 1992; Irvine, 1994; Tenner, 1996). Rationalization has led to advancements in technology that have probably allowed (and perhaps encouraged) participants to engage in physical activity with much greater intensity. Accompanying this increased intensity is often a dramatic rise in physical injuries.

Attitudinal Changes About Physical Activity. As pointed out earlier, one element of rationalization is the increased emphasis on prediction, quantification, and efficiency. Has this element of rationalization influenced our attitudes about participation in physical activity? As we learned in Chapter 7 (Psychological Foundations), our long-term involvement in physical activity has much to do with our intrinsic motivation, or our desire to participate in the activity for its own sake without the expectation or desire for external rewards. Have we, as a society, increased our reliance on technology, perhaps at the expense of enjoyment, intrinsic satisfaction, or appreciation of the physical activities engaged in? For example, in sailing, global positioning systems (GPS), once exclusively used by the military, are taking the place of the sailor's use of traditional navigation skills. These devices are capable of establishing the latitude and longitude of one's vessel within meters of accuracy. Some sailing advocates worry that over-reliance on these systems may, at best, diminish the intrinsic satisfaction of the activity, and at worst, put the sailor at risk if the GPS fails. One sailor friend of mine told me that some sailors believe the best back-up system for a failed GPS device is *another* GPS device!

In summary, rationalization has had a tremendous effect on our participation in physical activity. It is clear that many benefits can be gained from advances in technology. But it is equally clear that there can be negative consequences in using new technology in physical activity settings. From a sociocultural perspective, we must be aware of these costs and benefits and make personal and social decisions as to whether increased technology is in our best interest.

The Concept of Flow—Revisited

From a sociocultural viewpoint, it is possible to describe cultures on the basis of their technological advancements. Csikszentmihalyi (1990) has argued that there may be another way to compare cultures other than technological advancements. First, how well does a given culture provide access to its people's experiences that are in line with their goals? Second, how well do these experiences lead to personal growth of the individual? And third, how well does a given culture allow as many of its people to develop complex skills? With these criteria, it is possible that the state of the culture's technology does not necessarily influence the quality of life for its people. Despite improvements in life span and technology, the Industrial Revolution in Europe in the 18th century did not convincingly improve the life experiences of the English people (Thompson, 1963). The same can be said of the German people during the Nazi regime in the 1930s and 1940s. Technology and longer life does not necessarily ensure happiness or an increased quality of life. Conversely, the people in many so-called "primitive" cultures that lack high technology, such as the pygmies of the Uturi forest, live within their means, respect each other and their environment, and generally lead productive and challenging lives (Turnbull, 1961).

We will return again to the concept of flow and how it applies to other social conditions, such as parental upbringing of their children in the family life situation. In the next section, we examine several sociocultural factors that influence participation in physical activity.

SUMMARY

- Rationalization is the trend in a society of increasing use and dependency on technological advancements
- The positive aspects of technology are often offset by unexpected negative consequences, called revenge effects.
- Revenge effects have been shown to negatively influence health care and participation in physical activity.

SOCIOCULTURAL FACTORS IN PARTICIPATION PATTERNS OF PHYSICAL ACTIVITY

From child to adult, there are many sociocultural factors influencing our participation in physical activity. In this section, we examine some of these factors. First, the process of socialization is briefly discussed. Next, the importance of the family, peer groups, and the school to the socialization process into physical activity are discussed. Gender and racial factors in participation in sports and physical activity are examined. Adult physical activity patterns are then examined. Finally, a discussion of how sociocultural influences can affect physical activity patterns is presented.

The Process of Socialization

Socialization can be defined as "a complex developmental learning process that teaches the knowledge, values and norms essential to participation in social life" (McPherson, Curtis, and Loy, 1989, p. 37). In the socialization process, we learn how to be a family member and a member of various social groups and the large community. Our involvement in physical activity, sport, and exercise is influenced by social systems. In Figure 9.1, the various social systems that contribute to socialization are depicted. These systems are: the nuclear family, the extended family, our peer groups, the mass media, minor sport organizations, churches, and schools. As depicted in Figure 9.1, each of these social systems influences the socialization of an individual. There are several theoretical accounts of how one learns social roles. One particularly important one is called **social learning theory** (see Leonard, 1988, for a discussion). There are three major elements of social learning theory: reinforcement, coaching, and observational learning. Social learning theory acknowledges the influence of *reinforcement* on social learning, namely, the rewards and punishments one receives from significant others. *Coaching* refers to the deliberate teaching of a behavior by a significant other such as by a parent or teacher. Finally, social learning can come about as a result of *observational learning* or watching someone else perform the behavior to be learned. Also referred to as modeling, observational learning has been shown to be a powerful socializing device (Bandura, 1971; McCullagh, Weiss, and Ross, 1989). These three elements (reinforcement, coaching, and observational learning) provide the mechanisms that allow for the learning of social roles in the various social systems.

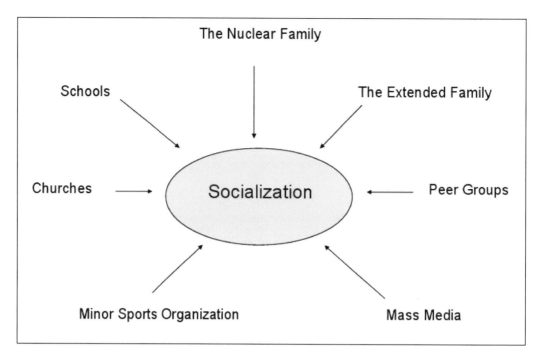

Figure 9.1

The significant others within social systems in which a child can interact (from McPherson, 1982).

Importance of the Family, Peer Groups, and the School

Of the several factors contributing to the socialization of the child into sport and physical activity, the family, peer groups, and the school have been found to be the most influential (Lewko and Greendorfer, 1988). Therefore, we will concentrate on these three factors in this section. The family, as a social system, is thought to have a profound influence on the child's participation in sport and physical activity. On the surface, this view would seem to make sense, given that children spend most of their early years at home or in the presence of their parents. Considerable research suggests that parents, in particular, exert strong influences on their children's interest in physical activity (Brustad, 1993; 1996; Eccles and Harold, 1991; Freedson and Evenson, 1991; Moore, Lombardi, White, Campbell, Oliveria, and Ellison, 1991; Weiss and Hayashi, 1995). For example, it has been shown that there is a relationship between the physical activity level of children and their parents. Moore et al. studied the physical activity patterns of children from four to seven years of age, as well as their parents. To directly measure physical activity, the subjects in the study wore Caltrec accelerometers that indicate the amount of body movement during a day. The parents and the children were then classified as being active or inactive based on established norms. The researchers found that children of two active parents were *six times* more likely to be active than children whose parents were classified as inactive.

These results do not tell us why active parents have active kids, and researchers have followed up with some studies for explanations (e.g., Brustad, 1996; Kimiecik, Horn, and Shurin, 1996). One possibility is that children may tend to imitate their parents as in observational learning (see above). Another possibility is that parents may help influence the child's belief that moderate to vigorous physical activity is good for the child. Some researchers have hypothesized that the father is more influential in the child's involvement in sport and physical activity, particularly for males. But complicating the matter are the possible interactive effects of the child's gender, age, and cultural background (Lewko and Greendorfer, 1988). One type of interesting interaction is between the level of competition and the families' attitudes toward that activity. Purdy, Eitzen, and Haufler (1982) found that parents of competitive swimmers reported higher expectations for their child's involvement and achievement in swimming, compared to parents of recreational swimmers. This finding suggests that parental attitudes can have a strong bearing on the child's choice of physical activity. Interestingly, it has been found that sons of former professional baseball players are *50 times* more likely to become professional baseball players than sons of fathers who were not professional players (Laband and Lentz, 1985)! Because of these complex interactions between the gender of the parent and the attributes of the child, it is not possible to identify a single cause of the child's socialization into sport and physical activity. This issue is of great importance in our understanding of the socialization process, but it requires further research.

According to Csikszentmihalyi (1990), the family can be structured in certain ways to promote "flow" and personal growth in the child. The family context is thought to provide an ideal training ground for the general enjoyment of life of the child's cognitive, emotional, and physical development. There are five family influences emphasized by Csikszentmihalyi: clarity, centering, choice, commitment, and challenge. *Clarity* refers to the child knowing what the parents expect of him/her. *Centering* is the acknowledgment of the importance of the present and the taking of opportunities that currently exist to promote personal growth. Often, parents overemphasize the importance of the future (i.e., graduating from school) and forget to stress the importance of getting the most out of the present, regardless of the activity. *Choice* is related to allowing the child to freely choose the activities they are interested in. The belief is that the child's intrinsic motivation of freely chosen activities will promote flow-like growth. *Commitment* on the part of family creates an atmosphere of trust that makes the child feel more comfortable pursuing certain tasks. Finally, parents in a family should be responsible for providing the child with *challenge* or increasingly complex opportunities for the child to allow for further personal growth. These suggestions from Csikszentmihalyi (1990) stress the importance of the family in helping the child develop cognitive, emotional, and physical skills.

While it is clear that the family can be influential in affecting the child's attitude about physical activity and sport, most sociocultural kinesiologists believe that the family is not the only factor. The child's peer group is also likely to affect the child's attitudes and participation in sports and physical activity. Especially during adolescence, a child's peer group becomes increasingly important in the socialization process. A peer group is defined as "a friendship group of children of a similar age" (Giddens, 1991, p. 65). Research is beginning to show the importance of the peer group in the child's attitudes and participation in sport and physical activity. One way that peer groups are important is serving as a source of comparison to judge one's competencies in skill development. A series of studies by Horn has indicated that peer groups serve as a source of comparison and evaluation beginning in early childhood and continuing throughout the teenage years (Horn and Hasbrook, 1986; Horn and Weiss, 1991; Horn, Glenn, and Wentzell, 1993). Many peer group comparisons occur within the context of sport teams. Within a sport team, it is relatively easy to evaluate how one is performing relative to others. Another important element of peer groups is that they may provide opportunities for friendship which, in turn, may influence one's affiliation with the sport or physical activity (Weiss, 1993; Weiss, Smith, and Theeboom, 1996).

The school may also influence one's participation in sport and physical activity. Some research suggests that the school's influence is more pronounced in secondary schools compared to elementary school. At the elementary level, the family and one's peers seem to have more influence, with the exception of private schools, where sport participation may be required (Armstrong, 1984). Within schools, teachers and coaches may have a strong effect on one's attitudes and participation in sport. In fact, Higginson (1985) found that teachers and coaches provided the most encouragement for participation in sport for girls over the age of 13.

Having said all this, it must be made clear that there are alternative approaches to understanding socialization into sports and physical activity other than examining outside influences (family, peers, etc.). As pointed out by Coakley (1998), socialization is an active process that represents both the external influences on the individual, as well the individual's thoughts, feelings, behaviors, and values while engaged in this process. Taking into account *both* the external influences and the individual's contribution are necessary to fully understand socialization into various types of physical activity.

Gender Differences in Sport and Physical Activity Participation

We have already noted in Chapter 7 that there is much evidence for socialization differences between developing young boys and girls. Research has suggested that boys are more encouraged to engage in vigorous physical activity while girls may even be punished for doing so. Even when allowed to participate, girls' participation in physical activity is often more regulated by their parents. For example, a young girl might be given permission to play as long as "she gets back in time to set the table" or "if she takes her little brother or sister with her" (Coakley, 1998, p. 242). In addition, it has been found that fathers spend considerably more shared physical activity participation time with their sons than their daughters (Ross and Pate, 1987). All of these factors have negatively influenced girls' participation patterns in physical activity. But in the words of songwriter Bob Dylan, "the times, they are a-changin'." In spite of these negative influences, girls' participation in sport and physical activity has increased over the last several decades. According to sport sociologist Jay Coakley (1998), there are several factors that have contributed to this increase.

One factor is *increased opportunities for girls' participation in sports*. Today there are sports teams and clubs that did not exist before the 1970s. Another factor is *government legislation*, such as the passage of Title IX in the United States in 1972. In 1988, the Civil Rights Restoration Act made it clear that all educational programs, including physical activity programs, must be equally accessible to males and females. Finally, in 1992, the U.S. Supreme court ruled that women athletes and coaches could sue if violations occurred. In Canada, the Royal Commission on the Status of Women was founded in 1970, and the Fitness and Amateur Sport Women's Program was established in 1980. Canada and the United States now have strong federal legislation in place protecting women's equal access to sport and physical activity programs. Since the late 1960s, the so-called *women's movement* has resulted in an increased sensitivity among the public regarding women's rights. The *health and fitness movement* over the last 25 years has encouraged both men and women to participate in physical activity to combat a variety of health problems, including obesity, cardiovascular disease, diabetes, etc. Finally, female athletes are provided much *more media coverage* today compared to 25 years ago. Increased media coverage has helped create female athlete role models, which encourage more female participation in sports and other physical activities. Research has indicated that the rate of participation of girls in sport activities has dramatically increased over the last 25 years. In 1971 only 7 percent of high school athletes in the United States were female. By 1982 this percentage had increased to 35 percent (Leonard, 1988, p. 276). By 1995 girls' sport team participation in schools increased to 42.4 percent, compared to 57.8 percent for boys.

While these increases in athletic sport activity participation should be viewed as contributing to youth physical fitness, the numbers on the general youth population in the United States gathered by several organizations and summarized in a report by the Surgeon General are discouraging (U.S. Department of Health and Human Services, 1996). Below are the major conclusions in this report on adolescent and young adults:

- Only about one half of U.S. young people (ages 12–21) regularly participate in vigorous physical activity. One fourth report no vigorous physical activity.
- Approximately one fourth of young people walk or bicycle (i.e., engage in light to moderate activity) nearly every day.
- About 14 percent of young people report no recent vigorous or light to moderate physical activity. This indicator of inactivity is higher among females than males and among black females than white females.

- Males are more likely than females to participate in vigorous physical activity, strengthening activities, and walking or bicycling.
- Participation in all types of physical activity declines strikingly as age or grade in school increases.
- Among high school students, enrollment in physical education remained unchanged during the first half of the 1990s. However, daily attendance in physical education declined from approximately 42 percent to 25 percent.
- The percentage of high school students who were enrolled in physical education and who reported being physically active for at least 20 minutes in physical education classes declined from approximately 81 percent to 70 percent during the first half of this decade.
- Only 19 percent of all high school students report being physically active for 20 minutes or more in daily physical education classes.
- (U.S. Department of Health and Human Services, 1996, pp. 200–201)

These rather dismal physical activity patterns and equally concerning dietary concerns of children (Popkin, et al., 1996) are making obesity, and its associated risks, a high priority health concern of the United States.

Racial Factors in Sport and Physical Activity

When discussing the influence of race on sport and physical activity participation, it is impossible to exclude the issue of prejudice and discrimination. **Prejudice** is defined as unfavorable feelings or thoughts exhibited toward a person or group. **Discrimination** is defined as unfavorable treatment or actions toward a person a group (Leonard, 1988). We are all aware of these types of unfortunate practices against minority groups and women in our society. In this brief section, we focus on prejudice and discrimination against minority groups such as African-Americans and Hispanics in sports and physical activity settings.

Both prejudice and discrimination are rampant in our society. Sport and other physical activity settings are microcosms of larger society and perpetuate these same unfortunate practices (Leonard, 1988). There are three major aspects of sports participation discrimination extensively researched by sport sociologists: *position allocation, performance differentials*, and *rewards and authority structure* (Eitzen and Sage, 1978). In position allocation research, sport sociologists have found that certain positions on different sports teams are underrepresented by minority group players. In Figures 9.2 and 9.3, illustrations are provided that show the percentages of whites, Hispanics, and African-Americans in professional football and baseball in the United States during the 1995–1996 seasons. In professional football, African-Americans are more underrepresented in offensive positions, particularly at quarterback, but also in the offensive line positions. There are few Hispanics playing professional football. In professional baseball, most of the African-American representation is in the outfield positions, with negligible representation at the pitcher and catcher positions. Hispanics also are poorly represented at these positions, but show an overrepresentation at second base and shortstop. In contrast to baseball and football, African-Americans dominate most of the positions in professional basketball, with the exception of the center position. Few Hispanics currently play professional basketball.

There are many hypotheses that attempt to explain differences in position allocation among whites and minorities. One view is that African-Americans and Hispanics are recruited for positions that require speed and quickness, while whites are recruited for positions that require leadership, decision-making skills, and dependability (Coakley, 1998). This view might explain position allocation in baseball and football, but does not explain why there are virtually no Hispanics in professional basketball. In addition, this view does not explain why there is such a large percentage of African-Americans in professional basketball.

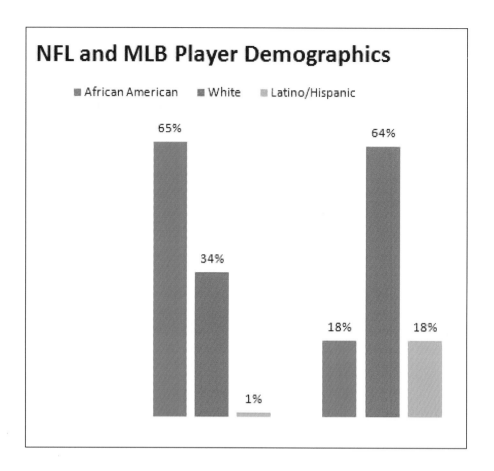

Figures 9.2 and 9.3
The percentages of African-American (in blue), Hispanic (in green) and white (in red) players in professional football and professional baseball in the United States during the 1995–1996 season. Data from Leonard (1997) and Branch (2007).

Another hypothesis argues that the concept of "centrality" is key to understanding position discrimination (Loy and McElvogue, 1970). **Centrality** is related to an individual's interaction with the rest of the group (or team) and the degree which the individual must coordinate his or her task with other members or players. For example, in baseball, much of the action of the players on the field revolves around the pitcher. In football, the quarterback position requires a high degree of centrality. In basketball, centrality is more distributed among the different positions. Loy and McElvogue argued that position allocation differences can be explained by the concept of centrality.

The **stereotype hypothesis** argues that whites stereotype other minority groups as possessing certain characteristics or skills. Thus, coaches or team management may have certain biases that will affect decisions about position allocation.

The **prohibitive cost hypothesis** states that the high cost of training athletes for certain positions, combined with the socioeconomic status of the athlete, determines position allocation differences (Medoff, 1986). If the training costs are higher than can be afforded for a given sport or position, as is the case with many minorities, it is unlikely that this sport or position will be pursued. For example, there are very few competitive-style swimming pools and swim programs in economically deprived areas. Thus, it is unlikely that young children living in this area will become competitive swimmers.

The **role-modeling hypothesis** suggests that young minorities seek to play positions successfully performed by other high-caliber, minority players. Tiger Woods, an African-American (with Asian heritage as well), burst onto the professional golfing world in 1996 and was chosen "Sportsman of the Year" by the magazine *Sports Illustrated* and "Player of the Year" by his peers in 1999. It will be interesting to see whether his presence will influence other young minorities into the sport of golf, a sport not known for its inclusion of minorities in the United States.

All of the above hypotheses have strengths and weaknesses in explaining position allocation differences among whites and minorities in sports (see Leonard, 1988, for an excellent review). It is likely that not any one hypothesis can explain all the racial differences associated with position allocation. Substantial work needs to be done to better our understanding of this important issue.

Another racial problem to be briefly addressed is **performance differentials** between whites and minorities. It is well known by sport sociologists that African-Americans in many of the major professional sports of baseball, football, and basketball have exhibited superior performance over whites. In professional baseball, African-Americans have generally shown superiority in a number of statistical categories such as batting average, home runs, runs batted in, etc. (e.g., Eitzen and Yetman, 1977; Leonard, 1977; Pascal and Rapping, 1970). Superior performance of African-Americans over whites has been shown in professional football and basketball (Scully, 1973). Given this superiority, how can it be claimed that discrimination exists? One possibility is that African-Americans may have to perform *better* than whites to receive an equal chance to play and continue playing their sport. Some research has suggested that less productive African-Americans may be released earlier (dropped from the team) than their white counterparts (Johnson and Marple, 1973). Furthermore, social scientist John Hoberman has recently questioned whether the remarkable athleticism displayed by African-Americans or their possession of high salaries in the professional ranks (see below) has done much to reduce discrimination or prejudice (Hoberman, 1997). His view is that the high numbers of successful African-Americans in many sports have fostered an "anti-intellectual" attitude that encourages young African-Americans to participate in sports at the expense of academic achievement. If Hoberman is right, the wonderful achievements of African-American athletes may have spurred a type of "revenge effect"—a shunning of interest in academics. More research is needed to verify Hoberman's claims.

A final discrimination issue to be addressed is **rewards and authority structures**. The focus here will primarily be at the professional sports level. The first question to be addressed is, are African-Americans and Hispanics discriminated against economically? It turns out that African-Americans are among the highest paid athletes in professional sports. If one compares the average paid salaries, there is no significant difference between African-Americans and whites. However, there is some evidence of discrimination at different positions in professional baseball (Leonard,1980; Christiano, 1986), but there is little comparative data in other professional sports. Thus, more research is needed to clarify this issue.

Leonard (1988) documented that there is a paucity of administration positions in professional sports. The following discussion represents some of his major findings. African-Americans and Hispanics occupy much fewer administrative positions within professional sports compared to whites. For example, as of 1989, of the 879 top administration positions in major league baseball, only 17 were held by African-Americans (or around 2 percent). Only 1½ percent were Hispanic or Asian. African-Americans held only 6 ½ percent and a little over 5 percent of the top administration positions in the National Football League and the National Basketball Association, respectively (*USA Today*, April 9, 1987, p. 1). No African-Americans are owners of any professional sports franchise. There are very few African-American coaches, sportscasters, and officials in professional sports.

African-Americans are underrepresented in Baseball's Hall of Fame. In addition, Roberto Clemente is the only Hispanic to be inducted in the Hall of Fame (at this writing).

What can account for all this underrepresentation of African-Americans and Hispanics in top administrative and authority positions? One possibility is that minorities who are ex-players may not be attracted to administrative positions at salaries far lower than they were earning as players. Another possibility is that they are not interested in working their way up through administrative systems largely controlled by whites. Related to this hypothesis is that minorities are simply discriminated against and prevented from attaining administrative positions, a view consistent with conflict or critical theories.

In summary, although progress has been made in improving minority representation in sports, prejudice and discrimination remain major obstacles to overcome in our society. Further research in sociocultural kinesiology is required to examine the many hypotheses related to discrimination discussed in the chapter. Hopefully, more research will lead to a better understanding of the problem and perhaps to some solutions on how discrimination can be reduced or eliminated in sports and other physical activity settings.

Adult Physical Activity Patterns

In this section we turn to physical activity patterns of adults. Over the years, major surveys have been periodically conducted to describe physical activity patterns in the United States such as the National Health Interview Survey (1985–991), the Behavioral Risk Factor Surveillance System (1986–1994), and the Third National Health and Nutrition Examination Survey (1988–1994). These surveys have generally shown a lack of physical activity in the adult population. Most recently, a report co-sponsored by the Centers for Disease Control and Prevention and the President's Council on Physical Fitness and Sports was published as part of the Healthy People 2010 Information Access Project. This report further indicates a lack of physical activity in most adults in the United States. Based on the results of this latest survey, the following conclusions were drawn:

- Only 15 percent of adults report physical activity for 5 or more days per week for 30 minutes or longer
- Only about 23 percent of adults in the United States report regular, vigorous physical activity that involves large muscle groups in dynamic movement for 20 minutes or longer 3 or more days per week.
- Roughly 40 percent of adults report no physical activity in their leisure time.
- Physical inactivity is more prevalent among women than men, among blacks and Hispanics than whites, among older than younger adults, and among the less affluent than the more affluent.

Clearly, these results represent a rather disappointing view of a nation supposedly embraced by a health-concerned conscience and living in a so-called "Fitness Boom" era!

Improving Participation in Physical Activity from a Sociocultural Perspective

A puzzling question that needs to be more effectively addressed is, "Why are physical activity patterns so low in both children and adults, in spite of increased awareness of the importance of physical activity for health and fitness?" Everyone knows that changing personal habits and routines is one of the most difficult things to do. What can society do to help individuals adopt and sustain a more physically active life?

A report from the U.S. Department of Health and Human Services (1996) provided insight into this problem. The report indicated that a major barrier to physical activity participation is that advances in technology have decreased people's need to be physically active in the workplace and in many areas of everyday life (a revenge effect!). In the past, earning a living was commonly a labor-intensive endeavor. For instance, in the past, farming required considerable energy expenditure, but technological improvements in farm equipment have greatly reduced physical activity in this profession. Walking, an excellent physical activity, has been significantly reduced since the advent of elevators, motor transportation, and parking garages near the workplace. Sedentary activities such as television viewing, video game playing, and computer use have largely replaced more physically demanding pursuits. Another reason for reduced physical activity, particularly in the cities and suburbs, is the increase in crime and traffic that prevents or discourages people walking or bicycling in their communities (U.S. Department of Transportation, 1993; 1994). Thus, the low levels of physical activity in the populace have been influenced by a number of factors, and correcting this problem will not be easy.

As pointed out by the U.S. Department of Health and Human Services (1996) report, there are several factors that appear to be important influences on physical activity participation:

- *Having confidence or "self-efficacy" in one's ability to be physically active.* Confidence can be increased in several ways such as providing clear directions to follow. You are more confident in pursuing a physical activity if you know exactly what it is you are supposed to do. Therefore, providing quality instructional programs in the schools, recreation centers, etc., that serve to improve skill level is one way to improve confidence in participation. Instructors who serve as good role models, who evoke trust, admiration, and respect, may be particularly effective.

- *Enjoying physical activity.* Enjoyment has been strongly associated with increased physical activity participation. The work discussed earlier by Csikszentmihalyi (1990) told us that to be enjoyable, an activity must provide challenge that is matched by our ability. Thus, instructional programs could incorporate this philosophy to maximize enjoyment of the participants. As individuals, we should probably participate in those physical activities that are the most enjoyable, whether they be gardening and yard work, bicycling to work, or swimming laps with friends during the lunch hour, for example. It is likely that we will discontinue those activities that are not enjoyable, even though we may know that participating in them would be good for our fitness or health.

- *Receiving support for family, friends, and peers.* For many people, it is either difficult or undesirable to exercise alone. The U.S. Department of Health and Human Services (1996) report summarized a number of research studies that have shown high relations between social support and physical activity participation. The family can provide support and encouragement for its individual members to participate in physical activity. Parents can help organize physical activities for their children and provide transportation to sport and exercise situations. Exercising or other types of physical activity can be experienced with friends, and new social relationships can be developed in various physical activity settings. Some research has shown that social support can even lead to greater adherence to exercise programs (Duncan and McAuley, 1993; Elward, Larson, and Wagner, 1992).

- *Intervention strategies.* The U.S. Department of Health and Human Services (1996) report also summarized a number of studies that have used intervention strategies to facilitate physical activity participation in children and adults. Intervention strategies have been used in a number of different ways or settings such as:

  ◊ requiring the individual to self-monitor their physical activity throughout the day;
  ◊ requiring primary care physicians to counsel their patients on exercise;
  ◊ manipulating the type of physical education instruction in schools (i.e., sports skills, exercise, lifetime skills) ;
  ◊ using mass media to improve awareness;
  ◊ construction of bike and walking trails in communities;
  ◊ developing work-site exercise and health programs for employees of a company; and
  ◊ targeting specific groups of people (racial, ethnic, overweight, people with disabilities, and older adults) for participation in exercise programs.

Intervention strategies have recently come under attack for not only being ineffective, but for also being based on questionable assumptions. Exercise interventions are typically designed to alter the type of exercise, duration, intensity, and frequency of exercise workouts based on an initial assessment of the individual. We have already learned in Chapter 4 that physiological benefits are gained when an individual partakes in a properly designed exercise program. But the question to be asked is, do people comply with an exercise program over the long term, and what should a properly designed exercise program look like? Of course, we know from our previous discussions that many people do poorly in adhering to an exercise program. In a review and position statement, Kimiecik and Lawson (1996) discussed why this may be the case. They argued that exercise interventions or prescriptions often ignore the individual's psychological readiness for incorporating and maintaining an exercise program in their lives. Sometimes exercise programs are overly restrictive and do not motivate the individual to exercise regularly. Also, some exercise prescriptions are too difficult, particularly for beginners. Finally, exercise interventions are not typically designed to promote self-responsibility and empowerment, characteristics required for long-term adherence.

One of the assumptions underlying exercise prescription, based on what Kimiecik and Lawson call the **human capital model**, is that people pursue a rational and regulated life. Because people are generally regarded as weak and in need of "protection" from disease and maligned lifestyles, the role of the health professional or exercise specialist is to provide all the expert information and whatever is required to cure the problem. Many individuals believe it is the responsibility of the health expert to cure them, rather than empowering themselves to take responsibilities for their own lives.

Kimiecik and Lawson propose a different model of exercise behavior change, called the **human development-potential perspective**. They propose that health professionals need to know more about the individual, the family and other sociocultural aspects affecting the individual. They also suggest that health professionals need to become more collaborative with their clients, and help their clients become more responsible in making decisions related to exercise, nutrition, and other related behaviors. In addition, there has to be a realization that all individuals are not alike. A change in lifestyle for one individual is likely to require different actions compared to another. Finding out more about the individual, their family interactions, and the sociocultural influences on the individual may require a more "phenomenological approach" (discussed in the next chapter) on the part of the health professional. Thus, a strong grounding in sociocultural studies on the part of the health professional is important in addressing the exercise adherence problem, if Kimiecik and Lawson are correct. An important research question to be addressed in the future is whether the human capital model or the human development-potential perspective is more effective in promoting the self-reliance and responsibility required for long-term physical activity changes in children and adults around the world.

Viewing the Human Body from a Sociocultural Perspective

As we learned earlier in Chapter One, there are a number of levels of analysis in the study of the human body and in the study of human physical activity. In this chapter, we have examined human physical activity from a sociocultural perspective or level of analysis. It is also possible to examine the human body itself from this perspective. Duncan (2007) discussed six ways in which society or people within a society have viewed the human body. One way is the 'imagined' body in which a type of idealized body is emphasized by society. For example, the notion of an idealized or 'perfect' body is constructed by the populace, fashion industry and media of a society. For example, in the 1960's, the model Twiggy, whose pencil-thin figure sharply constrasted with the fashion norms of the day, became a popular icon and helped to influence society's views of what a female body should look like. Some of the questions addressed by imagined body perspective are:

- Are there cultural pressures to look a certain way for women and men?

- Has the 'ideal' body image changed over the years?
- What historical forces shape this image?
- Are 'fat' women and men perceived differently?
- Are expectations different?
- Is it important how we look to others?

One of the consequences of trying to obtain the 'imagined' body, and in particular, the emphasis on a thin female body, are the economic costs associated with the diet industry which are over $60 billion a year, just in the United States alone!

The second perspective on viewing the human body from a sociocultural perspective is called the 'consumer body'. In this perspective, our identity is defined by our consumption of products and symbolic consumption. For example, we help define who we are by the types of products we consume, such as designer water, health club memberships, cosmetic surgery, etc. Duncan (2007) speculates that males play more video games because it helps to reinforce their masculine characteristics.

The third perspective is the 'transgressive body'. In this perspective, social norms are defied because the body of the individual does not appear to be socially acceptable. For example, extremely obese individuals are oftentimes not viewed favorably by others and by society in general. Transgender individuals who do not feel comfortable with the gender assigned at their birth are often viewed negatively and can face physical harm by others. Intersexed people are those born with both male and female anatomical characteristics and face many obstacles as they grow up. People with tattoos and body piercings are often scrutinized by others.

The fourth perspective is the 'disciplined body'. For various reasons, we may subject our bodies to disciplinary regimes in different settings. To improve our health or appearance, we subject ourself to exercise or dieting regimes. We may put ourselves under the regime of a physician's care to deal with a health issue.

The fifth perspective is the 'practiced body' that relates to the sociology of becoming skilled. To become better at a skill we must not only devote many hours of practice to the skill but often must learn and play by the rules of the skill and obey authority (i.e., coach, trainer).

The final perspective discussed by Duncan (2007) is the 'discursive body'. In this perspective, the body is described through language reflecting the biases of the individual or social group. This perspective is similar to symbolic interactionism, discussed earlier in the chapter. For example, women's athletic teams often have names that symbolize certain stereotypes such as the Wild Kittens or the Tigerettes. In describing the men's college basketball tournament, sports announcers and the media may refer to the NCAA Final Four, whereas the 'Women's NCAA Final Four', sometimes is used to qualify the women's competition. Is this qualification used to make the women's competition sound inferior to the men's?

In summary, the human body not only has anatomical and physiological descriptions, it can also be understood from a sociological perspective, as suggested by Duncan (2007). How society views the human body and how we portray our bodies can affect the perceptions, emotions and behaviors of others.

SUMMARY

- Socialization is the complex developmental learning process that teaches the knowledge, values, and norms essential to participation in social life.
- The family, peer groups, and the school are thought to be important factors in the socialization into sports and other physical activities.
- Until recently, boys have been more encouraged to engage in vigorous physical activity than girls.

- Government legislation, the women's movement, the health and fitness movement, and more media coverage have greatly influenced the increase in physical activity participation of girls and women in the United States.
- However, in spite of increased awareness, physical activity participation patterns for boy and girls are low in the United States and across the world.
- Prejudice and discrimination are still thought to negatively affect African-American and Hispanic participation in sports.
- A number of hypotheses have been developed to help explain how prejudice and discrimination practices in sports are manifested.
- Physical activity participation patterns of adults in the United States are low.
- A number of sociocultural strategies have been identified to improve participation in physical activity.
- The human body can be understood from a sociological perspective.

CHAPTER SUMMARY
- Sociocultural kinesiology is the study of social and cultural factors in all types of physical activity settings, such as play, recreation, games, exercise and sports.
- There are several theories of sociology that have been used to understand the participation in physical activity.
- Rationalization, the increasing use and dependency of technology in a society, has had a pronounced effect on participation in physical activity.
- To improve participation in physical activity, it is important that the activity be enjoyed (i.e., the challenge of the activity must be matched with the skill level of the participant).
- Socialization into sports and other physical activities is strongly influenced by the family, peer groups, and the school.
- Physical activity participation patterns of girls and boys in the United States and other countries around the world is low.
- Prejudice and discrimination against minority groups negatively affects their participation in sports.
- A number of social strategies have been formulated to improve participation in physical activity.
- A number of sociological perspectives can be used to describe the human body as discussed by Duncan (2007).

IMPORTANT TERMS

Adult physical activity patterns – the percentage of adults engaged in regular physical activity

Centering - the acknowledgment of the importance of the present and the taking of opportunities that currently exist to promote personal growth

Centrality - an individual's interaction with the rest of the group (or team) and the degree which the individual must coordinate his or her task with other members or players

Challenge – the level of difficulty of a task

Choice - allowing the child to freely choose the activities they are interested in

Clarity - refers to the child knowing what the parents expect of him/her

Commitment – a family atmosphere of trust that makes the child feel more comfortable pursuing certain tasks

Competitive behaviors - those behaviors that attempt to outperform or defeat another individual

Conflict theory – by Karl Mark argued that certain groups of people will achieve economic power and wealth at the expense of others

Cooperative behaviors - those behaviors that help or assist others for the purpose of achieving some goal

Critical theory – in addition to economic struggles, this theory argues that there are other struggles within a society that create conflict

Cultural relativism – not judging other cultures relative to one's own

Cultural universals – common elements shared by all cultures

Culture - the fundamental values and beliefs of a given society or social system as well as the norms followed and the material goods created by its members

Discrimination - unfavorable treatment or actions toward a person a group

Dissassociation – a lack of understanding in technology

Ethnocentrism - to judge other cultures relative to one's own

Functionalism – a theory that focusing on how elements of a society contribute to the continuation of the society as a whole

Human capital model – a view that health care is largely the responsibility of health professionals

Human development-potential perspective – a view that the patient, with the aid of the health professional should take more responsibility of their own health care

Observational learning – watching someone else perform the behavior to be learned

Performance differentials – differences in performance between whites and non-whites

Position allocation - certain positions on different sports teams that underrepresented by minority group players

Prejudice – unfavorable feelings or thoughts exhibited towards a person or group

Progressive differentiation - a compartmentalization of life's activities as a society evolves

Prohibitive cost hypothesis - states that the high cost of training athletes for certain positions, combined with the socioeconomic status of the athlete, determines position allocation differences

Rationalization - defined as a society's increasing reliance on the scientific method and technology

Reinforcement - the rewards and punishments one receives from significant others

Revenge effects - .the unanticipated consequences of new technology

Rewards and authority structure – related to the issue of underrepresentation of minorities in administrative positions and salary differences between whites and non-whites

Role-modeling hypothesis - suggests that young minorities seek to play positions successfully performed by other high-caliber, minority players.

Social learning theory – argues that three elements (reinforcement, coaching, and observational learning) provide the mechanisms that allow for the learning of social roles in various social systems.

Social system - a group, large or small, of individuals who interact

Socialization - a learning process through which a person acquires the necessary values, norms, roles, and skills necessary to function in a society

Society - a large, relatively permanent group of people such as the group of people within a country, who share common values and beliefs

Sociocultural kinesiology – a sub-field of kinesiology that examines the sociocultural influences on the participation in physical activity

Sociology - the scientific study of human behavior in group (or social) settings

Sport sociology – an area of kinesiology that examine the social influences in sport participation

Stereotype hypothesis - argues that whites stereotype other minority groups as possessing certain characteristics or skills

Symbolic interactionism – a theory arguing that much human behavior is driven by symbols

Women's movement – since the 1960's, this movement is largely responsible for increasing the public's awareness of women's rights

INTEGRATING KINESIOLOGY: PUTTING IT ALL TOGETHER

1. If you are planning to develop a fitness program for yourself (Chapter 4), what sociocultural issues should you consider?

2. Pick two countries you are familiar with, and compare and contrast their sociocultural differences. In what ways do you think these sociocultural factors influence people's participation in physical activity in these countries?

3. Using one of the theoretical frameworks discussed in this chapter (functionalism, conflict theory, critical theory, or symbolic interactionism), what kind of questions would you ask in addressing:
   - Gender differences in physical activity participation
   - Racial differences in physical activity participation
   - The influence of the immediate family on children's participation in physical activity.

4. Provide three additional examples (not discussed in this chapter) of the effects of rationalization on the participation in physical activity.

5. If you decide to learn a new skill or physical activity, such as ballroom dancing or scuba, how might your family, peer groups, and school (or academic setting) influence your participation.

6. If you become a parent, how might you use the five family influences emphasized by Csikszentmihalyi (1990) to help develop your child's cognitive, emotional, and physical skills?

KINESIOLOGY ON THE WEB

- www.healthypeople.gov/document/html/volume2/22physical.htm#_Toc490380803—This Web site contains the Healthy People 2010 Report on Physical Activity and Fitness.
- www.nasss.org/—This is the Web site for the North American Society for the Sociology of Sport. It contains resources for researchers and other professionals interested in the sociology of physical activity and important links to other organizations.

REFERENCES

Anders, G. (1993). Million-dollar M.D.: A top physician earns a fortune repairing knees in Vail, Colorado. *Wall Street Journal*, April 8, A1.

Arthur, C. (1992). Anyone for slower tennis? *New Scientist, 134*, May 2, 24–28.

Armstrong, C. F. (1984). The lessons of sport: Class socialization in British and American boarding schools. *Sociology of Sport Journal, 1, 314–31.*

Bandura, A. (1971). *Social Learning Theory*. Morristown, N.J.: General Learning.

Branch, J. (2007). Among Hispanics, N.F.L. mania hits cultural wall. *New York Times*, Feb. 3.

Brennan, T. A., et al. (1991). Incidence of adverse events and negligence in hospitalized patients: results of the Harvard Medical Practice Study 1. *New England Journal of Medicine, 324*, 370–76.

Brezina, P. B., Selengut, C., and Weyer, R. A. (1990). *Seeing Society: Perspectives on Social Life.* Boston: Allyn and Bacon.

Brustad, R. J. (1993). Who will go out and play? Parental and psychological influences on children's attraction to physical activity. *Pediatric Exercise Science, 5*, 210–23.

Brustad, R. J. (1996). Attraction to physical activity in urban schoolchildren: Parental socialization and gender influences. *Research Quarterly for Exercise and Sport, 67*, 316–23.

Chambers, M. (1996). Face masks may enhance injuries, not reduce them. *Denver Post*, November 6.

Christiano, K. J. (1986). Salary discrimination in major league baseball: The effect of race. *Sociology of Sport Journal, 13*, 144–53.

Coakley, J. J. (1998). *Sport in Society: Issues and Controversies*. St. Louis: Mosby.

Couzens, G. S. (1992). Football: A painful legacy for players? *Physician and Sportsmedicine, 20,* October 1, 146ff.

Csikszentmihalyi, M. (1990). *Flow: The Psychology of Optimal Experience*. New York: Harper and Row.

Duncan, M.C. (2007). Bodies in motion: The sociology of physical activiy. *Quest, 59,* 55-66.

Duncan, T. E., and McAuley, E. (1993). Social support and efficacy cognitions in exercise adherence: a latent growth curve analysis. *Journal of Behavioral Medicine, 16,* 199–218.

Eccles, J. S., and Harold, R. D. (1991). Gender differences in sport involvement: applying the Eccles expectancy-value model. *Journal of Applied Sport Psychology, 3,* 7–35.

Edell Health Letter (1991). Cushy shoes cause sprain. Vol. 10, September.

Elward, K., Larson, E., and Wagner, E. (1992). Factors associated with regular aerobic exercise in an elderly population. *Journal of the American Board of Family Practice, 5,* 467–74.

Eitzen, D. S., and Sage, G. H. (1978). *Sociology of American Sport*. Dubuque, IA: William C. Brown.

Eitzen, D. S., and Yetman, N. R. (1977). Immune from racism? *Civil Rights Digest, 9,* 3–13.

Freedson, P. S., and Evenson, S. (1991). Familial aggregation in physical activity. *Research Quarterly for Exercise and Sport, 62,* 384–89.

Freund, J. (1969). *The Sociology of Max Weber*. New York: Vintage.

Giddens, A. (1991). *Introduction to Sociology*. New York: W. W. Norton.

Gorn, E. J. (1986). *The Manly Art: Bare-Knuckle Prize Fighting in America*. Ithaca, NY: Cornell University Press.

Grady, D. (1996). Usual tests rarely aid doctors. From the *New York Times* cited in the *Denver Post*, Nov. 20.

Greendorfer, S. (1981). Emergence of and future prospects for sociology of sport. In G. A. Brooks (ed.), *Perspectives on the Academic Discipline of Physical Education: A Tribute to G. Lawrence Rarick*. Champaign, IL: Human Kinetics.

Guttman, A. (1982). *From Ritual to Record: The Nature of Modern Sports*. New York: Columbia Press.

Health News (1990). Bicycle helmets a must. *Vol. 8.,* No. 5, 5ff.

Healthy People (2010). Physical activity and fitness. A report co-sponsored by the Centers for Disease Control and Prevention and the President's Council on Physical Fitness and Sports.

Higginson, D. (1985). The influence of socializing agents in the sport-participation process. *Adolescence, 20,* 73–82.

Hoberman, J. (1997). *Darwin's Athletes: How Sport Has Damaged Black America and Preserved the Myth of Race*. Boston: Houghton-Mifflin.

Hoeberigs, J. H. (1992). Factors related to the incidence of running injuries: a review. *Sports Medicine, 13,* 408—22.

Horn, T. S., Glenn, S. D., and Wentzell, A. B. (1993). Sources of information underlying personal ability judgments in high school athletes. *Pediatric Exercise Science, 5,* 263–74.

Horn, T. S., and Hasbrook, C. A. (1986). Informational components influencing children's perceptions of physical competence. In M. R. Weiss and D. Gould (eds.). Sport for Children and Youths. Champaign, IL: Human Kinetics, pp. 81–88.

Horn, T. S., and Weiss, M. R. (1991). A developmental analysis of children's self-ability judgments in the physical domain. *Pediatric Exercise Science, 3,* 310–26.

Irvine, D. (1994). Sampras serves up short-changes final. *Manchester Guardian Weekly, 151,* July 10, 32.

Johnson, N. R., and Marple, D. P (1973). Racial discrimination in professional basketball. *Sociological Focus, 6,* 6–18.

Jones, B. H., Cowan, D.N ., and Knapik, J. J. (1994). Exercise training and injuries. *Sports Medicine, 18,* 202–14.

Kenyon, G. S., and Loy, J. W. (1965). Toward a sociology of sport. *Journal of Health, Physical Education, and Recreation, 36,* 24–25, 68–69.

Kimiecik, J. C., Horn, T. S., and Shurin, C. S. (1996). Relationships among children's beliefs, perceptions of their parents' beliefs, and their moderate-to-vigorous physical activity. *Research Quarterly for Exercise and Sport, 67,* 324–36.

Kimiecik, J. C., and Lawson, H. A. (1996). Toward new approaches for exercise behavior change and health promotion. *Quest, 48,* 102–25.

Laband, D. N., and Lentz, B. F. (1985). The natural choice. *Psychology Today*, August, pp. 37–39, 42–43.

Leape, L. L. (1991). The nature of adverse events in hospitalized patients. *New England Journal of Medicine*, February, *324*, 377–84.

Leonard, W. M. II (1977). Stacking and performance differentials of white,

Black, and Latin pro baseball players. *Review of Sport and Leisure II*, 77–106.

Leonard, W. M. II (1980). Social and performance characteristics of the pro baseball elite: a study of the 1977 starting lineups. *International*

Review of Sport Sociology, 2.

Leonard, W. M. II (1988). *A Sociological Perspective of Sport*. New York: Macmillan.

Leonard, W. M. II (1997). Racial composition of NBA, NFL and MLB teams and racial composition of franchise cities. *Journal of Sport Behavior, 20.*

Lewko, J. W., and Greendorfer, S. L. (1988). Family influences in sport socialization of children and adolescents. In F. L. Smoll, R. A. Magill, and M. J. Ash (eds.), *Children in Sport*, 3rd ed., pp. 257–302. New York: Academic.

Loy, J. W., and Ingham, A. G. (1973). Play, games, and sport in the psychosocial development of children and youth. In G.L. Rarick (ed.), *Physical Activity: Human Growth and Development*, New York: Academic, pp. 257–302.

Loy, J. W., and McElvogue, J. F. (1970). Racial segregation in American sport. *International Review of Sport Sociology, 5*, 5–23.

Lubell, A. (1989). Chronic brain injury in boxers: is it avoidable?. *Physician and Sportsmedicine, 17*, 126–32.

Malcolm, A. (1990). Canadian culture. In P. B. Brezina, C. Selengut, and R. A. Weyer (eds.). *Seeing Society: Perspectives on Social Life*. Boston: Allyn and Bacon, pp. 51–54.

McCullagh, P., Weiss, M. R., and Ross, D. (1989). Modeling considerations in motor skill acquisition and performance: an integrated approach. In K. Pandolf (ed.), *Exercise and Sport Science Reviews*, Vol. 17, 475–513, Baltimore: Williams and Wilkins.

McPherson, B. D. (1982). The child in competitive sport: Influence of the social milieu. In R. A. Magill, M. J. Ash, and F. L. Smoll (eds.), *Children in Sport* (2nd edition), Champaign, IL: Human Kinetics.

McPherson, B. D., Curtis, J. E., and Loy, J. W. (1989). *The Social Significance of Sport: An Introduction to the Sociology of Sport*. Champaign, IL: Human Kinetics.

Medoff, M. G. (1986). Positional segregation and the economic hypothesis. *Sociology of Sport Journal, 3*, 297–304.

Mitchell, R. G. (1982). Mountain experience: *The Psychology and Sociology of Adventure*. Chicago: University of Chicago Press.

Moody, S. (1988). Phantom of poliomyelitis still haunting ex-patients. *New Brunswick Home News*, September 12, A8.

Moore, L. L., Lombardi, D. A., White, M. J., Campbell, J. L., Oliveria, S. A., and Ellison, R. C. (1991). Influence of parents' physical activity levels on activity levels of young children. *Journal of Pediatrics, 118*, 215–19.

Munsat, T. L. (1991). Poliomyelitis: new problems with an old disease. *New England Journal of Medicine*, Vol. 324, No. 17, April 25, 1206–1207.

Murdock, G. P. (1945). The common denominator of cultures. In R. Linton (ed.), *The Science of Man in a World of Crisis*. New York: Columbia University Press.

Pascal, A., and Rapping, L. A. (1970). *Racial Discrimination in Organized Baseball*. Santa Monica, CA: RAND, 1970.

Popkin, B. M., Richards, M. K., and Montiero, C. A. (1996). Stunting is associated with overweight in children of four nations that are undergoing the nutrition transition. *The Journal of Nutrition, 126*, 3009–3016.

Powell, J. W., and Schootman, M. (1992). A multivariate risk analysis of selected playing surfaces in the National Football League: 1980–1989, an epidemiological study of knee injuries. *American Journal of Sports Medicine, 20*, 686–94.

Purdy, D., Eitzen, S., and Haufler, S. (1982). Age-group swimming: contributing factors and consequences. *Journal of Sport Behavior, 5*, 28–43.

Recer, P. (1996). Chronic illnesses cost nation $425 billion. *Associated Press*, October.

Rivara, F. P., et al. (1990). The Seattle children's bicycle helmet campaign: changes in helmet use and head injury admissions. *Pediatrics, 93*, 567–69.

Robinson, M. (1993). Snap, crackle, pop: climbing injuries to fingers and forearms. *Climbing,* June-July, 141–50.

Rolf, C. (1995). Overuse injuries of the lower extremity in runners. *Scandinavian Journal of Medicine and Science in Sports, 5,* 181–90.

Ross, J., and Pate, R. R. (1987). A summary of findings (for the National Children and Youth Fitness Study II). *Journal of Physical Education, Recreation and Dance, 58,* 51–56.

Schlucter, W. (1979). The paradox of rationalization: on the ethics and the world. In G. Roth and W. Schlucter (eds.). *Max Weber's Vision of History.* Berkeley: University of California Press, pp. 11–64.

Schmid, L., Hajek, E., Votipka, F., Teprik, O., and Blonstein, J. J. (1968). Experience with headgear in boxing. *Journal of Sports Medicine and Physical Fitness, 8,* 171–76.

Scully, G. (1973). Economic discrimination in professional sports. *Law and Contemporary Problems, 39,* 67–84.

Smith, A. (1776). *The Wealth of Nations.* ed. E. Cannon. London: Methuen, 1961.

Spenser, M. (1976). *Foundations of Modern Sociology.* Englewood Cliffs, NJ: Prentice-Hall.

Tenner, E. (1996). *Why Things Bite Back: Technology and the Revenge of Unintended Consequences.* New York: Alfred A. Knopf.

Thomas, D. (1994). Runners gain without pain? Researchers disagree on chronic injury risk. *Omaha World-Herald,* March 28.

Thompson, E. P. (1963). *The Making of the English Working Class.* New York: Viking.

Turnbull, C. M. (1961). *The Forest People.* Garden City, NY: Doubleday.

Underwood, J. (1995). It's pretty, it's trendy, but skiing is also much too dangerous. *New York Times,* February 25, sec. 8, 9.

U.S. Department of Health and Human Services (1996). *Physical Activity and Health: A Report of the Surgeon General.* Atlanta, GA: U.S. Department of Health and Human Services, Centers for Disease Control and Prevention, National Center for Chronic Disease Prevention and Health Promotion.

U.S. Department of Transportation (1993). *Measures to Overcome Impediments to Bicycling and Walking: The National Bicycling and Walking Study, Case Study No. 4.* Washington, D.C. Department of Transportation, Federal Highway Administration, Publication No. FHWA-PD-93-031.

U.S. Department of Transportation (1994). *Final Report: The National Bicycling and Walking Study: Transportation Choices for a Changing America.* Washington, D.C. Department of Transportation, Federal Highway Administration, Publication No. FHWA-PD-94-023.

U.S. National Institute for Occupational Safety and Health (1991). Division of Standards and Technology Transfer, *NIOSH Publications on Video Display Terminals,* 2nd ed., June.

Van Mechelen, W. (1992). Running injuries: a review of the epidemiological literature. *Sports Medicine, 14,* 320–35.

Weiss, M. R. (1993). Psychological effects of intensive sport participation in children and youth: self-esteem and motivation. In B. R. Cahill and A. J. Pearl (eds.), *Intensive Participation in Children's Sports,* Champaign, IL: Human Kinetics, pp. 39–69.

Weiss, M. R., and Hayashi, C. T. (1995). All in the family: parent-child influences in competitive gymnastics. *Pediatric Exercise Science, 7,* 36–48.

Weiss, M. R., Smith, A. L., and Theeboom, M. (1996). "That's what friends are for": Children's and teenagers perceptions of peer relationships in the sport domain. *Journal of Sport and Exercise Psychology, 18,* 347–79.

Chapter Ten
Epilogue

Chapter Ten

Epilogue

"One of the strengths of Kinesiology is its propensity for integrative scholarship."

– (Charles, 1996)

STUDENT OBJECTIVES

1. To be able to discuss interdisciplinary and cross-disciplinary approaches to science.
2. To understand how an integrative or cross-disciplinary approach can be applied to research and teaching within kinesiology.
3. To know the difference between two major empirical approaches: positivism and holism, and their application to the study of kinesiology.
4. To appreciate the contribution phenomenology can have on the study of human physical activity.
5. To understand the scope of human physical activity.
6. To appreciate some of the challenges facing the field of Kinesiology into the 21st century.

AN INTEGRATIVE APPROACH TO KINESIOLOGY

In this final chapter, I wish to present some ideas related to the nature of kinesiology as a field of study. In the previous chapters, the various foundations or subfields within kinesiology were presented. The emphasis was on the unique contribution that each subfield can make to kinesiology and other related life sciences. The various contributions of each subfield to the study of human movement and physical activity have been primarily provided by way of conceptual explanation and empirical investigation. It could be somewhat easy to conclude from reading the previous chapters that investigation in each of the subfields can proceed rather independently from each other. Indeed, there are unique journals and scholarly societies for each subfield (see Chapter 2), and it is common for scientists in one subfield today to do their work without much interaction with scientists in the other subfields. Even in those scholarly societies that contain more than one subfield (e.g., American College of Sports Medicine, North American Society for the Psychology of Sport and Physical Activity), interaction across the subfields is limited.

In this section, I wish to argue that in contrast to isolated scholarly activity in each subfield, much can be gained from an integrated approach to kinesiology. In fact, scholars differ on their choice of a name that best describes the field of human physical activity science (e.g., kinesiology, sport sciences, human movement science, etc.) and on the focus of inquiry (e.g., exercise, sports, physical activity). There does seem to be better unanimity among scholars that this type of *cross-disciplinary approach* best defines the potential uniqueness of the field of study that I have promoted in this book (Henry, 1964, 1978; Kroll, 1971; Newell, 1990; Rintala, 1991; Thomas, 1987). By integration I mean to bridge the gap between the subfields by conducting research that utilizes concepts and methodologies across more than one subfield. I fully agree with John Charles, who has written, "A challenge to kinesiology as it enters the 21st century is to develop scholarship that is integrative; that incorporates the interconnectedness of the humanities, social sciences, and life sciences into the study of human movement; and that incorporates the experiencing of physical activity into a meaningful program of study" (Charles, 1996, p. 162). Along these lines, I believe it is also possible to *teach* kinesiology in an integrative manner, such that students not only learn the content of each subfield of kinesiology, but also understand how the various subfields relate to one another. In this section, I want to provide some examples of both integrative research and teaching within kinesiology.

Integrative Research

As we have seen in the previous chapters, there has been much research in each of the various subfields within kinesiology. In addition, many theoretical concepts and experimental findings from *other* disciplines, such as physiology, psychology, and physics, have been used by researchers in the field of kinesiology. Psychology theories have always played an important part of research in the subfields of motor learning and sport and exercise psychology. Reinforcement theory developed by the late B. F. Skinner and other behaviorists in the discipline of psychology was incorporated into motor learning research in the 1960s to help develop this subfield within kinesiology. In addition, motivation theory developed largely in psychology and was later adopted by sport psychologists to understand both physical performance and exercise behavior. The sliding filament theory, developed by Huxley in the discipline of physiology, has helped exercise physiologists in kinesiology understand the role of muscle contraction in physical activity. Newton's laws and other concepts in the field of physics provided the backbone for the development of the subfield of biomechanics in kinesiology. All of these examples illustrate the importance of integrating knowledge from other scientific disciplines. This type of integration is called **interdisciplinary science**. It is somewhat ironic that many of my colleagues in kinesiology readily accept the importance of interdisciplinary science in their work but seem more reluctant to integrate their work with other subfields *within* kinesiology. Perhaps it is not clear to them why it is so important to do so.

I believe there are several reasons to adopt an integrative kinesiology approach to the study of human movement and physical activity. First and foremost is the realization that human physical activity defined in the broadest sense (see Newell, 1990) is the result of complex interactions of sociocultural, psychological, physiological, biomechanical, and anatomical factors. In other words, why and how we move in our environment depends on all of these factors. Focusing research on a single factor, while important in contributing to a given subfield, ultimately will fall well short of a comprehensive understanding of human movement within a given physical activity setting. Second is the acknowledgment that all of the subfields within kinesiology are equally important, and therefore it makes no sense to study one subfield in isolation of the others. The answer to the question of how and why we move will not come from research in only one subfield. Third, the *interactions* of the factors mentioned above are likely to have their own unique qualities that cannot be understood by focusing on a single factor. There are a number of interesting and important questions about human movement that require a working knowledge of more than one subfield. Fourth, and on a more practical note, the emphasis on only one or a few subfields within a department of kinesiology at the college or university level rids kinesiology of its unique nature. If a kinesiology department starts to look more like another psychology or biology department, university administrators may wonder why it should exist as a separate entity. This is a concern Franklin Henry noted as early as 1978, as many physical education departments were making transformations to a kinesiological emphasis. Unfortunately, we have seen this strategy played out at several departments of kinesiology, with often disastrous results (Scanlan, 1998; Wilmore, 1998; and the *Quest* 1998, *50*, no. 2 issue for further discussion).

While there are a number of examples of integrative kinesiology research in the literature, I have chosen only a few to illustrate my point. Observational learning, through the use of modeling, is one field of study that has been recognized as having the potential for integrative research. As early as 1989, McCullagh, Weiss, and Ross (1989) approached the topic from a learning, developmental, and social psychological perspective, and this approach has been re-emphasized in a recent review (McCullagh and Weiss, in press). An example of a specific integrative approach to kinesiology is a study by McCullagh (1987) that attempted to answer the question, does the type of model who demonstrates a skill influence the learning of that skill? In Chapter 9 we pointed out that demonstrating a skill or activity to someone is one important way to facilitate the learning process. There was some suggestion from other studies that certain types of models might be more effective for performance than others (e.g., Gould and Weiss, 1981). For example, watching a model who shared similar characteristics to the learner appeared to result in better performance than watching a dissimilar model. In McCullagh's study, subjects attempted to learn a complex motor task after watching a similar model (i.e., the subjects were told that the model was a college student just like the subjects) or a dissimilar model (the subjects were told that the model was a dancer/gymnast unlike the subjects). In actuality, the model was the same person in both conditions, so the similarity of model was manipulated by altering the *perceptions* of the type of model subjects were watching. The subjects received several demonstrations of the model performing the task before attempting the skill themselves across 20 trials of practice. After the practice phase, the subjects attempted to perform the skill for an additional 20 trials without the model or any feedback from the experiment to determine how much they had learned from the initial practice. The results showed that subjects viewing the similar model performed better than subjects who viewed the dissimilar model during the initial 20 practice trials, but no differences were found on the last 10 trials. These results suggested that model type affects initial practice of a skill but not long-term learning. McCullagh's study is an example of integrative kinesiology research that combines the sport psychology and motor learning subfields. The variable of model type is a social psychological variable, and the use of practice and learning/retention trials is a common methodology in motor learning research.

Another example of integrative kinesiology research is a study performed by Hatfield, Spalding, Mahon, Slater, Brody, and Vaccaro (1992). In this study, Hatfield et al. addressed the question of whether a certain type of psychological strategy can improve physiological efficiency during a running exercise. The psychological strategy used was a form of biofeedback, a technique that allows the performer to monitor certain physiological parameters during the activity. A group of competitive runners was required to run on a treadmill for a total of 36 minutes at a fairly intense level (around 70 percent of their maximal oxygen uptake) under three separate conditions of 12 minutes each. In one 12-minute set, the runners received biofeedback of their ventilation rate and information about how much upper body muscle activity was occurring during the running exercise. In another 12-minute set that served as a type of control condition, the runners were asked to perform a distracting hand-eye coordination task that required them to press a button using their left hand when a visual stimulus arrived in front of them. This condition was thought to distract the runners from monitoring their physiological behaviors while running on the treadmill. In the remaining 12-minute segment, subjects ran without either biofeedback or the attentional distracting task. Hatfield et al. hypothesized that biofeedback could help to improve some parameters related to ventilatory, respiratory, and muscular efficiency such as reduced oxygen consumption, ventilation rates, and upper body muscle activity. The results partially confirmed their hypothesis. Hatfield et al. found that one index of running efficiency, ventilation efficiency (or the ratio between the amount of oxygen consumed and the volume of pulmonary ventilation) was significantly improved in the biofeedback condition relative to the other two conditions. This finding suggested that for the same workload, biofeedback improved ventilation efficiency. The results did not show differences in muscle activity or total oxygen uptake during the running exercises between the three conditions, but they are still encouraging in showing that psychological strategies can influence physiological performance. Others have attempted to follow up on this finding by using different psychological strategies with some positive effects (Smith, Gill, Crews, Hopewell, and Morgan, 1995). These kinds of integrative kinesiology studies demonstrate that important physical activity phenomenon can be investigated by using theory and experimental techniques in more than one subfield.

In this last example of integrative kinesiology research, we turn to a study that combines techniques in biomechanics and motor control with a developmental emphasis. In a study by Jensen, Thelen, Ulrich, Schneider, and Zernicke (1995), interest was centered on the differences in lower limb control across infants of different ages and adults. The authors used a cross-sectional design (see Chapter 8) to determine differences between the different age groups studied—two-week-old infants, three months, seven months, and an adult group that comparisons were made to. Using sophisticated biomechanical techniques, the contributions of the hip, knee, and ankle joints were determined in kicking behavior of the lower limbs. The infants were held upright from behind by the experimenter while the parent stood in front to attract the infant's attention and to keep the infant's arousal level high. The adults supported themselves on one foot so that the unweighted leg hung freely. In this position, the adults performed air steps by lifting the knee of the unweighted leg to closely simulate the behavior of the infants. One of the main issues examined in this study was how the control of the kick changed across the different age groups. For the two-week-old infants, much of the kick was controlled by the muscle power generated in the hip and knee. However, the dominance of the hip reduced as a function of age. Older infants and adults showed much greater evidence of more distal control in the knee and the ankle. It was concluded that kicking in the young infant is a highly constrained and proximally controlled behavior. As development continues, the hip, knee, and ankle become much more adaptive and flexible. This particular study is a good example of integrative kinesiology research because concepts, methodology, and experimental techniques from several subfields were used to answer a specific problem in motor behavior: How is kicking controlled?

Integrative (cross-disciplinary) research helps us better understand how and why we move. One of the important challenges kinesiology faces is the development of theories that help link the major subfields, such as sport and exercise psychology to motor learning, biomechanics to development, and sociocultural to exercise physiology. We are beginning to see more examples of this type of research, which is promising for the kinesiology field of study. We must remember that physical activity is a result of the interaction among the various levels of analysis that span the subfields. Many more interesting links among the subfields are yet to be discovered, and theoretical advancement will only serve to expedite a richer explanation and understanding of human physical activity.

Roberta Park (1987) said, "At a time when a growing number of thoughtful individuals from many segments of higher education are beginning to doubt the wisdom of the excessive specialization that emerged in the 1960's and 1970's, the cross-disciplinary nature of our field offers the potential, still rarely realized, for new and insightful ways to deal with critical questions." (p. 197)

Integrative Teaching

How might this cross-disciplinary approach be imparted to students in kinesiology? Integrative teaching in kinesiology is the offering of courses and experiences that demand the use and integration of more than one subfield within kinesiology. Cross-disciplinary experiences are not necessarily achieved by only taking separate courses in the various subfields and expecting the student to integrate this knowledge on their own, although this strategy has its merits. While a foundation of study in each of the subfields is crucial for knowing the content of subfields, the *integration* of this content requires additional attention on the part of the teacher and student. With few exceptions, most college and university curricula at the undergraduate and graduate levels are arranged using a predominantly fragmented strategy. Namely, the student is required to take separate courses in each of the subfields (biomechanics, exercise physiology, sport psychology, etc.). One way to promote integration is to provide students with courses *specifically designed* for the integration of knowledge across the subfields. This kind of course is offered at San Francisco State University, in which upper-division undergraduate students are required to apply knowledge in the various subfields to a research problem. Students are broken into groups that represent the various subfields and then attempt to formulate a research question that can be examined from a number of perspectives (e.g., biomechanical, social-psychological, physiological). Applying different perspectives to address a specific problem is common in the training of physical therapists (Weeks, 1997) and osteopathic medical students (Hendryx, 1997), for example. At the graduate level, students could also be required to rotate among the various laboratories in the department, working with different professors in the various subfields within kinesiology. In addition, professors could sit on thesis committees in areas outside of their specific expertise, particularly if the research problem requires a more integrative approach. Today, it is probably more common for a biologist outside the kinesiology department to sit on an exercise physiology thesis committee, rather than a sport psychologist from the same department, for instance. This practice is typically adopted to provide the student with *interdisciplinary* input to their research problem. But when the problem is integrative in nature, cross-disciplinary input from other faculty within the kinesiology field of study would be desirable. A quote from Roberta Park (1987) illustrates the need for this type of student training:

> "Following the general tendency in graduate education to prepare narrowly specialized students, many graduate degree programs in physical education have become degree programs in exercise physiology, biomechanics, sport psychology, sport history, and the like. While the move away from vacuous generalist programs is to be commended, substituting these with programs of such narrow scope that students and faculty who choose to emphasize the social sciences have no understanding of or appreciation for the work of those in the biological area is to be lamented. Equally disturbing is the failure of students and faculty in the biological side of the field of physical education to inform themselves of relevant insights from the social sciences and humanities." (Park, 1987, p. 192)

Basing a physical activity curriculum around a problem-oriented approach has been proposed (Ingham and Lawson, 1985; see also Newell, 1990, for a discussion), but unfortunately this has not been frequently adopted by kinesiology departments.

It is my view that students within kinesiology who have more cross-disciplinary training will possess unique skills and insights into the study of human movement and physical activity. This unique perspective will not only allow them to pose important research questions, but will also likely make them more marketable in future career pursuits. Narrow training will result in a kinesiology student who cannot be distinguished from a physiologist or a psychologist.

SUMMARY

- A cross-disciplinary approach to kinesiology implies the interaction of concepts and methodologies among the various subfields within kinesiology.
- Integrating knowledge between scientific disciplines is called interdisciplinary science.
- Both cross-disciplinary research and teaching are advocated for the field of kinesiology.

TOWARD A PHILOSOPHY OF KINESIOLOGY

Types of Empiricism: Positivism and Holism

It should be apparent from the previous chapters that a strong scientific approach has been emphasized to serve as the primary vehicle for moving the kinesiology field of study forward. As indicated in Chapter 2, science, as we know it today, began in the Enlightenment period and continued its course through the Renaissance period and into the Modern Age. But one of the points that I would like to make in this section is that there are *different* approaches to science. In other words, there are different ways of knowing about the content within a field of study. Ways of knowing about this content is called **epistemology**. In most fields of science, an empirical approach is used that is grounded in a philosophy called **positivism.** An assumption of positivism is that all of nature is knowable, that given enough knowledge or facts based on scientific investigation, it is possible to describe, explain, predict, and even control naturally occurring phenomena. Furthermore, positivism is based on the principle that the only reliable scientific information is that which can be reduced to specific patterns of sensation (Harre, 1981, p. 3). Because the most reliable sense data are that which can isolated from outside or surrounding influences, an atomistic (or reductionistic) approach to science has invaded most fields of study. Atomism or reductionism relies heavily on another positivist assumption that the ultimate understanding of nature (or the human body) is found in its smallest elements. Molecular biologists discovering new genes and physicists identifying the elemental forces in nature are examples of the reductionistic approach.

But the importance of understanding the limitations of reductionism was well recognized by Albert Einstein, the world renowned physicist of the early 20th century. Speaking of physics, Einstein (1954) said, the scientist,

"… must content himself with describing the most simple events which can be brought within the domain of our experience; all events of a more complex order are beyond the power of the human intellect to reconstruct with the subtle accuracy and logical perfection which the theoretical physicist demands. Supreme purity, clarity, and certainty at the cost of completeness. But what can be the attraction of getting to know such a tiny section of nature thoroughly, while one leaves everything subtler and more complex shyly and timidly alone?" (pp. 225–26)

In another example, a little closer to home on the limitation of reductionism in understanding human motor control, psychologist Erwin Straus said some time ago,

"If I intend to go from here to there, I must take a certain number of steps. The path is one, my action is one, and I myself am one in my attitude toward the world. Now, if we follow the procedure of the physicist [using a reductionist strategy, my insert] and allow the identification of lived movement with the performed motion, what can we adopt as the basic unit? Where do we stop in this progressive division? Should we use the shortening of a single muscle as the basic unit or even the single tetanic contraction? Should we take, as the unit of a thousand different motions, the action of one whole muscle, a single muscle fiber, or an even more elementary unit? Should our subdivision stop at the microscopic structures of anatomy or at the elementary particles of nuclear physics?"

(Straus, 1980, p. 51)

Even with its limitations, the reductionist approach, influenced heavily by the works of Descartes and Newton (McKinney, 1988), has become known as *modern or classical science* (Prigogine and Stengers, 1984). But a different type of science, somewhat obscured by the so-called successes of classical science, has been developing over the last two centuries—a science that focuses not on the individual elements of complex natural and biological systems, but on how the patterns among the elements behave within the systems. We briefly alluded to this approach, referred to as chaos theory (Gleick, 1987) in Chapter 6. This theory subscribes to a more holistic account of nature's workings. This **holistic** science views nature as different layers of complexity, each layer with its own set of interacting elements, and with all the layers interacting among themselves. Applied to the human body and physical activity, there is a recognition that all the layers (molecular, biochemical, systemic [e.g., muscular, nervous, skeletal, etc.], psychological, sociocultural) contribute in important ways to human movement, with no one layer more important than another. This approach embraces a truly "mind-body" connection in attempting to explain how human movement is actualized. The holistic approach has gained respectability, not only because of new mathematical, statistical, and research methodology techniques designed to study complex behavior, but also because of greater realization of the limits of the reductionistic approach (see Kelso, 1996; Prigogine and Stengers, 1984).

In my view, a reductionist strategy, while important, must be accompanied by an equally demanding attempt to understand how the fundamental elements of a system work together. I must admit, this opinion is similar to the Chinese philosophers' view that nature is an interplay between two polar opposites, Yin and Yang (see McKinney, 1988; Capra, 1982). Yin is an integrating, cooperative activity, while Yang is an aggressive, rational activity. According to the Chinese, reality is not one activity or another but a constant oscillation between them. I envision the holistic scientific strategy likened to Yin and reductionism to Yang. I believe that if a philosophy of kinesiology is to be forwarded, it must include the recognition of both the reductionistic and the holistic approaches in advancing our knowledge of the human engaged in physical activity. Reductionism is useful for identifying the important elements of components at the various layers or levels. However, a holistic approach is required to describe how these elements interact and how the various levels influence one another. This type of philosophy can help guide both integrative teaching and research in kinesiology.

Phenomenology

Our previous discussion of positivism and holism essentially focused on how we view the nature of reality. That is, do we view the human being engaged in physical activity as a collection of independent elements, or as a collection of elements interacting with one another at a given level and between levels? My answer to that question was somewhat of a compromise. Positivism, with its reductionistic emphasis, provides a way to identify the important components at any level, but holism provides a means to understand the complex interactions of the components. In either case, the focus of inquiry is on *matter*, whether it be molecules, blood flow, neuronal transmission, muscular contractions, or psychological motives. The definition of matter could also be extended to include the *relationships* among the components, such as the relationship between heart rate and blood flow, between mass and acceleration in the definition of force, or between motivation and adherence to exercise. Whether we are talking about individual components or their relationship to other components, we are still talking about matter and how it is organized. Matter is the content of empiricism, a scientific, methodological approach shared by both reductionism and holism. Is there another level of inquiry relevant to the study of kinesiology?

While perhaps in its infancy as applied to kinesiology, there is another level of inquiry that has been used to investigate important elements of human existence, namely human consciousness, interpretation, and meaning. **Phenomenology**, advanced by a number of philosophers, including Edmund Husserl (1859–1938) and Martin Heidegger (1889–1976), is a type of inquiry that focuses on the essence of lived experiences rather than the behavior of matter (Polkinghorne, 1989). Its ambition is to learn more about an individual's consciousness and how existence is experienced. Phenomenological research usually depends on interviewing techniques that require a dialogue between the subject (or participant) and the investigator (see Dale, 1996, for an excellent introduction). A type of phenomenological inquiry is **hermeneutics,** the study of the *meaning* in one's existence that takes into account the social-cultural context and the history of the individual (Allen and Jensen, 1990; Harris, 1983; Hekman, 1986; Park, 1986). Hermeneutics, a term possibly stemming from the mythological Greek god, Hermes (Grondin, 1994) or an actual ancient Egyptian, Hermes Trismegistus (Sardello, 1992, p. 170), is concerned with the meaning associated with one's behaviors. Applied to kinesiology, a hermeneutics research investigation might be concerned with the meaning of a physical activity to an individual rather than the biomechanical, physiological, or even psychological changes that accompany human physical activity. The focus of inquiry is on meaning, not matter.

Why are phenomenology and hermeneutics relevant to kinesiology? For starters, I believe these types of epistemologies could be used to help us better understand an anomaly that seems to escape reason, namely, given the clear benefits of exercise for both physiological and psychological health, why don't more people exercise (see Kimiecik and Lawson, 1996)? Could it be that if we knew more about the "meaning" of exercise and physical activity to people, we would be better able to understand why they choose not to exercise, even though they know it is beneficial?

In another example related to kinesiology taken from Fahlberg and Fahlberg (1994), we might observe a young woman jogging and ask, How would we study this behavior from either an empirical or a phenomenological perspective? From an empirical perspective, we might be concerned with the mechanical efficiency of her gait pattern and measure a variety of angles, torques, and electromyographic activity from her muscles. From a motor control perspective, we could study the coordination between her upper and lower extremities by measuring the phasing between the limbs. Physiologically, we could determine her cardiorespiratory endurance and measure her maximum oxygen uptake. Psychologically, we might be interested in whether she is intrinsically or extrinsically motivated to jog, and give her appropriate questionnaires to fill out to verify her motivational orientation. But phenomenologically, we would be more interested in understanding the meaning or importance of jogging in the woman's life. By engaging in a phenomenological and hermeneutic dialogue with a jogger, Fahlberg (1990) was able to uncover interesting insight into how jogging fit in with the young woman's life. During the dialogue, the jogger said,

> "Many people have other ways to feel good about themselves. They're married, they've got a job or something like that. They've got something in their lives that's important to them … and as far as I'm concerned, I don't have that … you know, I live for my workout. If I didn't work out, I mean it's that strong with me that, basically, if I wasn't working out like this, there wouldn't be any point in even being on this earth, as far as being alive. I mean, I might as well go out and commit suicide or something like that. As long as I got my workout and I can work out, I'm fine. But don't take it away. *Don't* take it away."
>
> (Fahlberg, 1990, p. 75)

This quote provides an example of the uniqueness of the phenomenology and hermeneutic approach. With empiricism, we might find details of how her jogging is coordinated, whether it is mechanically efficient, and how much oxygen is consumed during a workout. We might even be able to determine the young woman's motivational state (in this case, she might very well be intrinsically motivated). But the fact that jogging plays such a central role in the woman's life could help us better understand something about her existence and the meaning of "working out" in relation to other activities in which she might or might not participate. One might see why an injury could be devastating to her—having clear implications for any type of rehabilitation she might need to do. It could tell us something about the social interactions she has with others, or how she goes about setting priorities in her life. In short, phenomenology tells us more about the *person*, not about the person's physiological or anatomical condition (see Estes, 1994; and Kleinman, 1979, for further discussion about phenomenology).

Empiricism helps us identify and describe the matter of the human body and how that matter is organized. But the human being is more than just matter. We possess qualities that uniquely define us as a distinct species, and these qualities may likely play a critical role in our further understanding of humans engaged in physical activity. A phenomenological strategy is required to uncover these unique qualities. Borrowing from Rintala (1991), human movement science should be more than just the study of human *movement*. It should also be about the study of the *human* moving.

The Future of Kinesiology

As a field of study, kinesiology has grown in interest all over the world. Part of this popularity is the recognition that kinesiology can prepare an individual for a number of careers or professions. I would also hope that kinesiology will continue to be recognized as a legitimate *scholarly* enterprise as well. Without continued scholarly development, the field will be unable to meet the rigorous demands of the scientific community into the 21st century. For kinesiological science to thrive, we must continue to attract the bright and motivated student whose allegiance to the field will be strong. We must provide cross-disciplinary experiences in our curricula that will help develop true scholars of kinesiology, and not just biologists, psychologists, or sociologists with an interest in physical activity. We must continue to support our journals and professional organizations. As scientists, that means publishing and presenting some of our best work in scholarly journals and organizations that help promote kinesiology as a cross-disciplinary field of study. The American Kinesiology Association (AKA) was founded to help promote the field of kinesiology. I am hopeful that the AKA will help expose kinesiology to the public, and provide political influence that will open more funding possibilities for physical activity research. In 2009 the AKA made recommendations as to the desired content of the undergraduate core curriculum in kinesiology. According to the AKA, the undergraduate degree in kinesiology should include principles and experiences focused on physical activity across the life span. These include:

- Physical activity in health, wellness, and quality of life
- Scientific foundations of physical activity
- Cultural, historical, and philosophical dimensions of physical activity
- The practice of physical activity.

I believe that it is important for the AKA to continue its efforts to provide leadership in the development and evaluation of undergraduate curricula in kinesiology throughout the United States.

Efforts should also be made to reunite the professional and the performance elements of physical activity with the scholarly enterprise of kinesiology, as suggested by Newell (1990). The break of the academic elements from physical education and dance into what we now call the scholarly enterprise of kinesiology occurred several years ago. This break might just as easily be called a divorce. It is time to heal the wounds, make amends, and bring the academic, professional, and performance elements of physical activity together into one integrated framework. This task will not be easy; some say it might be impossible. But there is no question in my mind that if it can occur, the field of study called "kinesiology" will blossom and thrive into the 21st century.

SUMMARY

- An epistemology is a way of knowing about nature.
- Empiricism is an epistemology that relies on theory, predictions, laws, and the scientific method.
- One type of empiricism is positivism, which states that it is possible to describe, explain, predict, and control naturally occurring phenomena.
- Atomism or reductionism assumes that the ultimate understanding of nature is found in its smallest elements.
- Another type of empiricism is called holism, the study of how complex matter is organized.
- Both positivism and holism are helpful in advancing the state of knowledge within kinesiology.
- Phenomenology is an epistemology that focuses on the essence and meaning of lived experiences in humans.
- Hermeneutics is a type of phenomenology that attempts to understand the meaning of lived experiences and takes into account the sociocultural and developmental aspects of the individual.
- Phenomenology may help us understand the human engaged in physical activity.
- The scope of human physical activity is large.
- Much effort must be made to advance the field of kinesiology into the 21st century.

CHAPTER SUMMARY

- An integrative or cross-disciplinary approach to research and teaching within kinesiology is advocated.
- Two major epistemologies to be used to advance the study of kinesiology are reductionism and holism.
- The importance of lived physical activity experiences can be studied using phenomenology and hermeneutics.
- The study of kinesiology should not be restricted to a small set of physical activity types or situations.
- A number of important issues need to be addressed for continued growth of kinesiology into the 21st century.

IMPORTANT TERMS

Classical science – considered to be largely based on Newtonian physics

Epistemology – ways of knowing about the content in science

Hermeneutics - the study of the meaning in one's existence that takes into account the social-cultural context and the history of the individual

Holistic science – the science of complex systems made up of many interacting elements

Integrative research – the type of research that utilizes the knowledge and techniques from more than one sub-field within a discipline

Integrative teaching - the offering of courses and experiences that demand the use and integration of more than one subfield within kinesiology

Interdisciplinary science – the type of science that utilizes knowledge and techniques from more than one discipline

Phenomenology - a type of inquiry that focuses on the essence of lived experiences rather than the behavior of matter

Positivism – a scientific philosophy that all of nature is knowable, that given enough knowledge or facts based on scientific investigation, it is possible to describe, explain, predict, and even control naturally occurring phenomena.

Reductionism – a scientific philosophy holding that the ultimate understanding of nature (or the human body) is found in its smallest elements

Yin and Yang – an ancient Chinese view that nature is an interplay between two polar opposites, Yin and Yang where Yin is integrating, cooperative activity, while Yang is aggressive, rational activity.

INTEGRATING KINESIOLOGY: PUTTING IT ALL TOGETHER

1. Read a scientific article from a journal in one of the subfields of kinesiology. In the introduction, identify the purpose of the study. How might the purpose be changed to include a more integrative approach to the problem being studied?
2. What are some advantages and disadvantages of inter- and cross- disciplinary training in kinesiology?
3. Identify some differences between reductionistic and holistic approaches to science. What is your opinion on the importance of each approach?
4. In your opinion, should kinesiology be the study of human movement, or the study of the human moving (or both)? Defend your answer.
5. Pick three physical activities that you participate in. Identify the "meaning" or importance of each activity in your life. Why do you think you feel this way about each activity?

KINESIOLOGY ON THE WEB

- http://en.wikipedia.org/wiki/Phenomenology_(philosophy)—This Web site contains information on the philosophy of phenomenology.
- http://en.wikipedia.org/wiki/Yin_and_yang—Visit this Web site for a further discussion of the concept of Yin and Yang.
- http://en.wikipedia.org/wiki/Holism—This Web site contains various links to the concept of holism.

REFERENCES

Allen, M., and Jensen, L. (1990). Hermeneutical inquiry: meaning and scope. *Western Journal of Nursing Research, 12,* 241–53.

Capra, F. (1982). *The Turning Point: Science, Society, and the Rising Culture.* New York: Simon & Schuster.

Charles, J. M. (1996). Scholarship reconceptualized: the connectedness of kinesiology, *Quest, 48,* 152–64.

Dale, G. A. (1996). Existential phenomenology: emphasizing the experience of the athlete in sport psychology research. *The Sport Psychologist, 10,* 307–21.

Estes, S. (1994). Knowledge and kinesiology. *Quest, 46,* 392–409.

Einstein, A. (1954). *Ideas and Opinions.* New York: Crown.

Fahlberg, L.L. (1990). A hermeneutical study of meaning in health behavior. Women and exercise. Doctoral dissertation, University of Utal, Salt Lake City.

Fahlberg, L.L., and Fahlberg, L.A. (1994). A human science for the study of movement: An integration of multiple ways of knowing. Research Quarterly for Exercise and Sport, 65, 100=109.

Gleick, J. (1987). *Chaos: Making a New Science.* New York: Viking Penguin.

Gould, D., and Weiss, M. (1981). The effects of model similarity and model talk on self-efficacy and muscular endurance. *Journal of Sport Psychology, 3,* 17–29.

Groudin, J. (1994). *Introduction to Philosophical Hermeneutics.* New Haven and London: Yale University Press. Originally published by Wissenschaftliche Buchgesellschaft, Darmstadt, Germany, 1991.

Harris, J. (1983). Broadening horizons: interpretive cultural research, hermeneutics, and scholarly inquiry in physical education. *Quest, 35,* 82–96.

Hatfield, B. D., Spalding, T. W., Mahon, A. D., Slater, B. A., Brody, E. B., and Vaccaro, P. (1992). The effect of psychological strategies upon cardiorespiratory and muscular activity during treadmill running. *Medicine and Science in Sports and Exercise, 24,* 218–25.

Hekman, S. (1986). *Hermeneutics and the Sociology of Knowledge.* Notre Dame, IN: University of Notre Dame Press.

Hendryx, J. (1997). Osteopathic physician. Personal communication, Boulder, Colorado.

Henry, F. M. (1964). Physical education: An academic discipline. *Journal of Health, Physical Education and Recreation, 35,* 32–33; 69.

Henry, F. M. (1978). The academic discipline of physical education. *Quest, Monograph 29,* 13–29.

Kimiecik, J. C., and Lawson, H. A. (1996). Toward new approaches for exercise behavior change and health promotion. *Quest, 48,* 102–25.

Kelso, J. A. S. (1996). *Dynamic Patterns: The Self-Organization of Brain and Behavior.* Cambridge, MA: MIT Press.

Kleinman, S. (1979). The significance of human movement: a phenomenological approach. In E. Gerber and W. Morgan (eds.), *Sport and the Body: A Philosophical Symposium* (pp. 177–80). Philadelphia: Lea and Feiberger.

Kroll, W. (1971). *Perspectives in Physical Education.* New York: Academic.

McCullagh, P. (1987). Model similarity on motor performance. *Journal of Sport Psychology, 9,* 249–60.

McCullagh, P., Weiss, M. R., and Ross, D. (1989). Modeling considerations in motor skill acquisition and performance: an integrated approach. In K. Pandolf (ed.), *Exercise and Sport Science Reviews,* Vol. 17, 475–513, Baltimore: Williams and Wilkins.

McCullagh, P., and Weiss, M. R. (in press). Modeling considerations for motor skill performance and psychological responses. In R. N. Singer, H. Hausenblas, and C. Janelle (eds.). *Handbook on Research in Sport Psychology.* New York: John Wiley and Sons.

McKensie, D. S. (1970). Functional replacement of the upper extremity today. In G. Murdoch (ed.), *Prosthetic and Orthotic Practice.* London: Edward Arnold, 331–35.

McKinney, R. H. (1988). Toward a resolution of paradigm conflict: holism versus postmodernism. *Philosophy Today,* Winter, 299–311.

Newell, K. M. (1990). Physical activity, knowledge types, and degree programs. *Quest, 42,* 243–68.

Park, R. (1986). Hermeneutics, semiotics, and the 19th-century quest for a corporeal self. *Quest, 38,* 33–49.

Park, R. J. (1987). The future of graduate education in the sociocultural foundations: history. *Quest, 39,* 191–200.

Park, R. J. (1998). Critical issues for the future: A house divided. *Quest, 50,* 213–24.

Polkinghorne, D. (1989). Phenomenological research methods. In R. Valle and S. Halling (eds.), *ExistentialPhenomenological Perspectives in Psychology* (pp. 41–60). New York: Plenum.

Prigogine, I., and Stengers, I. (1984). *Order Out of Chaos: Man's New Dialogue with Nature.* New York: Bantam.

Rintala, J. (1991). The mind-body revisited. *Quest, 43,* 260–79.

Sardello, R. (1992). *Facing the World with Soul.* Hudson, NY: Lindisfarne.

Scanlan, T. K. (1998). Thriving versus surviving. *Quest, 50,* 1126–33.

Smith, A. L., Gill, D. L., Crews, D. J., Hopewell, R., and Morgan, D. W. (1995). Attentional strategy use by experienced distance runners: physiological and psychological effects. *Research Quarterly for Exercise and Sport, 66,* 142–50.

Straus, E. W. (1980). *Phenomenological Psychology.* New York and London: Garland. Reprint of the 1966 edition published by New York: Basic Books.

Thomas, J. R. (1987). Are we already in pieces or just falling apart? *Quest, 39,* 114–21.

Weeks, D. (1997). Department of Physical Therapy, Regis University. Personal communication.

Wilmore, J. H. (1998). Building strong academic programs for our future. *Quest, 50,* 103–107.

CPSIA information can be obtained at www.ICGtesting.com
Printed in the USA
LVIW01n2102060916
503467LV00005B/7